# Multistate Systems Reliability Theory with Applications

WILEY SERIES IN PROBABILITY AND STATISTICS

Established by WALTER A. SHEWHART and SAMUEL S. WILKS



A complete list of the titles in this series appears at the end of this volume.

# Multistate Systems Reliability Theory with Applications

**Bent Natvig**

*University of Oslo, Norway*

WILEY

A John Wiley and Sons, Ltd., Publication

This edition first published 2011


*Registered office*
John Wiley & Sons Ltd, The Atrium, Southern Gate, Chichester, West Sussex, PO19 8SQ, United Kingdom

For details of our global editorial offices, for customer services and for information about how to apply for permission to reuse the copyright material in this book please see our website at www.wiley.com.



*Library of Congress Cataloging-in-Publication Data*

Natvig, Bent, 1946-
Multistate systems reliability theory with applications / Bent Natvig.
p. cm.
Includes bibliographical references and index.
ISBN 978-0-470-69750-4 (cloth)
1. Reliability (Engineering) 2. Stochastic systems. I. Title.
TA169.N38 2011
620′.00452 – dc22

2010034245

A catalogue record for this book is available from the British Library.

Print ISBN: 978-0-470-69750-4
ePDF ISBN: 978-0-470-97707-1
oBook ISBN: 978-0-470-97708-8
ePub ISBN: 978-0-470-97713-2

Typeset in 11/13 Times-Roman by Laserwords Private Limited, Chennai, India
Printed and bound in Singapore by Markono Print Media Pte Ltd.

*To my Mother, Kirsten,*

*for giving me jigsaws instead of cars,*
*and to Helga*
*for letting me (at least for the last 40 years)*
*run in the sun*
*and work when normal people relax.*

# Contents

# Preface

In the magazine *Nature* (1986) there was an article on a near catastrophe on the night of 14 April 1984 in a French pressurized water reactor (PWR) at Le Bugey on the Rhône river, not far from Geneva. This incident provides ample motivation for multistate reliability theory.

'The event began with the failure of the rectifier supplying electricity to one of the two separate 48 V direct-current control circuits of the 900 MW reactor which was on full power at the time. Instantly, a battery pack switched on to maintain the 48 V supply and a warning light began to flash at the operators in the control room. Unfortunately, the operators ignored the light (if they had not they could simply have switched on an auxiliary rectifier).

What then happened was something that had been completely ignored in the engineering risk analysis for the PWR. The emergency battery now operating the control system began to run down. Instead of falling precipitously to zero, as assumed in the "all or nothing" risk analysis, the voltage in the control circuit steadily slipped down from its nominal 48 V to 30 V over a period of three hours. In response a number of circuit breakers began to trip out in an unpredictable fashion until finally the system, with the reactor still at full power, disconnected itself from the grid.

The reactor was now at full power with no external energy being drawn from the system to cool it. An automatic "scram" system then correctly threw in the control rods, which absorbed neutrons and shut off the nuclear reaction. However, a reactor in this condition is still producing a great deal of heat −300 MW in this case. An emergency system is then supposed to switch in a diesel generator to provide emergency core cooling (otherwise the primary coolant would boil and vent within a few hours). But the first generator failed to switch on because of the loss of the first control circuit. Luckily the only back-up generator in the system then switched on, averting a serious accident.'

Furthermore, *Nature* writes: 'The Le Bugey incident shows that a whole new class of possible events had been ignored – those where electrical systems fail gradually. It shows that risk analysis must not only take into account a yes or no, working or not working, for each item in the reactor, but the possibility of working with a slightly degraded system.'

This book is partly a textbook and partly a research monograph, also covering research on the way to being published in international journals. The first two chapters, giving an introduction to the area and the basics, are accompanied by exercises. In a course, these chapters should at least be combined with the two first and the three last sections of Chapter 3. This will cover basic bounds in a time interval for system availabilities and unavailabilities given the corresponding component availabilities and unavailabilities, and how the latter can be arrived at. In addition come applications to a simple network system and an offshore electrical power generation system. This should be followed up by Chapter 4, giving a more in-depth application to an offshore gas pipeline network.

The rest of Chapter 3 first gives improved bounds both in a time interval and at a fixed point of time for system availabilities and unavailabilities, using modular decompositions, leaving some proofs to Appendix A. Some of the results here are new. Strict and exactly correct bounds are also covered in the rest of this chapter.

In Chapter 5 component availabilities and unavailabilities in a time interval are considered uncertain and a new theory for Bayesian assessment of system availabilities and unavailabilities in a time interval is developed including a simulation approach. This is applied to the simple network system. Chapter 6 gives a new theory for measures of importance of system components covering generalizations of the Birnbaum, Barlow–Proschan and Natvig measures from the binary to the multistate case both for unrepairable and repairable systems. In Chapter 7 a corresponding numerical study is given based on an advanced discrete event simulation program. The numerical study covers two three-component systems, a bridge system and an offshore oil and gas production system.

Chapter 8 is concerned with probabilistic modeling of monitoring and maintenance based on a marked point process approach taking the dynamics into account. The theory is illustrated using the offshore electrical power generation system. We also describe how a standard simulation procedure, the data augmentation method, can be implemented and used to obtain a sample from the posterior distribution for the parameter vector without calculating the likelihood explicitly.

Which of the material in the rest of Chapter 3 and in Chapters 5–8 should be part of a course is a matter of interest and taste.

A reader of this book will soon discover that it is very much inspired by the Barlow and Proschan (1975a) book, which gives a wonderful tour into the world of binary reliability theory. I am very much indebted to Professor Richard Barlow for, in every way, being the perfect host during a series of visits, also by some of my students, to the University of California at Berkeley during the 1980s. I am also very thankful to Professor Arnljot Høyland, at what was then named The Norwegian Institute of Technology, for letting me, in the autumn of 1976, give the first course in moderny reliability theory in Norway based on the Barlow and Proschan (1975a) book.

Although I am the single author of the book, parts of it are based on important contributions from colleagues and students to whom I am very grateful. Government grant holder Jørund Gåsemyr is, for instance, mainly responsible for the new results in Chapter 3, for the simulation approach in Chapter 5 and is the first author of the paper on which Chapter 8 is based. Applying the data augmentation approach was his idea. Associate Professor Arne Bang Huseby, my colleague since the mid 1980s, has developed the advanced discrete event simulation program necessary for the numerical study in Chapter 7. The extensive calculations have been carried through by my master student Mads Opstad Reistadbakk. Furthermore, the challenging calculations in Chapter 5 were done by post. doc Trond Reitan. The offshore electrical power generation system case study in Chapter 3 is based on a suggestion by Professor Arne T. Holen at The Norwegian Institute of Technology and developed as part of master theses by my students Skule Sørmo and Gutorm Høgåsen. Finally, Chapter 4 is based on the master thesis of my student Hans Wilhelm Mørch, heavily leaning on the computer package MUSTAFA (MUltiSTAte Fault-tree Analysis) developed by Gutorm Høgåsen as part of his PhD thesis.

Finally I am thankful to the staff at John Wiley & Sons, especially Commissioning Editor, Statistics and Mathematics Dr Ilaria Meliconi for pushing me to write this book. I must admit that I have enjoyed it all the way.

Bent Natvig
Oslo, Norway

# Acknowledgements

We acknowledge *Applied Probability* for giving permission to include Tables 2.1, 2.2, 2.4 and 2.15 from Natvig (1982a), Figure 1.1 and Tables 1.1 and 3.1 from Funnemark and Natvig (1985) and Figure 1.2 and Tables 2.7–2.10, 3.6 and 3.7 from Natvig *et al.* (1986). In addition we acknowledge World Scientific Publishing for giving permission to include Figure 4.1 and Tables 4.1–4.15 from Natvig and Mørch (2003).

# List of abbreviations

| | |
|---|---|
| BCS | Binary coherent system |
| BMS | Binary monotone system |
| BTMMS | Binary type multistate monotone system |
| BTMSCS | Binary type multistate strongly coherent system |
| MCS | Multistate coherent system |
| MMS | Multistate monotone system |
| MSCS | Multistate strongly coherent system |
| MUSTAFA | MUltiSTAte Fault-tree Analysis |
| MWCS | Multistate weakly coherent system |

# 1

# Introduction

In reliability theory a key problem is to find out how the reliability of a complex system can be determined from knowledge of the reliabilities of its components. One inherent weakness of traditional binary reliability theory is that the system and the components are always described just as functioning or failed. This approach represents an oversimplification in many real-life situations where the system and their components are capable of assuming a whole range of levels of performance, varying from perfect functioning to complete failure. The first attempts to replace this by a theory for multistate systems of multistate components were done in the late 1970s in Barlow and Wu (1978), El-Neweihi *et al.* (1978) and Ross (1979). This was followed up by independent work in Griffith (1980), Natvig (1982a), Block and Savits (1982) and Butler (1982) leading to proper definitions of a multistate monotone system and of multistate coherent systems and also of minimal path and cut vectors. Furthermore, in Funnemark and Natvig (1985) upper and lower bounds for the availabilities and unavailabilities, to any level, in a fixed time interval were arrived at for multistate monotone systems based on corresponding information on the multistate components. These were assumed to be maintained and interdependent. Such bounds are of great interest when trying to predict the performance process of the system, noting that exactly correct expressions are obtainable just for trivial systems. Hence, by the mid 1980s the basic multistate reliability theory was established. A review of the early development in this area is given in Natvig (1985a). Rather recently, probabilistic modeling of partial monitoring of components with applications to preventive system maintenance has been extended by Gåsemyr and Natvig (2005) to multistate monotone systems of multistate components. A newer review of the area is given in Natvig (2007).

The theory was applied in Natvig *et al.* (1986) to an offshore electrical power generation system for two nearby oilrigs, where the amounts of power that may possibly be supplied to the two oilrigs are considered as system states. This application is also used to illustrate the theory in Gåsemyr and Natvig (2005). In Natvig and Mørch (2003) the theory was applied to the Norwegian offshore gas pipeline network in the North Sea, as of the end of the 1980s, transporting gas to Emden in Germany. The system state depends on the amount of gas actually delivered, but also to some extent on the amount of gas compressed, mainly by the compressor component closest to Emden. Rather recently the first book (Lisnianski and Levitin, 2003) on multistate system reliability analysis and optimization appeared. The book also contains many examples of the application of reliability assessment and optimization methods to real engineering problems. This has been followed up by Lisnianski *et al.* (2010).

Working on the present book a series of new results have been developed. Some generalizations of bounds for the availabilities and unavailabilities, to any level, in a fixed time interval given in Funnemark and Natvig (1985) have been established. Furthermore, the theory for Bayesian assessment of system reliability, as presented in Natvig and Eide (1987) for binary systems, has been extended to multistate systems. Finally, a theory for measures of component importance in nonrepairable and repairable multistate strongly coherent systems has been developed, and published in Natvig (2011), with accompanying advanced discrete simulation methods and an application to a West African production site for oil and gas.

## 1.1 Basic notation and two simple examples

Let $S = \{0, 1, \ldots, M\}$ be the set of states of the system; the $M + 1$ states representing successive levels of performance ranging from the perfect functioning level $M$ down to the complete failure level 0. Furthermore, let $C = \{1, \ldots, n\}$ be the set of components and $S_i$, $i = 1, \ldots, n$ the set of states of the $i$th component. We claim $\{0, M\} \subseteq S_i \subseteq S$. Hence, the states 0 and $M$ are chosen to represent the endpoints of a performance scale that might be used for both the system and its components. Note that in most applications there is no need for the same detailed description of the components as for the system.

Let $x_i$, $i = 1, \ldots, n$ denote the state or performance level of the $i$th component at a fixed point of time and $\boldsymbol{x} = (x_1, \ldots, x_n)$. It is assumed that the state, $\phi$, of the system at the fixed point of time is a deterministic function of $\boldsymbol{x}$, i.e. $\phi = \phi(\boldsymbol{x})$. Here $\boldsymbol{x}$ takes values in $S_1 \times S_2 \times \cdots \times S_n$ and $\phi$ takes values in $S$. The function $\phi$ is called the structure function of the system. We often denote a multistate system by $(C, \phi)$. Consider, for instance, a system of $n$ components in parallel where $S_i = \{0, M\}, i = 1, \ldots, n$. Hence, we have a binary description of component states. In binary theory, i.e. when $M = 1$, the system state is 1 iff at least one component is functioning. In multistate theory we may let the state of the system be the number of components functioning, which is far more informative. In this case, for $M = n$,

$$\phi(\boldsymbol{x}) = \sum_{i=1}^{n} x_i / n. \tag{1.1}$$

As another simple example consider the network depicted in Figure 1.1. Here component 1 is the parallel module of the branches $a_1$ and $b_1$ and component 2 the parallel module of the branches $a_2$ and $b_2$. For $i = 1, 2$ let $x_i = 0$ if neither of the branches work, 1 if one branch works and 3 if two branches work. The states of the system are given in Table 1.1.

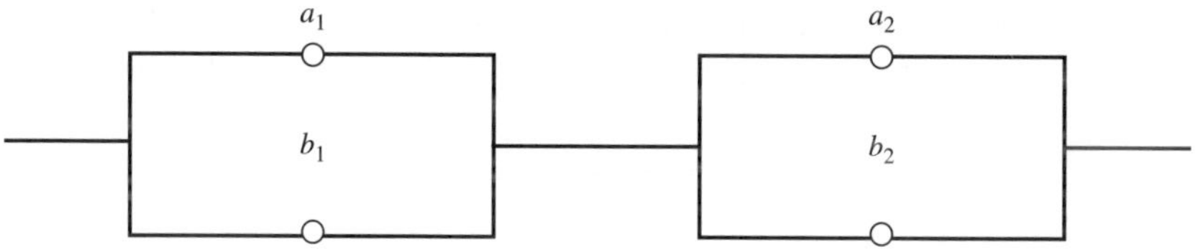


*Figure 1.1* *A simple network.*

Note, for instance, that the state 1 is critical both for each component and the system as a whole in the sense that the failing of a branch leads

Table 1.1 States of the simple network system of Figure 1.1.

| | | | | |
|---|---|---|---|---|
| | 3 | 0 | 2 | 3 |
| Component 2 | 1 | 0 | 1 | 2 |
| | 0 | 0 | 0 | 0 |
| | | 0 | 1 | 3 |
| | | Component 1 | | |

to the 0 state. In binary theory the functioning state comprises the states $\{1, 2, 3\}$ and hence only a rough description of the system's performance is possible. It is not hard to see that the structure function is given by

$$\phi(\boldsymbol{x}) = x_1x_2 - I(x_1x_2 = 3) - 6I(x_1x_2 = 9), \tag{1.2}$$

where $I(\cdot)$ is the indicator function.

The following notation is needed throughout the book.

$$(\cdot_i, \boldsymbol{x}) = (x_1, \ldots, x_{i-1}, \cdot, x_{i+1}, \ldots, x_n).$$

$\boldsymbol{y} < \boldsymbol{x}$ means $y_i \leq x_i$ for $i = 1, \ldots, n$, and $y_i < x_i$ for some $i$.

Let $A \subset C$. Then

$\boldsymbol{x}^A$ = vector with elements $x_i, i \in A$,

$A^c$ = subset of $C$ complementary to $A$.

## 1.2 An offshore electrical power generation system

In Figure 1.2 an outline of an offshore electrical power generation system, considered in Natvig *et al.* (1986), is given. The purpose of this system is to supply two nearby oilrigs with electrical power. Both oilrigs have their own main generation, represented by equivalent generators $A_1$ and $A_3$ each having a capacity of 50 MW. In addition, oilrig 1 has a standby generator $A_2$ that is switched into the network in case of outage of $A_1$ or $A_3$, or may be used in extreme load situations in either of the two oilrigs. The latter situation is, for simplicity, not treated in this book. $A_2$ is in cold standby, which means that a short startup time is needed before it is switched into the network. This time is neglected in the following model. $A_2$ also has a capacity of 50 MW. The control unit, $U$, continuously supervises the supply from each of the generators with automatic control of the switches. If, for instance, the supply from $A_3$ to oilrig 2 is not sufficient, whereas the supply from $A_1$ to oilrig 1 is sufficient, $U$ can activate $A_2$ to supply oilrig 2 with electrical power through the standby subsea cables $L$.

The components to be considered here are $A_1$, $A_2$, $A_3$, $U$ and $L$. We let the perfect functioning level $M$ equal 4 and let the set of states of all components be $\{0, 2, 4\}$. For $A_1$, $A_2$ and $A_3$ these states are interpreted as

0: The generator cannot supply any power;

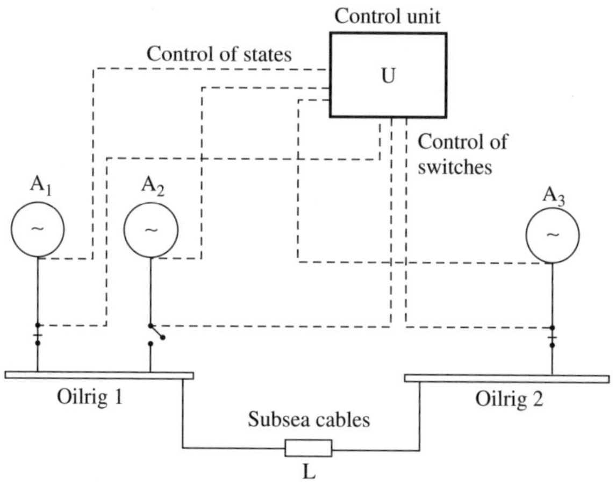


*Figure 1.2 Outline of an offshore electrical power generation system.*

2: The generator can supply a maximum of 25 MW;

4: The generator can supply a maximum of 50 MW.

Note that as an approximation we have, for these generators, chosen to describe their supply capacity on a discrete scale of three points. The supply capacity is not a measure of the actual amount of power delivered at a fixed point of time. There is continuous power-frequency control to match generation to actual load, keeping electrical frequency within prescribed limits.

The control unit $U$ has the states

0: $U$ will, by mistake, switch the main generators $A_1$ and $A_3$ off without switching $A_2$ on;

2: $U$ will not switch $A_2$ on when needed;

4: $U$ is functioning perfectly.

The subsea cables $L$ are actually assumed to be constructed as double cables transferring half of the power through each simple cable. This leads to the following states of $L$

0: No power can be transferred;

2: 50% of the power can be transferred;

4: 100% of the power can be transferred.

Let us now, for simplicity, assume that the mechanism that distributes the power from $A_2$ to platform 1 or 2 is working perfectly. Furthermore, as a start, assume that this mechanism is a simple one either transferring no power from $A_2$ to platform 2, if $A_2$ is needed at platform 1, or transferring all power from $A_2$ needed at platform 2. Now let $\phi_1(A_1, A_2, U) =$ the amount of power that can be supplied to platform 1, and $\phi_2(A_1, A_2, A_3, U, L) =$ the amount of power that can be supplied to platform 2. $\phi_1$ will now just take the same states as the generators whereas $\phi_2$ can also take the following states

1: The amount of power that can be supplied is a maximum of 12.5 MW;

3: The amount of power that can be supplied is a maximum of 37.5 MW.

Number the components $A_1, A_2, A_3, U, L$ successively 1, 2, 3, 4, 5. Then it is not too hard to be convinced that $\phi_1$ and $\phi_2$ are given respectively by

$$\phi_1(\boldsymbol{x}) = I(x_4 > 0) \min(x_1 + x_2 I(x_4 = 4), 4) \tag{1.3}$$

$$\phi_2(\boldsymbol{x}) = I(x_4 > 0) \min(x_3 + x_2 I(x_4 = 4) I(x_1 = 4) x_5/4, 4). \tag{1.4}$$

Let us still assume that the mechanism that distributes the power from $A_2$ to platform 1 or 2 is working perfectly. However, let it now be more advanced, transferring excess power from $A_2$ to platform 2 if platform 1 is ensured a delivery corresponding to state 4. Of course in a more refined model this mechanism should be treated as a component. The structure functions are now given by

$$\phi_1^*(\boldsymbol{x}) = \phi_1(\boldsymbol{x}) \tag{1.5}$$

$$\phi_2^*(\boldsymbol{x}) = I(x_4 > 0) \min(x_3 + \max(x_1 + x_2 I(x_4 = 4) - 4, 0) x_5/4, 4), \tag{1.6}$$

noting that $\max(x_1 + x_2 I(x_4 = 4) - 4, 0)$ is just the excess power from $A_2$ which one tries to transfer to platform 2.

## 1.3 Basic definitions from binary theory

Before going into the specific restrictions that we find natural to claim on the structure function $\phi$, it is convenient first to recall some basic definitions from the traditional binary theory. This theory is nicely introduced in Barlow and Proschan (1975a).

**Definition 1.1:** A system is a binary monotone system (BMS) iff its structure function $\phi$ satisfies:

(i) $\phi(\boldsymbol{x})$ is nondecreasing in each argument

(ii) $\phi(\mathbf{0}) = 0$ and $\phi(\mathbf{1}) = 1 \quad \mathbf{0} = (0, \ldots, 0), \mathbf{1} = (1, \ldots, 1)$.

The first assumption roughly says that improving one of the components cannot harm the system, whereas the second says that if all components are in the failure state, then the system is in the failure state and, correspondingly, that if all components are in the functioning state, then the system is in the functioning state.

We now impose some further restrictions on the structure function $\phi$.

**Definition 1.2:** A binary coherent system (BCS) is a BMS where each component is relevant, i.e. the structure function $\phi$ satisfies: $\forall i \in \{1, \ldots, n\}$, $\exists(\cdot_i, \boldsymbol{x})$ such that $\phi(1_i, \boldsymbol{x}) = 1$, $\phi(0_i, \boldsymbol{x}) = 0$.

A component which is not relevant is said to be irrelevant. We note that an irrelevant component can never directly cause the failure of the system. As an example of such a component consider a condenser in parallel with an electrical device in a large engine. The task of the condenser is to cut off high voltages which might destroy the electrical device. Hence, although irrelevant, the condenser can be very important in increasing the lifetime of the device and hence the lifetime of the whole engine. The limitation of Definition 1.2 claiming relevance of the components, is inherited by the various definitions of a multistate coherent system considered in this book.

Let $C_0(\boldsymbol{x}) = \{i | x_i = 0\}$ and $C_1(\boldsymbol{x}) = \{i | x_i = 1\}$.

**Definition 1.3:** Let $\phi$ be the structure function of a BMS. A vector $\boldsymbol{x}$ is said to be a path vector iff $\phi(\boldsymbol{x}) = 1$. The corresponding path set is $C_1(\boldsymbol{x})$. A minimal path vector is a path vector $\boldsymbol{x}$ such that $\phi(\boldsymbol{y}) = 0$ for all $\boldsymbol{y} < \boldsymbol{x}$. The corresponding minimal path set is $C_1(\boldsymbol{x})$.

**Definition 1.4:** Let $\phi$ be the structure function of a BMS. A vector $\boldsymbol{x}$ is said to be a cut vector iff $\phi(\boldsymbol{x}) = 0$. The corresponding cut set is $C_0(\boldsymbol{x})$. A minimal cut vector is a cut vector $\boldsymbol{x}$ such that $\phi(\boldsymbol{y}) = 1$ for all $\boldsymbol{y} > \boldsymbol{x}$. The corresponding minimal cut set is $C_0(\boldsymbol{x})$.

We also need the following notation

$$\coprod_{i \in A} x_i = 1 - \prod_{i \in A}(1 - x_i) \quad x_1 \amalg x_2 = 1 - (1 - x_1)(1 - x_2).$$

We then have the following representations for the series and parallel systems respectively

$$\min_{1\le i\le n} x_i = \prod_{i=1}^{n} x_i \qquad \max_{1\le i\le n} x_i = \coprod_{i=1}^{n} x_i. \tag{1.7}$$

Consider a BCS with minimal path sets $P_1, \ldots, P_p$ and minimal cut sets $K_1, \ldots, K_k$. Since the system is functioning iff for at least one minimal path set all the components are functioning, or alternatively, iff for all minimal cut sets at least one component is functioning, we have the two following representations for the structure function

$$\phi(\boldsymbol{x}) = \coprod_{j=1}^{p} \prod_{i\in P_j} x_i = \max_{1\le j\le p} \min_{i\in P_j} x_i, \tag{1.8}$$

$$\phi(\boldsymbol{x}) = \prod_{j=1}^{k} \coprod_{i\in K_j} x_i = \min_{1\le j\le k} \max_{i\in K_j} x_i. \tag{1.9}$$

**Definition 1.5:** The monotone system $(A, \chi)$ is a module of the monotone system $(C, \phi)$ iff

$$\phi(\boldsymbol{x}) = \psi[\chi(\boldsymbol{x}^A), \boldsymbol{x}^{A^c}],$$

where $\psi$ is a monotone structure function and $A \subseteq C$.

Intuitively, a module is a monotone subsystem that acts as if it were just a supercomponent. Consider again the example where a condenser is in parallel with an electrical device in a large engine. The parallel system of the condenser and the electrical device is a module, which is relevant.

**Definition 1.6:** A modular decomposition of a monotone system $(C, \phi)$ is a set of disjoint modules $\{(A_k, \chi_k)\}_{k=1}^{r}$ together with an organizing monotone structure function $\psi$, i.e.

(i) $C = \cup_{i=1}^{r} A_i$ where $A_i \cap A_j = \emptyset \quad i \neq j$,

(ii) $\phi(\boldsymbol{x}) = \psi[\chi_1(\boldsymbol{x}^{A_1}), \ldots, \chi_r(\boldsymbol{x}^{A_r})] = \psi[\boldsymbol{\chi}(\boldsymbol{x})]$.

Making a modular decomposition of a system is a way of breaking it into a collection of subsystems which can be dealt with more easily.

**Definition 1.7:** Given a BMS structure function $\phi$, its dual structure function $\phi^D$ is given by

$$\phi^D(\boldsymbol{x}^D) = 1 - \phi(\boldsymbol{x}),$$

where $\boldsymbol{x}^D = (x_1^D, \ldots, x_n^D) = \mathbf{1} - \boldsymbol{x} = (1 - x_1, \ldots, 1 - x_n)$.

**Definition 1.8:** The random variables $T_1, \ldots, T_n$ are associated iff $\text{Cov}[\Gamma(\boldsymbol{T}), \Delta(\boldsymbol{T})] \geq 0$ for all pairs of nondecreasing binary functions $\Gamma, \Delta$. $\boldsymbol{T} = (T_1, \ldots, T_n)$.

We list some basic properties of associated random variables.

$P_1$ Any subset of a set of associated random variables is a set of associated random variables.

$P_2$ The set consisting of a single random variable is a set of associated random variables.

$P_3$ nondecreasing and nonincreasing functions of associated random variables are associated.

$P_4$ If two sets of associated random variables are independent of each other, then their union is a set of associated random variables.

$P_5$ Independent random variables are associated.

## 1.4 Early attempts to define multistate coherent systems

We now return to multistate reliability theory and begin by discussing the structure function considered by Barlow and Wu (1978).

**Definition 1.9:** Let $P_1, \ldots, P_p$ be nonempty subsets of $C = \{1, \ldots, n\}$ such that $\cup_{i=1}^n P_i = C$ and $P_j \not\subseteq P_i, i \neq j$. Then

$$\phi(\boldsymbol{x}) = \max_{1 \leq j \leq p} \min_{i \in P_j} x_i. \tag{1.10}$$

If the sets $\{P_1, \ldots, P_p\}$ are considered minimal path sets in a binary system, they uniquely determine a BCS $(C, \phi_0)$ where $\phi_0$ is defined by Equation (1.8). On the other hand, starting out with a BCS $\phi_0$, its minimal path sets $\{P_1, \ldots, P_p\}$ are uniquely determined. Hence, what Barlow and Wu (1978) essentially do when defining their structure function is just to extend the domain and range of Equation (1.8) from $\{0, 1\}$ to $\{0, 1, \ldots, M\}$. It is hence a one-to-one correspondence between the binary structure function $\phi_0$ and the multistate structure function $\phi$. Furthermore, if $\{K_1, \ldots, K_k\}$ are the minimal cut sets of $(C, \phi_0)$ it follows from Theorem 3.5, page 12 of Barlow and Proschan (1975a) that for $\phi(\boldsymbol{x})$ of Equation (1.10) we have

$$\phi(\boldsymbol{x}) = \min_{1 \leq j \leq k} \max_{i \in K_j} x_i. \tag{1.11}$$

Setting $p = 1$ in Equation (1.10) and $k = 1$ in Equation (1.11), noting that $P_1 = K_1 = C$, we respectively get what are naturally called the multistate series and parallel systems.

El-Neweihi *et al.* (1978) suggest the following definition of a multistate coherent system.

**Definition 1.10:** Let $S_i = S, i = 1, \ldots, n$. A system is a multistate coherent system iff its structure function $\phi$ satisfies:

(i) $\phi(\boldsymbol{x})$ is nondecreasing in each argument

(ii) $\forall i \in \{1, \ldots, n\}, \forall j \in \{0, 1, \ldots, M\}, \exists(\cdot_i, \boldsymbol{x})$ such that, $\phi(j_i, \boldsymbol{x}) = j$ and $\phi(\ell_i, \boldsymbol{x}) \neq j \quad \ell \neq j$

(iii) $\forall j \in \{0, 1, \ldots, M\} \quad \phi(\boldsymbol{j}) = j \quad \boldsymbol{j} = (j, \ldots, j)$.

It is easy to see that the structure function of Definition 1.9 is just a special case of the one in Definition 1.10. Furthermore, note that (i) and (ii) of Definition 1.10 are generalizations of the claims of Definition 1.2. In the binary case, (iii) of Definition 1.10 is implied by the corresponding (i) and (ii). This is not true in the multistate case.

## 1.5 Exercises

1.1 Verify Equation (1.1).

1.2 Verify Equation (1.2).

1.3 Verify Equations (1.3) and (1.4).

1.4 Verify Equations (1.5) and (1.6).

1.5 Prove Property $P_5$ of associated random variables by applying Properties $P_2$ and $P_4$.

1.6 Show that the structure function of Definition 1.9 is just a special case of the one in Definition 1.10.

# 2

# Basics

## 2.1 Multistate monotone and coherent systems

For $j \in \{1, \dots, M\}$ define the indicators

$$I_j(x_i) = \begin{cases} 1 & \text{if } x_i \geq j \\ 0 & \text{if } x_i < j \end{cases}, \tag{2.1}$$

and the indicator vector

$$\boldsymbol{I}_j(\boldsymbol{x}) = (I_j(x_1), \dots, I_j(x_n)). \tag{2.2}$$

We then have the following representations for the multistate series and parallel systems generalizing Equation (1.7)

$$\min_{1 \leq i \leq n} x_i = \sum_{j=1}^{M} \prod_{i=1}^{n} I_j(x_i) \quad \max_{1 \leq i \leq n} x_i = \sum_{j=1}^{M} \coprod_{i=1}^{n} I_j(x_i). \tag{2.3}$$

We now give the obvious generalization, suggested in Block and Savits (1982) for $S_i = S$, $i = 1, \dots, n$, of a BMS, given in Definition 1.1, to the multistate case for which a series of results can be derived.

**Definition 2.1:** A system is a multistate monotone system (MMS) iff its structure function $\phi$ satisfies:

(i) $\phi(\boldsymbol{x})$ is nondecreasing in each argument

(ii) $\phi(\boldsymbol{0}) = 0$ and $\phi(\boldsymbol{M}) = M \quad \boldsymbol{0} = (0, \dots, 0)$, $\boldsymbol{M} = (M, \dots, M)$.

*Multistate Systems Reliability Theory with Applications* Bent Natvig

The multistate series and parallel systems are obviously very special cases of an MMS. The same is true for the $k$-out-of-$n$ system having structure function $\phi(\boldsymbol{x}) = x_{(n-k+1)}$, where $x_{(1)} \leq x_{(2)} \leq \cdots \leq x_{(n)}$ is an increasing rearrangement of $x_1, \ldots, x_n$. Note that the $n$-out-of-$n$ system is the series system and that the 1-out-of-$n$ system is the parallel system. Also the structure functions given by Equations (1.1)–(1.6) are examples of an MMS. The first assumption roughly says, as in the binary case, that improving one of the components cannot harm the system, whereas the second says that if all components are in the complete failure state, then the system is in the complete failure state and, correspondingly, that if all components are in the perfect functioning state, then the system is in the perfect functioning state. Ross (1979) showed that it is possible to obtain interesting results by just imposing the first assumption on $\phi$.

The obvious generalization of the binary dual structure function $\phi^D$ given by Definition 1.7 is as follows

**Definition 2.2:** Given an MMS structure function $\phi$, its dual structure function $\phi^D$ is given by

$$\phi^D(\boldsymbol{x}^D) = M - \phi(\boldsymbol{x}),$$

where $\boldsymbol{x}^D = (x_1^D, \ldots, x_n^D) = \boldsymbol{M} - \boldsymbol{x} = (M - x_1, \ldots, M - x_n)$.

It can easily be checked that the dual system of an MMS is an MMS. Furthermore, it should be remarked that Definitions 1.5 and 1.6 defining, respectively, a module and a modular decomposition for a BMS are valid also for an MMS. For a module $(A, \chi)$, if $S_\chi$ denotes the set of states of $\chi$, we assume $S_i \subseteq S_\chi \subseteq S$ for $i \in A$. As for a BMS, if an MMS $(C, \phi)$ has a modular decomposition $\{(A_k, \chi_k)\}_{k=1}^r$, $\phi^D$ has a modular decomposition $\{(A_k, \chi_k^D)\}_{k=1}^r$ with organizing structure function $\psi^D$. In the simple network system of Figure 1.1 the two components can be considered as modules and the four branches as components. In the offshore electrical power generation system of Figure 1.2 we have the following module of $\phi_2$

$$\chi(x_1, x_2, x_5) = x_2 I(x_1 = 4) x_5/4, \tag{2.4}$$

which is the amount of power that can be transferred from $A_2$ to platform 2 if the control unit $U$ works perfectly. The organizing structure function is given by

$$\psi(x_3, x_4, \chi) = I(x_4 > 0) \min(x_3 + I(x_4 = 4)\chi, 4). \tag{2.5}$$

We now impose some further restrictions on the structure function $\phi$ of an MMS starting out from Definition 1.10 of a multistate coherent system given by El-Neweihi *et al.* (1978). We begin by questioning whether (iii) of Definition 1.10 must come into play when defining a multistate coherent system. The answer seems to be yes, because of the following theorem, the main part of which is given as Theorem 3.1 in El-Neweihi *et al.* (1978). This result is also noted by Griffith (1980).

**Theorem 2.3:** Let $\phi(\boldsymbol{x})$ be a multistate structure function which is nondecreasing in each argument and let $S_i = S, i = 1, \ldots, n$. Then (iii) of Definition 1.10 is equivalent to

$$\min_{1 \le i \le n} x_i \le \phi(\boldsymbol{x}) \le \max_{1 \le i \le n} x_i. \tag{2.6}$$

**Proof:** Assume (iii) of Definition 1.10 holds and let $m = \max_{1 \le i \le n} x_i$. Then $\phi(\boldsymbol{x}) \le \phi(\boldsymbol{m}) = m$, since $\phi$ is nondecreasing in each argument. Hence, the upper bound of Equation (2.6) is proved. The proof of the lower bound is completely similar. Now assume on the contrary that Equation (2.6) holds. By choosing $\boldsymbol{x} = \boldsymbol{j}$, (iii) of Definition 1.10 follows immediately.

Hence, when $\phi(\boldsymbol{x})$ is nondecreasing in each argument and $S_i = S$, $i = 1, \ldots, n$, (iii) of Definition 1.10 holds iff $\phi(\boldsymbol{x})$ is bounded below by the multistate series system and above by the multistate parallel system. Now choose $j \in \{1, \ldots, M\}$ and let the states $\{0, \ldots, j-1\}$ correspond to the failure state and $\{j, \ldots, M\}$ to the functioning state if a binary approach had been applied. Equation (2.6) says that for any $j$, i.e. for any way of distinguishing between the binary failure and functioning state, if all components are in the binary failure state, the system itself is in the binary failure state and correspondingly if all components are in the binary functioning state, the system itself is in the binary functioning state. This is consistent with claiming that (iii) of Definition 1.10 holds in the binary case.

Following the binary approach above it seems natural, for any way of distinguishing between the binary failure and functioning state, to claim each component to be relevant. More precisely, for any $j$ and any component $i$ there should exist a vector $(\cdot_i, \boldsymbol{x})$ such that if the $i$th component is in the binary failure state, the system itself is in the binary failure state and, correspondingly, if the $i$th component is in the binary functioning state, the system itself is in the binary functioning state. This motivates the following definition of a multistate strongly coherent system, which for the case $S_i = S$, $i = 1, \ldots, n$ is introduced as a multistate coherent system of type 1 in Natvig (1982a). For this case the following definitions of a multistate coherent system and a multistate weakly coherent system are presented in Griffith (1980).

To deal with the case where $S_i = S, i = 1, \ldots, n$ is not true, the following notation is needed

$$S_{i,j}^0 = S_i \cap \{0, \ldots, j-1\} \quad \text{and} \quad S_{i,j}^1 = S_i \cap \{j, \ldots, M\}. \qquad (2.7)$$

**Definition 2.4:** Consider an MMS with structure function $\phi$ satisfying

(i) $\min_{1 \le i \le n} x_i \le \phi(\boldsymbol{x}) \le \max_{1 \le i \le n} x_i$.

If in addition $\forall i \in \{1, \ldots, n\}$, $\forall j \in \{1, \ldots, M\}$, $\exists\ (\cdot_i, \boldsymbol{x})$ such that

(ii) $\phi(k_i, \boldsymbol{x}) \ge j$, $\phi(\ell_i, \boldsymbol{x}) < j$, $\forall k \in S_{i,j}^1$, $\forall \ell \in S_{i,j}^0$, we have a multistate strongly coherent system (MSCS),

(iii) $\phi(k_i, \boldsymbol{x}) > \phi(\ell_i, \boldsymbol{x})$ $\forall k \in S_{i,j}^1$, $\forall \ell \in S_{i,j}^0$, we have a multistate coherent system (MCS),

(iv) $\phi(M_i, \boldsymbol{x}) > \phi(0_i, \boldsymbol{x})$, we have a multistate weakly coherent system (MWCS).

Note that

$$\forall j \in \{1, \ldots, M\}, \exists\ (\cdot_i, \boldsymbol{x}) \text{ such that (iii) holds} \Leftrightarrow$$
$$\exists\ (\cdot_i, \boldsymbol{x}) \text{ such that } \phi(k_i, \boldsymbol{x}) > \phi(\ell_i, \boldsymbol{x})\ \forall k, \ell \in S_i \text{ such that } k > \ell.$$

When $M = 1$, all these systems reduce to the established binary coherent system (BCS) given by Definition 1.2. All are generalizations of the multistate coherent system suggested by El-Neweihi *et al.* (1978) given in Definition 1.10. Consider, for instance, the MMS of two components having structure function tabulated in Table 2.1. This is obviously an MSCS but not of the latter type. On the other hand the MMS of Table 2.2 is neither an MSCS nor an MCS, but an MWCS.

Loosely speaking, modifying Block and Savits (1982), condition (ii) says that every level of each component is relevant to the same level

Table 2.1 An MSCS, but not an MCS suggested by El-Neweihi *et al.* (1978).

| | | | | |
|---|---|---|---|---|
| Component 2 | 2 | 1 | 1 | 2 |
| | 1 | 1 | 1 | 1 |
| | 0 | 0 | 1 | 1 |
| | | 0 | 1 | 2 |
| | | Component 1 | | |

Table 2.2 An MWCS, but neither an MSCS nor an MCS.

| | | | | |
|---|---|---|---|---|
| Component 2 | 2 | 1 | 1 | 2 |
| | 1 | 0 | 1 | 1 |
| | 0 | 0 | 1 | 1 |
| | | 0 | 1 | 2 |
| | | Component 1 | | |

of the system, condition (iii) says that every level of each component is relevant to the system, whereas condition (iv) simply says that every component is relevant to the system. If we in (ii) had just claimed that at least one level of each component was relevant to the same level of the system, the MMS of Table 2.2 would have been an MSCS.

For a BCS one can prove the following practically very useful principle: redundancy at the component level is superior to redundancy at the system level except for a parallel system where it makes no difference. Assuming $S_i = S$ $(i = 1, \ldots, n)$ this is also true for an MCS, but not for an MWCS, as shown by Griffith (1980). To prove the full theorem we need the following notation

$$x \vee y = \max(x, y)$$

$$\boldsymbol{x} \vee \boldsymbol{y} = (x_1 \vee y_1, \ldots, x_n \vee y_n)$$

$$x \wedge y = \min(x, y)$$

$$\boldsymbol{x} \wedge \boldsymbol{y} = (x_1 \wedge y_1, \ldots, x_n \wedge y_n).$$

**Theorem 2.5:** Let $\phi$ be the structure function of an MCS and assume $S_i = S$, $i = 1, \ldots, n$. Then for all $\boldsymbol{x}$ and $\boldsymbol{y}$

(i) $\phi(\boldsymbol{x} \vee \boldsymbol{y}) \geq \phi(\boldsymbol{x}) \vee \phi(\boldsymbol{y})$

(ii) $\phi(\boldsymbol{x} \wedge \boldsymbol{y}) \leq \phi(\boldsymbol{x}) \wedge \phi(\boldsymbol{y})$.

Equality holds for all $\boldsymbol{x}$ and $\boldsymbol{y}$ in (i) iff the structure function is parallel, i.e. $\phi(\boldsymbol{x}) = \max_{1 \leq i \leq n} x_i$, and in (ii) iff the structure function is series, i.e. $\phi(\boldsymbol{x}) = \min_{1 \leq i \leq n} x_i$.

**Proof:** Since $\phi(\boldsymbol{x})$ is nondecreasing in each argument $\phi(\boldsymbol{x} \vee \boldsymbol{y}) \geq \phi(\boldsymbol{x})$ and $\phi(\boldsymbol{x} \vee \boldsymbol{y}) \geq \phi(\boldsymbol{y})$. Hence, (i) follows. (ii) is proved similarly. Now assume that the structure function is parallel. Then both sides of (i) equal $\max_{1 \leq i \leq n} \max(x_i, y_i)$ and one of the implications is proved. Assume, on the other hand, that $\phi(\boldsymbol{x} \vee \boldsymbol{y}) = \phi(\boldsymbol{x}) \vee \phi(\boldsymbol{y})$ for all $\boldsymbol{x}$ and $\boldsymbol{y}$. From (iii) of Definition 2.4 $\forall i \in \{1, \dots, n\}$, $\forall j \in \{1, \dots, M\}$, $\exists$ $(\cdot_i, \boldsymbol{x})$ such that $\phi(j_i, \boldsymbol{x}) > \phi((j-1)_i, \boldsymbol{x})$. Since

$$(j_i, \boldsymbol{x}) = (j_i, \boldsymbol{0}) \vee (0_i, \boldsymbol{x}) \quad ((j-1)_i, \boldsymbol{x}) = ((j-1)_i, \boldsymbol{0}) \vee (0_i, \boldsymbol{x}),$$

it follows from our assumption that

$$\phi(j_i, \boldsymbol{x}) = \phi(j_i, \boldsymbol{0}) \vee \phi(0_i, \boldsymbol{x})$$
$$\phi((j-1)_i, \boldsymbol{x}) = \phi((j-1)_i, \boldsymbol{0}) \vee \phi(0_i, \boldsymbol{x}).$$

Hence, $\phi(j_i, \boldsymbol{0}) > \phi((j-1)_i, \boldsymbol{0})$. Since, from Definition 2.1, $\phi(\boldsymbol{0}) = 0$, this implies that $\phi(j_i, \boldsymbol{0}) = j$ for $j = 0, \dots, M$. Now finally

$$\phi(\boldsymbol{x}) = \phi(x_1, 0, \dots, 0) \vee \phi(0, x_2, 0, \dots, 0) \vee \cdots \vee \phi(0, \dots, x_n)$$
$$= x_1 \vee x_2 \vee \cdots \vee x_n = \max_{1 \le i \le n} x_i,$$

and the reverse implication is proved. Hence, the equivalence involving the parallel system is established. The equivalence linked to the series system is proved similarly.

## 2.2 Binary type multistate systems

We now discuss a special type of an MMS remembering Equations (2.1) and (2.2).

**Definition 2.6:** An MMS is said to be a binary type multistate monotone system (BTMMS) iff there exist binary monotone structures $\phi_j$, $j = 1, \dots, M$ such that its structure function $\phi$ satisfies $\phi(\boldsymbol{x}) \ge j \Leftrightarrow \phi_j(\boldsymbol{I}_j(\boldsymbol{x})) = 1$ for all $j \in \{1, \dots, M\}$ and all $\boldsymbol{x}$. If the binary monotone structure functions are coherent, we have a binary type multistate strongly coherent system (BTMSCS).

The BTMSCS was introduced as a multistate coherent system of type 2 in Natvig (1982a). Choose again $j \in \{1, \dots, M\}$ and, remembering Equation (2.7), let the states $S_{i,j}^0$ correspond to the failure state and $S_{i,j}^1$ to the functioning state for the $i$th component if a binary approach is applied. By the definition above, $\phi_j$ will, from the binary states of the components, uniquely determine the corresponding binary state of the system.

The MMS represented by the simple network in Figure 1.1, with structure function tabulated in Table 1.1, is an MSCS, but not a BTMMS. Now consider the MMS with structure function tabulated in Table 2.3.

Table 2.3 A BTMMS, but not an MWCS.

| Component 2 \ Component 1 | 0 | 1 | 2 |
|---|---|---|---|
| 2 | 2 | 2 | 2 |
| 1 | 1 | 1 | 1 |
| 0 | 0 | 0 | 0 |

This is given by $\phi(x_1, x_2) = x_2$, which is nondecreasing in each argument. Since $\phi(0,0) = 0$ and $\phi(2,2) = 2$, it is an MMS. Furthermore, since $\phi_j(z_1, z_2) = z_2,\ j = 1, 2$ it is a BTMMS. However, since $\phi(2_1, x_2) = x_2 = \phi(0_1, x_2)$, it is not an MWCS.

The binary structures $\phi_j,\ j = 1, \ldots, M$ of a BTMMS cannot be chosen arbitrarily, as is demonstrated in the following theorem.

**Theorem 2.7:** The binary structures $\phi_j,\ j = 1, \ldots, M$ of a BTMMS satisfy

$$\phi_j(\boldsymbol{z}) \geq \phi_{j+1}(\boldsymbol{z})$$

for all $j \in \{1, \ldots, M-1\}$ and all binary $\boldsymbol{z}$.

**Proof:** Let $\boldsymbol{z}$ be a binary vector such that $\phi_{j+1}(\boldsymbol{z}) = 1$. Define the vector $\boldsymbol{x}$ by

$$x_i = \begin{cases} j+1 & \text{if } z_i = 1 \\ 0 & \text{if } z_i = 0 \end{cases}.$$

Obviously $\boldsymbol{I}_j(\boldsymbol{x}) = \boldsymbol{I}_{j+1}(\boldsymbol{x}) = \boldsymbol{z}$. Since then $\phi_{j+1}(\boldsymbol{I}_{j+1}(\boldsymbol{x})) = 1$, $\phi(\boldsymbol{x}) \geq j+1 > j$ and consequently $\phi_j(\boldsymbol{z}) = \phi_j(\boldsymbol{I}_j(\boldsymbol{x})) = 1$.

The following theorem, which follows immediately from Definition 2.6, gives a unique correspondence between the structure function $\phi$ and the binary structures $\phi_j,\ j = 1, \ldots, M$ of a BTMMS.

**Theorem 2.8:** For a BTMMS we have the following unique correspondence between the structure function $\phi$ and the binary structures $\phi_j$, $j = 1, \ldots, M$

$$\phi(\boldsymbol{x}) = 0 \Leftrightarrow \phi_1(\boldsymbol{I}_1(\boldsymbol{x})) = 0 \tag{2.8}$$

$$\phi(\boldsymbol{x}) = j \Leftrightarrow \phi_j(\boldsymbol{I}_j(\boldsymbol{x})) - \phi_{j+1}(\boldsymbol{I}_{j+1}(\boldsymbol{x})) = 1,\ j \in \{1, \ldots, M-1\} \tag{2.9}$$

$$\phi(\boldsymbol{x}) = M \Leftrightarrow \phi_M(\boldsymbol{I}_M(\boldsymbol{x})) = 1. \tag{2.10}$$

Note especially that by applying Equations (2.8) and (2.9) we obtain

$$\exists\, j \in \{0, \ldots, M-1\} \text{ such that } \phi(\boldsymbol{x}) = j \Leftrightarrow$$
$$\exists\, j \in \{1, \ldots, M\} \text{ such that } \phi_j(I_j(\boldsymbol{x})) = 0.$$

Since, by applying Theorem 2.7

$$\phi_j(I_j(\boldsymbol{x})) \geq \phi_j(I_M(\boldsymbol{x})) \geq \phi_M(I_M(\boldsymbol{x})),$$

it follows that

$$\exists\, j \in \{0, \ldots, M-1\} \text{ such that } \phi(\boldsymbol{x}) = j \Leftrightarrow \phi_M(I_M(\boldsymbol{x})) = 0,$$

which is equivalent to Equation (2.10). Hence, the latter relation represents nothing new. Starting out with $\phi_1, \ldots, \phi_M, \phi$ is uniquely determined by Equations (2.8) and (2.9). On the other hand, starting out with $\phi, \phi_1$ is uniquely determined by Equation (2.8). Then $\phi_2, \ldots, \phi_M$ is uniquely determined by applying Equation (2.9) for $j = 1, \ldots, M-1$.

In the following theorem we show that a BTMMS actually is a special type of an MMS, that a BTMSCS is a special type of an MSCS and the new result that an MSCS that is a BTMMS, is a BTMSCS.

**Theorem 2.9:** A BTMMS is also an MMS and a BTMSCS also an MSCS. An MSCS that is a BTMMS, is a BTMSCS.

**Proof:** First we will show that a BTMMS is an MMS as defined by Definition 2.1. The claim (i) of this definition is seen to be satisfied since $\phi_j(I_j(\boldsymbol{x}))$ is nondecreasing in each argument for $j \in \{1, \ldots, M\}$ having used Equations (2.1) and (2.2), Definition 2.6 and (i) of Definition 1.1. The claim (ii) of Definition 2.1 follows from Equations (2.8) and (2.10) since $\phi_1(\mathbf{0}) = 0$ and $\phi_M(\mathbf{1}) = 1$ by Definition 2.6 and (ii) of Definition 1.1. This completes the proof of the first part of the theorem. (i) of Definition 2.4 is, for a BTMMS, equivalent to

$$\min_{1 \leq i \leq n} I_j(x_i) \leq \phi_j(I_j(\boldsymbol{x})) \leq \max_{1 \leq i \leq n} I_j(x_i)$$

for all $j \in \{1, \ldots, M\}$. This again follows by Definition 2.6 since Equation (2.6) holds for a BMS. What remains to prove the second part of the theorem is (ii) of Definition 2.4. This is equivalent to showing $\forall i \in \{1, \ldots, n\}$, $\forall j \in \{1, \ldots, M\}$, $\exists\, (\cdot_i, \boldsymbol{x})$ such that

$$\phi_j(1_i, I_j(\boldsymbol{x})) = 1 \text{ and } \phi_j(0_i, I_j(\boldsymbol{x})) = 0.$$

By applying Definition 1.2 on $\phi_j$ for $j = 1, \ldots, M$, the above statement is seen to be true. To prove the last part of the theorem, assume that (ii) of Definition 2.4 holds. Since the system is assumed to be a BTMMS, the equivalent representation above is valid. Again by applying Definition 1.2 on $\phi_j$ for $j = 1, \ldots, M$, it follows from Definition 2.6 that the system is a BTMSCS.

As demonstrated in the previous section, the MMS of two components having the structure function tabulated in Table 2.1 is not of the type suggested by El-Neweihi *et al.* (1978). It is, however, a BTMSCS with $\phi_1$ and $\phi_2$ being, respectively, a parallel and series structure. The structure function tabulated in Table 2.4, however, belongs to an MCS suggested by El-Neweihi *et al.* (1978), but not to a BTMMS.

We now show that the BTMSCS is a generalization of the multistate system suggested by Barlow and Wu (1978).

**Theorem 2.10:** A BTMSCS where all the binary coherent structure functions $\phi_j$ are identical, reduces to the one suggested by Barlow and Wu (1978) in Definition 1.9.

**Proof:** Denote the common binary coherent structure function by $\phi_0$ and the corresponding minimal path sets by $P_1, \ldots, P_p$. Then we have, from Definition 2.6 and Equation (1.8), for all $j \in \{1, \ldots, M\}$ and all $\boldsymbol{x}$

$$\phi(\boldsymbol{x}) \geq j \Leftrightarrow \phi_0(\boldsymbol{I}_j(\boldsymbol{x})) = 1 \Leftrightarrow \max_{1 \leq j \leq p} \min_{i \in P_j} I_j(x_i) = 1$$

$$\Leftrightarrow \max_{1 \leq j \leq p} \min_{i \in P_j} x_i \geq j.$$

Hence, Equation (1.10) is satisfied and the proof is completed.

Pooling all relevant results we can, as in Figure 2.1, illustrate the relationships between the different multistate systems presented in this book.

Table 2.4 An MCS suggested by El-Neweihi *et al.* (1978), but not a BTMMS.

| | | | | |
|---|---|---|---|---|
| Component 2 | 2 | 1 | 2 | 2 |
| | 1 | 0 | 1 | 2 |
| | 0 | 0 | 0 | 1 |
| | | 0 | 1 | 2 |
| | | Component 1 | | |

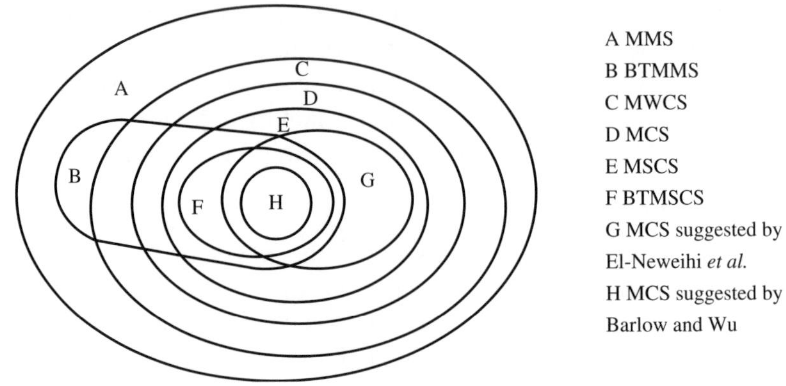


*Figure 2.1 Relationships between the different multistate systems.*

We conclude this section by demonstrating that the structure function of a BTMSCS is far more general than the one suggested by Barlow and Wu (1978) in Definition 1.9.

**Theorem 2.11:** For two-component systems there are $M+1$ different BTMSCSs, whereas there are just two of the type suggested by Barlow and Wu (1978). The corresponding numbers for three-component systems are

$$9+(M-1)30+\binom{M-1}{2}46+\binom{M-1}{3}33+\binom{M-1}{4}9$$

and 9.

**Proof:** First consider the two-component systems. The only possible binary coherent structures are

$$\psi_1(x_1,x_2)=x_1 \amalg x_2 \qquad \psi_2(x_1,x_2)=x_1x_2,$$

i.e. the parallel and series structure. Since

$$\psi_1(\boldsymbol{x}) \geq \psi_2(\boldsymbol{x}) \forall \boldsymbol{x} \text{ and } \exists\ \boldsymbol{x} \text{ such that } \psi_1(\boldsymbol{x}) > \psi_2(\boldsymbol{x}),$$

which we abbreviate to $\psi_1 > \psi_2$, it is, according to Theorem 2.7, impossible to choose the binary coherent structures $\phi_1, \ldots, \phi_M$ of a BTMSCS such that $\phi_k = \psi_2$ and $\phi_{k+1} = \psi_1, k = 1, \ldots, M-1$. From Theorem 2.8 for a BTMSCS there is a unique correspondence between the structure function $\phi$ and the binary coherent structures $\phi_j,\ j = 1, \ldots, M$. Hence, applying Theorem 2.10, the results for the two-component systems follow immediately.

Now consider the three-component systems. From Barlow and Proschan (1975a), page 7, the only possible binary coherent structures are

$$\psi_1(\boldsymbol{x}) = x_1 \amalg x_2 \amalg x_3 \quad \psi_2(\boldsymbol{x}) = x_1 \amalg (x_2x_3) \quad \psi_3(\boldsymbol{x}) = x_2 \amalg (x_1x_3)$$

$$\psi_4(\boldsymbol{x}) = x_3 \amalg (x_1x_2) \quad \psi_5(\boldsymbol{x}) = (x_1x_2) \amalg (x_1x_3) \amalg (x_2x_3)$$

$$\psi_6(\boldsymbol{x}) = x_1(x_2 \amalg x_3) \quad \psi_7(\boldsymbol{x}) = x_2(x_1 \amalg x_3) \quad \psi_8(\boldsymbol{x}) = x_3(x_1 \amalg x_2)$$

$$\psi_9(\boldsymbol{x}) = x_1x_2x_3.$$

From Theorems 2.8 and 2.10 there are just nine different three-component systems of the type suggested by Barlow and Wu (1978).

The structure functions above can be ordered in the following way

$$\psi_1 > \begin{pmatrix} \psi_2 \\ \psi_3 \\ \psi_4 \end{pmatrix} > \psi_5 > \begin{pmatrix} \psi_6 \\ \psi_7 \\ \psi_8 \end{pmatrix} > \psi_9.$$

Among $\psi_2, \psi_3, \psi_4$ and among $\psi_6, \psi_7, \psi_8$ there is no ordering. This ordering divides the structure functions into five natural groups. Now, for $i = 1, \ldots, 5$ let

$a_i =$ the number of ways to choose $i$ structure functions coming from different groups

$b_i =$ the number of ways to choose $M$ elements from $i$ groups such that all groups are represented

By Theorems 2.7 and 2.8 the number of different three-component BTM-SCSs equals $\sum_{i=1}^{5} a_i b_i$. It is not hard to see that

$$a_1 = \binom{9}{1} = 9$$

$$a_2 = \binom{3}{2} + \binom{2}{1}\binom{3}{1}\binom{3}{1} + \binom{3}{1}\binom{3}{1} = 30$$

$$a_3 = \binom{3}{3} + \binom{3}{2}\binom{2}{1}\binom{3}{1} + \binom{3}{1}\binom{3}{1}\binom{3}{1} = 46$$

$$a_4 = \binom{2}{1}\binom{3}{1} + \binom{3}{2}\binom{3}{1}\binom{3}{1} = 33$$

$$a_5 = \binom{3}{1}\binom{3}{1} = 9.$$

If we can show for $i = 1, 2, 3, 4, 5$ that

$$b_i = \binom{M-1}{i-1},$$

our proof is completed. This, however, follows from Feller (1968), page 38.

By applying Theorem 2.11 and choosing $M = 7$ we get, in addition to the nine different three-component systems of the type suggested by Barlow and Wu (1978), 1665 BTMSCSs.

## 2.3 Multistate minimal path and cut vectors

In this section we start by generalizing on the one hand the notions of path vector, path set, minimal path vector and minimal path set and on the other hand the notions of cut vector, cut set, minimal cut vector and minimal cut set given respectively by Definitions 1.3 and 1.4 for a BMS. $\boldsymbol{y} < \boldsymbol{x}$ still means that $y_i \leq x_i$ for $i = 1, \ldots, n$, and $y_i < x_i$ for some $i$.

**Definition 2.12:** Let $\phi$ be the structure function of an MMS and let $j \in \{1, \ldots, M\}$. A vector $\boldsymbol{x}$ is said to be a path vector to level $j$ iff $\phi(\boldsymbol{x}) \geq j$. The corresponding path set and binary type path set are respectively given by

$$C_\phi^j(\boldsymbol{x}) = \{i | x_i \geq 1\} \text{ and } C_{\phi BT}^j(\boldsymbol{x}) = \{i | x_i \geq j\}.$$

A minimal path vector to level $j$ is a path vector $\boldsymbol{x}$ such that $\phi(\boldsymbol{y}) < j$ for all $\boldsymbol{y} < \boldsymbol{x}$. The corresponding path set and binary type path set are also said to be minimal.

**Definition 2.13:** Let $\phi$ be the structure function of an MMS and let $j \in \{1, \ldots, M\}$. A vector $\boldsymbol{x}$ is said to be a cut vector to level $j$ iff $\phi(\boldsymbol{x}) < j$. The corresponding cut set and binary type cut set are respectively given by

$$D_\phi^j(\boldsymbol{x}) = \{i | x_i < M\} \text{ and } D_{\phi BT}^j(\boldsymbol{x}) = \{i | x_i < j\}.$$

A minimal cut vector to level $j$ is a cut vector $\boldsymbol{x}$ such that $\phi(\boldsymbol{y}) \geq j$ for all $\boldsymbol{y} > \boldsymbol{x}$. The corresponding cut set and binary type cut set are also said to be minimal.

Note that the binary type path and cut sets to level $j$ are, for a BTMMS, identical to the path and cut sets of $\phi_j$, $j = 1, \ldots, M$ given by Definitions 1.3 and 1.4. Note also that for $j \in \{1, \ldots, M\}$ the existence of a minimal path and cut set and a binary type minimal path and cut set to level $j$ is guaranteed by an MMS. Note finally that only components with states $\geq 1$ are of any relevance to the path sets and with states $< M$ of any relevance to the cut sets.

We now illustrate the Definitions 2.12 and 2.13 by finding the minimal path and cut vectors to all levels first for the simple network system given in Figure 1.1 and then for the offshore electrical power generation system presented in Section 1.2. First of all, from Table 1.1 it is easy to come up with the minimal path vectors and minimal cut vectors given respectively in Tables 2.5 and 2.6. Note that the same vector may be a minimal cut vector to more than one level. This is not true for the minimal path vectors.

For the offshore electrical power generation system presented in Section 1.2 we will, for simplicity, concentrate on the structure functions $\phi_1$ and $\phi_2$ given respectively by Equations (1.3) and (1.4). From these equations we come up with the minimal path vectors and minimal cut vectors for $\phi_1$ respectively in Tables 2.7 and 2.8 and for $\phi_2$ respectively in Tables 2.9 and 2.10. The minimal path vectors and minimal cut vectors for the module of $\phi_2$ given by Equation (2.4) are given respectively in Tables 2.11 and 2.12, whereas the minimal path vectors and minimal cut vectors for the organizing structure given by Equation (2.5) are given respectively in Tables 2.13 and 2.14. Note that the same vector may be a minimal path vector to more than one level. The same is true for a minimal cut vector.

Table 2.5 Minimal path vectors for the simple network system of Figure 1.1.

| Level | Component 1 | Component 2 |
|---|---|---|
| 1 | 1 | 1 |
| 2 | 1 | 3 |
| 2 | 3 | 1 |
| 3 | 3 | 3 |

Table 2.6 Minimal cut vectors for the simple network system of Figure 1.1.

| Level | Component 1 | Component 2 |
|---|---|---|
| 1, 2 | 0 | 3 |
| 1, 2 | 3 | 0 |
| 2 | 1 | 1 |
| 3 | 1 | 3 |
| 3 | 3 | 1 |

Table 2.7 Minimal path vectors of $\phi_1$.

| Level | $A_1$ | $A_2$ | $U$ |
|---|---|---|---|
| 2 | 0 | 2 | 4 |
| 2 | 2 | 0 | 2 |
| 4 | 0 | 4 | 4 |
| 4 | 2 | 2 | 4 |
| 4 | 4 | 0 | 2 |

Table 2.8 Minimal cut vectors of $\phi_1$.

| Level | $A_1$ | $A_2$ | $U$ |
|---|---|---|---|
| 2 | 0 | 0 | 4 |
| 2 | 0 | 4 | 2 |
| 2, 4 | 4 | 4 | 0 |
| 4 | 0 | 2 | 4 |
| 4 | 2 | 0 | 4 |
| 4 | 2 | 4 | 2 |

Table 2.9 Minimal path vectors of $\phi_2$.

| Level | $A_1$ | $A_2$ | $A_3$ | $U$ | $L$ |
|---|---|---|---|---|---|
| 1 | 4 | 2 | 0 | 4 | 2 |
| 1, 2 | 0 | 0 | 2 | 2 | 0 |
| 2 | 4 | 2 | 0 | 4 | 4 |
| 2 | 4 | 4 | 0 | 4 | 2 |
| 3 | 4 | 2 | 2 | 4 | 2 |
| 3, 4 | 0 | 0 | 4 | 2 | 0 |
| 3, 4 | 4 | 4 | 0 | 4 | 4 |
| 4 | 4 | 2 | 2 | 4 | 4 |
| 4 | 4 | 4 | 2 | 4 | 2 |

As examples of how to arrive at these tables note that

$$\phi_2(4, 4, 0, 4, 4) = 4,$$

whereas

$$\phi_2(4, 4, 0, 2, 4) = \phi_2(2, 4, 0, 4, 4) = 0$$

$$\phi_2(4, 4, 0, 4, 2) = \phi_2(4, 2, 0, 4, 4) = 2.$$

Hence, (4, 4, 0, 4, 4) is a minimal path vector both to level 3 and 4. Similarly,

$$\phi_2(4, 2, 0, 4, 4) = 2,$$

Table 2.10 Minimal cut vectors of $\phi_2$.

| Level | $A_1$ | $A_2$ | $A_3$ | $U$ | $L$ |
|---|---|---|---|---|---|
| 1, 2 | 2 | 4 | 0 | 4 | 4 |
| 1, 2 | 4 | 0 | 0 | 4 | 4 |
| 1, 2 | 4 | 4 | 0 | 2 | 4 |
| 1, 2 | 4 | 4 | 0 | 4 | 0 |
| 1, 2, 3, 4 | 4 | 4 | 4 | 0 | 4 |
| 2 | 4 | 2 | 0 | 4 | 2 |
| 3, 4 | 2 | 4 | 2 | 4 | 4 |
| 3, 4 | 4 | 0 | 2 | 4 | 4 |
| 3, 4 | 4 | 2 | 0 | 4 | 4 |
| 3, 4 | 4 | 4 | 0 | 4 | 2 |
| 3, 4 | 4 | 4 | 2 | 2 | 4 |
| 3, 4 | 4 | 4 | 2 | 4 | 0 |
| 4 | 4 | 2 | 2 | 4 | 2 |

Table 2.11 Minimal path vectors for the module of the power generation system.

| Level | $A_1$ | $A_2$ | $L$ |
|---|---|---|---|
| 1 | 4 | 2 | 2 |
| 2 | 4 | 2 | 4 |
| 2 | 4 | 4 | 2 |
| 4 | 4 | 4 | 4 |

Table 2.12 Minimal cut vectors for the module of the power generation system.

| Level | $A_1$ | $A_2$ | $L$ |
|---|---|---|---|
| 1, 2 | 4 | 0 | 4 |
| 1, 2 | 4 | 4 | 0 |
| 1, 2, 4 | 2 | 4 | 4 |
| 2 | 4 | 2 | 2 |
| 4 | 4 | 2 | 4 |
| 4 | 4 | 4 | 2 |

whereas

$$\phi_2(4, 4, 0, 4, 4) = \phi_2(4, 2, 2, 4, 4) = 4.$$

Hence, (4, 2, 0, 4, 4) is a minimal cut vector both to level 3 and 4.

Consider the MMS of two components having the structure function tabulated in Table 2.15.

Table 2.13 Minimal path vectors for the organizing structure of the power generation system.

| Level | $A_3$ | $U$ | $\chi$ |
|---|---|---|---|
| 1 | 0 | 4 | 1 |
| 1, 2 | 2 | 2 | 0 |
| 2 | 0 | 4 | 2 |
| 3 | 2 | 4 | 1 |
| 3, 4 | 0 | 4 | 4 |
| 3, 4 | 4 | 2 | 0 |
| 4 | 2 | 4 | 2 |

Table 2.14 Minimal cut vectors for the organizing structure of the power generation system.

| Level | $A_3$ | $U$ | $\chi$ |
|---|---|---|---|
| 1 | 0 | 4 | 0 |
| 1, 2 | 0 | 2 | 4 |
| 1, 2, 3, 4 | 4 | 0 | 4 |
| 2 | 0 | 4 | 1 |
| 3 | 2 | 4 | 0 |
| 3, 4 | 2 | 2 | 4 |
| 3, 4 | 0 | 4 | 2 |
| 4 | 2 | 4 | 1 |

Table 2.15 An MCS suggested by El-Neweihi *et al.*

| | | | | | |
|---|---|---|---|---|---|
| | 3 | 1 | 2 | 3 | 3 |
| Component 2 | 2 | 0 | 1 | 2 | 3 |
| | 1 | 0 | 1 | 1 | 3 |
| | 0 | 0 | 0 | 0 | 1 |
| | | 0 | 1 | 2 | 3 |
| | | Component 1 | | | |

This is obviously of the type suggested by El-Neweihi *et al.* (1978). In their paper $\boldsymbol{x}$ is said to be an upper critical connection vector to level $j$ iff $\phi(\boldsymbol{x}) = j$ and $\boldsymbol{y} < \boldsymbol{x}$ implies $\phi(\boldsymbol{y}) < j$. Hence (3, 1) is an upper critical connection vector to level 3. However, it is a minimal path vector not just to level 3, but also to level 2. This shows that the set of minimal path vectors to level $j$ in addition to including the set of upper critical connection vectors to level $j$ also might include some few upper critical connection vectors to a higher level. It is not true, as claimed in Griffith (1980), that $\phi(\boldsymbol{y}) \geq j$ iff $\boldsymbol{y} \geq \boldsymbol{x}$ for some upper critical connection vector $\boldsymbol{x}$ to level $j$.

We now present some results corresponding to ones given in El-Neweihi *et al.* (1978). Note that due to the above observations, our Theorem 2.15 is far more elegant and useful than their Theorem 3.3. Theorem 2.15 is given in Natvig (1982a), but equivalent results are given in independent work by Block and Savits (1982) and Butler (1982).

**Theorem 2.14:** For an MSCS and for all $j \in \{1, \ldots, M\}$ the union of all minimal path sets to level $j$ and of all binary type minimal path sets to level $j$ equals $C = \{1, \ldots, n\}$. The same is true for the union of all minimal cut sets to level $j$ and of all binary type minimal cut sets to level $j$.

**Proof:** For any $j \in \{1, \ldots, M\}$, any $i \in C$ is a member of the union of all minimal path sets to level $j$ and of all binary type minimal path sets to level $j$ since, according to (ii) of Definition 2.4, we can construct a minimal path vector to level $j$ starting out with $(k_i, \boldsymbol{x})$. The proof is similar for the cut sets starting out with $(l_i, \boldsymbol{x})$.

Note that Theorem 2.14 generalizes a well-known result from binary theory.

**Theorem 2.15:** Let $(C, \phi)$ be an MMS. Furthermore, for $j \in \{1, \ldots, M\}$ let $\boldsymbol{y}_k^j = (y_{1k}^j, \ldots, y_{nk}^j)$, $k = 1, \ldots, n_\phi^j$ be its minimal path vectors to level $j$ and $\boldsymbol{z}_k^j = (z_{1k}^j, \ldots, z_{nk}^j)$, $k = 1, \ldots, m_\phi^j$ be its minimal cut vectors to level $j$ and

$$C_\phi^j(\boldsymbol{y}_k^j), k = 1, \ldots, n_\phi^j \text{ and } D_\phi^j(\boldsymbol{z}_k^j), k = 1, \ldots, m_\phi^j$$

the corresponding minimal path and cut sets to level $j$. Furthermore, introduce the following generalizations of respectively the minimal path series structure and the minimal cut parallel structure from binary theory

$$J_{\boldsymbol{y}_k^j}(\boldsymbol{x}) = \begin{cases} 1 & \text{if } x_i \geq y_{ik}^j \text{ for } i \in C_\phi^j(\boldsymbol{y}_k^j) \\ 0 & \text{otherwise} \end{cases},$$

$$J_{\boldsymbol{z}_k^j}(\boldsymbol{x}) = \begin{cases} 1 & \text{otherwise} \\ 0 & \text{if } x_i \leq z_{ik}^j \text{ for } i \in D_\phi^j(\boldsymbol{z}_k^j) \end{cases}.$$

Then

(i) $\phi(\boldsymbol{x}) \geq j \Leftrightarrow \exists\ 1 \leq k \leq n_\phi^j$ such that $x_i \geq y_{ik}^j$ for $i \in C_\phi^j(\boldsymbol{y}_k^j)$

$\Leftrightarrow \max_{1\leq k\leq n_\phi^j} \boldsymbol{J}_{\boldsymbol{y}_k^j}(\boldsymbol{x}) = \coprod_{k=1}^{n_\phi^j} \boldsymbol{J}_{\boldsymbol{y}_k^j}(\boldsymbol{x}) = 1$

(ii) $\phi(\boldsymbol{x}) < j \Leftrightarrow \exists\ 1 \leq k \leq m_\phi^j$ such that $x_i \leq z_{ik}^j$ for $i \in D_\phi^j(\boldsymbol{z}_k^j)$

$\Leftrightarrow \min_{1\leq k\leq m_\phi^j} \boldsymbol{J}_{\boldsymbol{z}_k^j}(\boldsymbol{x}) = \prod_{k=1}^{m_\phi^j} \boldsymbol{J}_{\boldsymbol{z}_k^j}(\boldsymbol{x}) = 0.$

The proof is straightforward.

The following theorem gives the link between the structure function of a BTMMS and respectively the binary type minimal path and cut sets.

**Theorem 2.16:** For $j \in \{1, \ldots, M\}$ consider a BTMMS $(C, \phi)$ having binary type minimal path sets to level $j$ $C_{\phi BT}^j(\boldsymbol{y}_k^j)$, $k = 1, \ldots, n_\phi^j$ and binary type minimal cut sets to level $j$ $D_{\phi BT}^j(\boldsymbol{z}_k^j)$, $k = 1, \ldots,$ $m_\phi^j$. Then

$$\phi(\boldsymbol{x}) \geq j \Leftrightarrow \coprod_{k=1}^{n_\phi^j} \prod_{i \in C_{\phi BT}^j(\boldsymbol{y}_k^j)} I_j(x_i) = 1 \tag{2.11}$$

$$\phi(\boldsymbol{x}) < j \Leftrightarrow \prod_{k=1}^{m_\phi^j} \coprod_{i \in D_{\phi BT}^j(\boldsymbol{z}_k^j)} I_j(x_i) = 0. \tag{2.12}$$

**Proof:** The results follow immediately from Definitions 2.12, 2.13 and 2.6 and Equations (1.8) and (1.9).

Given Theorem 2.16 it is natural to believe that most of the theory for a BMS can be extended to a BTMMS.

We also need the following theorem, which is a corrected and extended version of a result proved as part of the proof of Theorem 3.1 in Butler (1982). These corrections and extensions are treated in depth in Gåsemyr (2010). Note that we have augmented the notation introduced in Theorem 2.15.

**Theorem 2.17:** Let $(C, \phi)$ be an MMS with modular decomposition given by Definition 1.6. Introduce the following MMS structure functions of the system's components defined by

$$I(\phi_\ell^j(\boldsymbol{x}) \geq j) = \boldsymbol{J}_{\boldsymbol{y}_{\psi\ell}^j}(\boldsymbol{\chi}(\boldsymbol{x})) \;\; j = 1, \ldots, M, \;\; \ell = 1, \ldots, n_\psi^j,$$

or alternatively by

$$I(\phi_\ell^j(\boldsymbol{x}) \geq j) = \boldsymbol{J}_{\boldsymbol{z}_{\psi\ell}^j}(\boldsymbol{\chi}(\boldsymbol{x})) \;\; j = 1, \ldots, M, \;\; \ell = 1, \ldots, m_\psi^j.$$

Then the minimal path vectors and minimal cut vectors to level $j$ of $(C, \phi)$ are respectively included in

$$\boldsymbol{y}^j_{\phi_\ell^j m} \quad m = 1, \ldots, n^j_{\phi_\ell^j}, \quad \ell = 1, \ldots, n_\psi^j, \tag{2.13}$$

and

$$\boldsymbol{z}^j_{\phi_\ell^j m} \quad m = 1, \ldots, m^j_{\phi_\ell^j}, \quad \ell = 1, \ldots, m_\psi^j. \tag{2.14}$$

Assume that for each module $(A_k, \chi_k)$, $k = 1, \ldots, r$ and $m \in S_{\chi_k} - \{0\}$ every minimal path vector to level $m$, $\boldsymbol{y}^m_{\chi_k}$, satisfies $\chi_k(\boldsymbol{y}^m_{\chi_k}) = m$, and every minimal cut vector to level $m, \boldsymbol{z}^m_{\chi_k}$, satisfies $\chi_k(\boldsymbol{z}^m_{\chi_k}) = \max\{s \in S_{\chi_k} : s < m\}$. Then the minimal path vectors and minimal cut vectors to level $j$ of $(C, \phi)$ are respectively exactly given by (2.13) and (2.14).

**Proof:** The proof goes along the same lines as the corresponding part of the proof of Theorem 4.1, page 44 in Barlow and Proschan (1975a). By construction

$$\boldsymbol{z}^j_{\phi_\ell^j m} \quad m = 1, \ldots, m^j_{\phi_\ell^j}, \quad \ell = 1, \ldots, m_\psi^j$$

are cut vectors to level $j$ of $(C, \phi)$ and also include all corresponding minimal cut vectors. However, these cut vectors to level $j$ of $(C, \phi)$ are not necessarily minimal and distinct as claimed by Butler (1982). Assume that for each module $(A_k, \chi_k), k = 1, \ldots, r$ and $m \in S_{\chi_k} - \{0\}$ every minimal cut vector to level $m, \boldsymbol{z}^m_{\chi_k}$, satisfies $\chi_k(\boldsymbol{z}^m_{\chi_k}) = \max\{s \in S_{\chi_k} : s < m\}$. Then, again by construction, the cut vectors are minimal and distinct. Hence, the minimal cut vectors to level $j$ of $(C, \phi)$ are exactly given by (2.14). A corresponding argument is applied for the minimal path vectors to level $j$ of $(C, \phi)$. This completes the proof.

Let us illustrate Theorem 2.17 by combining the minimal cut vector (0, 4, 0) to level 1 of the organizing structure of the power generation system given in Table 2.14, by the minimal cut vectors to level 1 for the corresponding module given in Table 2.12. This leads to the following minimal cut vectors to level 1 of $(C, \phi)$: (4, 0, 0, 4, 4), (4, 4, 0, 4, 0) and (2, 4, 0, 4, 4) according to Theorem 2.17. Hence, they are listed in Table 2.10. Starting out with the minimal cut vector (0, 4, 1) to level 2 of the organizing structure leads to the same minimal cut vectors to level 2 of $(C, \phi)$ in addition to (4, 2, 0, 4, 2). Although the given assumption does not hold, these are the correct ones listed in Table 2.10.

Remembering Definition 2.2 of the dual structure function $\phi^D$, we conclude this section by giving a theorem which is an extension of well-known results in binary theory. Since dealing both with $\phi$ and $\phi^D$, we have again augmented the notation for the minimal path and cut vectors introduced in Theorem 2.15.

**Theorem 2.18:** Let $(C, \phi)$ be an MMS with dual MMS $(C, \phi^D)$ given by Definition 2.2. Furthermore, for $j \in \{1, \ldots, M\}$ let $\boldsymbol{y}^j_{\phi k}, k = 1, \ldots, n^j_\phi$ be the minimal path vectors to level $j$ and $\boldsymbol{z}^j_{\phi k}, k = 1, \ldots, m^j_\phi$ the minimal cut vectors to level $j$ for $(C, \phi)$. Then $\boldsymbol{y}^{jD}_{\phi k} = \boldsymbol{z}^{j^D}_{\phi^D k}$ and $D^{j^D}_{\phi^D}(\boldsymbol{z}^{j^D}_{\phi^D k}) = C^j_\phi(\boldsymbol{y}^j_{\phi k}), k = 1, \ldots, m^{j^D}_{\phi^D} = n^j_\phi$ are respectively the minimal cut vectors and minimal cut sets, and $\boldsymbol{z}^{jD}_{\phi k} = \boldsymbol{y}^{j^D}_{\phi^D k}$ and $C^{j^D}_{\phi^D}(\boldsymbol{y}^{j^D}_{\phi^D k}) = D^j_\phi(\boldsymbol{z}^j_{\phi k}), k = 1, \ldots, n^{j^D}_{\phi^D} = m^j_\phi$ respectively the minimal path vectors and minimal path sets, to the dual level $j^D = M - j + 1$ for $(C, \phi^D)$.

Note that the special, but convenient, notation $j^D = M - j + 1$, used in the rest of the book, is not consistent with the notation $x_i^D = M - x_i, i = 1, \ldots, n$ introduced in Definition 2.2. This should, however, not lead to any confusion.

## 2.4 Stochastic performance of multistate monotone and coherent systems

In this section we concentrate on the relationship, at a fixed point of time, between the stochastic performance of the system $(C, \phi)$ and the stochastic performances of the components. Introduce the random state $X_i$ of the $i$th component, $i = 1, \ldots, n$ and the random vector $\boldsymbol{X} = (X_1, \ldots, X_n)$. Now, if $\phi$ is a multistate structure function, $\phi(\boldsymbol{X})$ is the corresponding random system state.

**Definition 2.19:** Let $i = 1, \ldots, n$, $j = 1, \ldots, M$. The reliability, $p_i^j$, and the unreliability, $q_i^j$, to level $j$ of the $i$th component are given by

$$p_i^j = P[X_i \geq j] \qquad q_i^j = P[X_i < j].$$

Furthermore, let

$$r_i^j = P[X_i = j] \quad i = 1, \ldots, n, j = 0, \ldots, M.$$

Let $j = 1, \ldots, M$. The reliability, $p_\phi^j$, and the unreliability, $q_\phi^j$, to level $j$ for an MMS with structure function $\phi$ are given by

$$p_\phi^j = P[\phi(\boldsymbol{X}) \geq j] \qquad q_\phi^j = P[\phi(\boldsymbol{X}) < j].$$

Furthermore, let

$$r_\phi^j = P[\phi(\boldsymbol{X}) = j] \quad j = 0, \ldots, M.$$

Finally, introduce the $n \times M$ component reliability and unreliability matrices

$$\boldsymbol{P}_\phi = \left\{p_i^j\right\}_{\substack{i=1,\ldots,n\\ j=1,\ldots,M}} \qquad \boldsymbol{Q}_\phi = \left\{q_i^j\right\}_{\substack{i=1,\ldots,n\\ j=1,\ldots,M}}.$$

We also introduce the performance function of the system, $h_\phi$, as the expected system state, given by

$$h_\phi = E\phi(\boldsymbol{X}) = \sum_{j=1}^{M} p_\phi^j. \tag{2.15}$$

The following theorem is a generalization of Theorem 1.2, page 22 in Barlow and Proschan (1975a) and Theorem 4.1 in El-Neweihi *et al.* (1978) given for their multistate coherent system.

**Theorem 2.20:** Let $S_i = S, i = 1, \ldots, n$ and assume $X_1, \ldots, X_n$ are independent. Furthermore, assume $0 < r_i^j < 1, i = 1, \ldots, n$, $j = 0, \ldots, M$. Then, if $\phi$ is the structure function of an MCS, the performance function $h_\phi(\boldsymbol{P}_\phi)$, being a function of just $\boldsymbol{P}_\phi$ since $X_1, \ldots, X_n$ are independent, is strictly increasing in $p_i^j$ and $r_i^j$ at the expense of $r_i^0, i = 1, \ldots, n$, $j = 1, \ldots, M$. The same is true for the reliability to level $j$ $p_\phi^j$ of the system, if $\phi$ is the structure function of an MSCS.

**Proof:** By using the fact that $\sum_{k=0}^{M} r_i^k = 1, i = 1, \ldots, n$, we have

$$h_\phi(\boldsymbol{P}_\phi) = \sum_{k=0}^{M} r_i^k E\phi(k_i, \boldsymbol{X}) = \sum_{k=1}^{M} r_i^k E[\phi(k_i, \boldsymbol{X}) - \phi(0_i, \boldsymbol{X})] + E[\phi(0_i, \boldsymbol{X})]$$

$$= \sum_{k=1}^{M} (p_i^k - p_i^{k+1}) E[\phi(k_i, \boldsymbol{X}) - \phi(0_i, \boldsymbol{X})] + E[\phi(0_i, \boldsymbol{X})]$$

$$= \sum_{k=1}^{M} p_i^k E[\phi(k_i, \boldsymbol{X}) - \phi((k-1)_i, \boldsymbol{X})] + E[\phi(0_i, \boldsymbol{X})].$$

Thus, for $i = 1, \ldots, n,\ j = 1, \ldots, M$

$$\frac{\partial h_\phi(\boldsymbol{P}_\phi)}{\partial p_i^j} = E[\phi(j_i, \boldsymbol{X}) - \phi((j-1)_i, \boldsymbol{X})].$$

Since $\phi$ is nondecreasing in each argument $E[\phi(j_i, \boldsymbol{X}) - \phi((j-1)_i, \boldsymbol{X})] \geq 0$. In addition $\phi(j_i, \boldsymbol{x}) - \phi((j-1)_i, \boldsymbol{x}) > 0$ for some $(\cdot_i, \boldsymbol{x})$ since the structure function is an MCS. Since $0 < r_i^j < 1$, $i = 1, \ldots, n,\quad j = 0, \ldots, M$, $P[(\cdot_i, \boldsymbol{X}) = (\cdot_i, \boldsymbol{x})] > 0$. This leads to $E[\phi(j_i, \boldsymbol{X}) - \phi((j-1)_i, \boldsymbol{X})] > 0$, completing the proof for $h_\phi(\boldsymbol{P}_\phi)$ for $p_i^j$. The result for $r_i^j$ follows similarly since

$$\frac{\partial h_\phi(\boldsymbol{P}_\phi)}{\partial r_i^j} = E[\phi(j_i, \boldsymbol{X}) - \phi(0_i, \boldsymbol{X})].$$

The proof for $p_\phi^j$ is completely parallel, placing $I_j$ in front of $\phi$.

Next we give Theorem 4.2 in El-Neweihi *et al.* (1978), which shows that $h_\phi(\boldsymbol{P}_\phi)$ is nondecreasing with respect to stochastic ordering.

**Theorem 2.21:** Assume $X_1, \ldots, X_n, \tilde{X}_1, \ldots, \tilde{X}_n$ to be mutually independent random variables having component reliability matrices $\boldsymbol{P}_\phi$ and $\tilde{\boldsymbol{P}}_\phi$. Assume $p_i^j \leq \tilde{p}_i^j$ for $i = 1, \ldots, n,\ j = 1, \ldots, M$. Then

(i) $p_\phi^j \leq \tilde{p}_\phi^j$ for $j = 1, \ldots, M$

(ii) $h_\phi(\boldsymbol{P}_\phi) \leq h_\phi(\tilde{\boldsymbol{P}}_\phi)$.

**Proof:** For $i = 1, \ldots, n$, $p_i^j \leq \tilde{p}_i^j$ for $j = 1, \ldots, M$ implies that $X_i$ is stochastically less than or equal to$\tilde{X}_i$. Since $\phi$ is nondecreasing, $\phi(\boldsymbol{X})$

is stochastically less than or equal to $\phi(\tilde{X})$. This immediately completes the proof.

The following theorem, based on state enumeration, gives expressions for $h_\phi$ without making any assumptions on the structure function $\phi$. The proof is immediate by the definition of expectation.

**Theorem 2.22:** Let $\phi$ be the structure function of a multistate system. Then

$$h_\phi = \sum_{\boldsymbol{x}} \phi(\boldsymbol{x}) P[\cap_{i=1}^{n}(X_i = x_i)]. \tag{2.16}$$

Assuming $X_1, \ldots, X_n$ to be independent, we get

$$h_\phi(\boldsymbol{P}_\phi) = \sum_{\boldsymbol{x}} \phi(\boldsymbol{x}) \prod_{i=1}^{n} r_i^{x_i}. \tag{2.17}$$

The number of addends in Equations (2.16) and (2.17) equals $(M+1)^n$ if $S_i = S, i = 1, \ldots, n$, which easily gets far too large for any computer. Hence, we have to find other ways of establishing $h_\phi$.

Our first attempt is generalizing what, in binary theory, is known as the inclusion–exclusion method, see Barlow and Proschan (1975a), page 25.

**Theorem 2.23:** Let $(C, \phi)$ be an MMS. Furthermore, for $j \in \{1, \ldots, M\}$ let $\boldsymbol{y}_k^j = (y_{1k}^j, \ldots, y_{nk}^j)$, $k = 1, \ldots, n_\phi^j$ be its minimal path vectors to level $j$ and $\boldsymbol{z}_k^j = (z_{1k}^j, \ldots, z_{nk}^j)$, $k = 1, \ldots, m_\phi^j$ be its minimal cut vectors to level $j$ and

$$C_\phi^j(\boldsymbol{y}_k^j), k = 1, \ldots, n_\phi^j \text{ and } D_\phi^j(\boldsymbol{z}_k^j), k = 1, \ldots, m_\phi^j$$

the corresponding minimal path and cut sets to level $j$. Then

$$p_\phi^j = \sum_{k=1}^{n_\phi^j} (-1)^{k-1} S_k^j \tag{2.18}$$

$$p_\phi^j = 1 - \sum_{k=1}^{m_\phi^j} (-1)^{k-1} T_k^j, \tag{2.19}$$

where

$$S_k^j = \sum_{1 \le i_1 < i_2 < \cdots < i_k \le n_\phi^j} P[\cap_{i \in \cup_{s=1}^{k} C_\phi^j(\boldsymbol{y}_{i_s}^j)} (X_i \ge \max_{1 \le s \le k} y_{ii_s}^j)] \tag{2.20}$$

$$T_k^j = \sum_{1 \le i_1 < i_2 < \cdots < i_k \le m_\phi^j} P[\cap_{i \in \cup_{s=1}^k D_\phi^j(\mathbf{z}_{i_s}^j)} (X_i \le \min_{1 \le s \le k} z_{ii_s}^j)]. \tag{2.21}$$

If $X_1, \ldots, X_n$ are independent, we have

$$S_k^j = \sum_{1 \le i_1 < i_2 < \cdots < i_k \le n_\phi^j} \prod_{i \in \cup_{s=1}^k C_\phi^j(\mathbf{y}_{i_s}^j)} p_i^{\max_{1 \le s \le k} y_{ii_s}^j}$$

$$T_k^j = \sum_{1 \le i_1 < i_2 < \cdots < i_k \le m_\phi^j} \prod_{i \in \cup_{s=1}^k D_\phi^j(\mathbf{z}_{i_s}^j)} q_i^{\min_{1 \le s \le k} z_{ii_s}^j + 1}.$$

**Proof:** For $j \in \{1, \ldots, M\}$, by applying (i) of Theorem 2.15 and the general addition law of probability theory, we get

$$p_\phi^j = P(\phi(\mathbf{X}) \ge j) = P[\cup_{k=1}^{n_\phi^j} \cap_{i \in C_\phi^j(\mathbf{y}_k^j)} (X_i \ge y_{ik}^j)]$$

$$= \sum_{k=1}^{n_\phi^j} (-1)^{k-1} \sum_{1 \le i_1 < i_2 < \cdots < i_k \le n_\phi^j} P[\cap_{s=1}^k \cap_{i \in C_\phi^j(\mathbf{y}_{i_s}^j)} (X_i \ge y_{ii_s}^j)],$$

leading to Equation (2.18). Equation (2.19) is proved similarly by applying (ii) of Theorem 2.15. The results for independent $X_1, \ldots, X_n$ follow immediately from Equations (2.20) and (2.21).

The total number of addends in Equation (2.18) is

$$\sum_{k=1}^{n_\phi^j} \sum_{1 \le i_1 < i_2 < \cdots < i_k \le n_\phi^j} 1 = \sum_{k=0}^{n_\phi^j} \binom{n_\phi^j}{k} - 1 = (1+1)^{n_\phi^j} - 1 = 2^{n_\phi^j} - 1.$$

Similarly, the number of addends in Equation (2.19) is $2^{m_\phi^j} - 1$. As for the state enumeration method these numbers may be too large for a computer. However, by the inclusion–exclusion principle of Feller (1968), pages 98–101 we have

$$1 - T_1^j \le p_\phi^j \le S_1^j$$

$$S_1^j - S_2^j \le p_\phi^j \le 1 - T_1^j + T_2^j$$

$$1 - T_1^j + T_2^j - T_3^j \le p_\phi^j \le S_1^j - S_2^j + S_3^j,$$

and so on, giving upper and lower bounds on $p_\phi^j$ for $j \in \{1, \ldots, M\}$. Applying Equation (2.15) we automatically get upper and lower bounds on $h_\phi$ too.

We now give a stochastic version of Theorem 2.5.

**Theorem 2.24:** Let $\phi$ be a structure function of an MCS and assume $S_i = S, i = 1, \ldots, n$. Let $\boldsymbol{X}$ and $\tilde{\boldsymbol{X}}$ be two random component state vectors. Then

(i) $P[\phi(\boldsymbol{X} \vee \tilde{\boldsymbol{X}}) \geq j] \geq P[(\phi(\boldsymbol{X}) \vee \phi(\tilde{\boldsymbol{X}})) \geq j] \quad j = 1, \ldots, M$

(ii) $P[\phi(\boldsymbol{X} \wedge \tilde{\boldsymbol{X}}) \geq j] \leq P[(\phi(\boldsymbol{X}) \wedge \phi(\boldsymbol{X})) \geq j] \quad j = 1, \ldots, M$

(iii) $E[\phi(\boldsymbol{X} \vee \tilde{\boldsymbol{X}})] \geq E[\phi(\boldsymbol{X}) \vee \phi(\tilde{\boldsymbol{X}})]$

(iv) $E[\phi(\boldsymbol{X} \wedge \tilde{\boldsymbol{X}})] \leq E[\phi(\boldsymbol{X}) \wedge \phi(\tilde{\boldsymbol{X}})]$.

Equality holds for all $P[(\boldsymbol{X} = \boldsymbol{x}) \cap (\tilde{\boldsymbol{X}} = \tilde{\boldsymbol{x}})]$ in (i) and (iii) iff the structure function is parallel, i.e. $\phi(\boldsymbol{x}) = \max_{1 \leq i \leq n} x_i$, and in (ii) and (iv) iff the structure function is series, i.e. $\phi(\boldsymbol{x}) = \min_{1 \leq i \leq n} x_i$.

**Proof:**

$$\begin{aligned}
&P[\phi(\boldsymbol{X} \vee \tilde{\boldsymbol{X}}) \geq j] - P[(\phi(\boldsymbol{X}) \vee \phi(\tilde{\boldsymbol{X}})) \geq j] \\
&\quad = E[I_j(\phi(\boldsymbol{X} \vee \tilde{\boldsymbol{X}})) - I_j(\phi(\boldsymbol{X}) \vee \phi(\tilde{\boldsymbol{X}}))] \\
&\quad = \sum_{\boldsymbol{x}} \sum_{\tilde{\boldsymbol{x}}} [I_j(\phi(\boldsymbol{x} \vee \tilde{\boldsymbol{x}})) - I_j(\phi(\boldsymbol{x}) \vee \phi(\tilde{\boldsymbol{x}}))] P[(\boldsymbol{X} = \boldsymbol{x}) \cap (\tilde{\boldsymbol{X}} = \tilde{\boldsymbol{x}})] \\
&\quad \geq 0,
\end{aligned}$$

having applied (i) of Theorem 2.5 and the fact that $I_j(x_i)$ is nondecreasing in $x_i$. Hence, (i) is established. The inequality of (iii) follows from (i) by applying Equation (2.15). Similar arguments applying (ii) of Theorem 2.5 give (ii) and (iv).

If equality holds in (i) and (iii) for all $P[(\boldsymbol{X} = \boldsymbol{x}) \cap (\tilde{\boldsymbol{X}} = \tilde{\boldsymbol{x}})]$, it especially holds for $P[(\boldsymbol{X} = \boldsymbol{x}) \cap (\tilde{\boldsymbol{X}} = \tilde{\boldsymbol{x}})] > 0$ for all $\boldsymbol{x}$ and $\tilde{\boldsymbol{x}}$. Hence, equality in (i) and (iii) for all $P[(\boldsymbol{X} = \boldsymbol{x}) \cap (\tilde{\boldsymbol{X}} = \tilde{\boldsymbol{x}})]$ is equivalent to

$$\phi(\boldsymbol{x} \vee \tilde{\boldsymbol{x}}) = \phi(\boldsymbol{x}) \vee \phi(\tilde{\boldsymbol{x}}) \text{ for all } \boldsymbol{x} \text{ and } \tilde{\boldsymbol{x}}.$$

Applying Theorem 2.5 this is again true iff the structure function is parallel. A similar proof establishes that equality holds in (ii) and (iv) for all $P[(\boldsymbol{X} = \boldsymbol{x}) \cap (\tilde{\boldsymbol{X}} = \tilde{\boldsymbol{x}})]$ iff the structure function is series. This completes the proof.

In Section 3 of Chapter 2 of Barlow and Proschan (1975a) a series of bounds on $p_\phi^1$ is given for a binary coherent system (BCS). Actually, it is only necessary that we have a binary monotone system (BMS). Some of these results will be generalized in the following. It is easy to check that all bounds given in the rest of this section lie inside the interval [0, 1], as is a minimum claim. This is not true for those given in Theorem 2.23 based on the inclusion–exclusion method.

**Theorem 2.25:** If $X_1, \ldots, X_n$ are associated random variables, for $j \in \{1, \ldots, M\}$, for the multistate series and parallel systems we respectively have

$$P[\min_{1\le i\le n} X_i \ge j] \ge \prod_{i=1}^{n} p_i^j \tag{2.22}$$

$$P[\max_{1\le i\le n} X_i \ge j] \le \coprod_{i=1}^{n} p_i^j. \tag{2.23}$$

**Proof:** We obviously have

$$P[\min_{1\le i\le n} X_i \ge j] = P[\prod_{i=1}^{n} I_j(X_i) = 1].$$

$I_j(X_i)$ is nondecreasing in $X_i, i = 1, \ldots, n$, and so, by Property $P_3$ of associated random variables, $I_j(X_1), \ldots, I_j(X_n)$ are associated. Equation (2.22) now follows from Theorem 3.1, page 32 in Barlow and Proschan (1975a), this theorem being just the present theorem in the binary case. Equation (2.23) is proved similarly.

**Corollary 2.26:** Let $\phi$ be a multistate structure function which is nondecreasing in each argument and such that Equation (2.6) holds. Assume $X_1, \ldots, X_n$ are associated random variables. Then we have

$$\prod_{i=1}^{n} p_i^j \le p_\phi^j \le \coprod_{i=1}^{n} p_i^j \quad j \in \{1, \ldots, M\} \tag{2.24}$$

$$\sum_{j=1}^{M}\prod_{i=1}^{n} p_i^j \le h_\phi \le \sum_{j=1}^{M}\coprod_{i=1}^{n} p_i^j. \tag{2.25}$$

**Proof:** Equation (2.24) follows directly from Equations (2.6), (2.22) and (2.23). Equation (2.25) follows from Equations (2.15) and (2.24).

For the special case where $X_1, \ldots, X_n$ are independent, this result is given by Theorem 4.4 in El-Neweihi *et al.* (1978). As a simple application of the crude bounds in Equation (2.25), consider the simple network system of Figure 1.1 with states given in Table 1.1. Let the probability of a branch working be $p$, and assume that branches within a component work independently whereas the two components are associated. Then

$$\{1-(1-p)^2\}^2 + 2p^4 \leq h \leq 3 - (1-p)^4 - 2(1-p^2)^2. \tag{2.26}$$

For $p = 0$ and $p = 1$, we get the obvious results, whereas for $p = \frac{1}{2}$, $\frac{11}{16} \leq h \leq \frac{29}{16}$.

The following generalization of what, in binary theory, are known as Min–Max bounds, see Barlow and Proschan (1975a) page 37, is given in the first part of Lemma 3.1 in Block and Savits (1982) and in Theorem 2.3 and Corollary 2.2 in Butler (1982).

**Theorem 2.27:** Let $(C, \phi)$ be an MMS. Furthermore, for $j \in \{1, \ldots, M\}$ let $\boldsymbol{y}_k^j = (y_{1k}^j, \ldots, y_{nk}^j)$, $k = 1, \ldots, n_\phi^j$ be its minimal path vectors to level $j$ and $\boldsymbol{z}_k^j = (z_{1k}^j, \ldots, z_{nk}^j)$, $k = 1, \ldots, m_\phi^j$ be its minimal cut vectors to level $j$ and

$$C_\phi^j(\boldsymbol{y}_k^j), k = 1, \ldots, n_\phi^j \text{ and } D_\phi^j(\boldsymbol{z}_k^j), k = 1, \ldots, m_\phi^j$$

the corresponding minimal path and cut sets to level $j$. Let

$$\begin{aligned} \ell_\phi^{\prime\prime j} &= \max_{1 \leq k \leq n_\phi^j} P[\cap_{i \in C_\phi^j(\boldsymbol{y}_k^j)} (X_i \geq y_{ik}^j)] \\ \bar{\ell}_\phi^{\prime\prime j} &= \max_{1 \leq k \leq m_\phi^j} P[\cap_{i \in D_\phi^j(\boldsymbol{z}_k^j)} (X_i \leq z_{ik}^j)]. \end{aligned}$$

Then

$$\ell_\phi^{\prime\prime j} \leq p_\phi^j \leq 1 - \bar{\ell}_\phi^{\prime\prime j}. \tag{2.27}$$

Furthermore, let

$$\ell_\phi^{\prime j}(\boldsymbol{P}_\phi) = \max_{1 \leq k \leq n_\phi^j} \prod_{i \in C_\phi^j(\boldsymbol{y}_k^j)} p_i^{y_{ik}^j} \qquad \bar{\ell}_\phi^{\prime j}(\boldsymbol{Q}_\phi) = \max_{1 \leq k \leq m_\phi^j} \prod_{i \in D_\phi^j(\boldsymbol{z}_k^j)} q_i^{z_{ik}^j + 1}.$$

If $X_1, \ldots, X_n$ are associated, then

$$\ell'^j_\phi(\boldsymbol{P}_\phi) \leq p^j_\phi \leq 1 - \bar{\ell}'^j_\phi(\boldsymbol{Q}_\phi). \qquad (2.28)$$

**Proof:** From Theorem 2.15 we have

$$I(\phi(\boldsymbol{X}) \geq j) = \max_{1 \leq k \leq n^j_\phi} \boldsymbol{J}_{\boldsymbol{y}^j_k}(\boldsymbol{X}) = \coprod_{k=1}^{n^j_\phi} \boldsymbol{J}_{\boldsymbol{y}^j_k}(\boldsymbol{X}) =$$

$$\prod_{k=1}^{m^j_\phi} \boldsymbol{J}_{\boldsymbol{z}^j_k}(\boldsymbol{X}) = \min_{1 \leq k \leq m^j_\phi} \boldsymbol{J}_{\boldsymbol{z}^j_k}(\boldsymbol{X}).$$

For $k = 1, \ldots, n^j_\phi$ and $r = 1, \ldots, m^j_\phi$ we then have

$$\boldsymbol{J}_{\boldsymbol{y}^j_k}(\boldsymbol{X}) \leq I(\phi(\boldsymbol{X}) \geq j) \leq \boldsymbol{J}_{\boldsymbol{z}^j_r}(\boldsymbol{X}).$$

Hence,

$$P[\boldsymbol{J}_{\boldsymbol{y}^j_k}(\boldsymbol{X}) = 1] \leq p^j_\phi \leq P[\boldsymbol{J}_{\boldsymbol{z}^j_r}(\boldsymbol{X}) = 1] = 1 - P[\boldsymbol{J}_{\boldsymbol{z}^j_r}(\boldsymbol{X}) = 0]$$

for $k = 1, \ldots, n^j_\phi$ and $r = 1, \ldots, m^j_\phi$. Equation (2.27) now follows by maximizing over $k$ and $r$. Remembering the indicators given by Equation (2.1), Equation (2.27) can be written

$$\max_{1 \leq k \leq n^j_\phi} P\left[\prod_{i \in C^j_\phi(\boldsymbol{y}^j_k)} I_{y^j_{ik}}(X_i) = 1\right] \leq p^j_\phi$$

$$\leq 1 - \max_{1 \leq k \leq m^j_\phi} P\left[\prod_{i \in D^j_\phi(\boldsymbol{z}^j_k)} (1 - I_{z^j_{ik}+1}(X_i)) = 1\right].$$

Since $X_1, \ldots, X_n$ are associated, by applying Theorem 3.1, page 32 of Barlow and Proschan (1975a) and Property $P_3$ of associated random variables, we arrive at Equation (2.28).

Note that we can apply the lower bounds in Equations (2.27) and (2.28) on the dual structure and dual level, applying Theorem 2.18, and arrive at the corresponding upper bounds. Let us illustrate this on the bounds in the former equation using the augmented notation of the latter theorem.

$$\bar{\ell}''^j_\phi = \max_{1 \leq k \leq m^j_\phi} P[\cap_{i \in D^j_\phi(\boldsymbol{z}^j_{\phi k})}(X_i \leq z^j_{\phi ik})]$$

$$= \max_{1 \le k \le m^j_\phi} P[\cap_{i \in D^j_\phi(\mathbf{z}^j_{\phi k})}(X^D_i \ge z^{jD}_{\phi ik})]$$

$$= \max_{1 \le k \le n^{jD}_{\phi D}} P[\cap_{i \in C^{jD}_{\phi D}(\mathbf{y}^{jD}_{\phi^D k})}(X^D_i \ge y^{jD}_{\phi^D ik})] = \ell''^{jD}_{\phi D} \le p^{jD}_{\phi D}$$

$$= P[\phi^D(\boldsymbol{X^D}) \ge j^D] = P[M - \phi(\boldsymbol{X}) \ge M - j + 1]$$

$$= P[\phi(\boldsymbol{X}) \le j - 1] = 1 - p^j_\phi.$$

Hence, we have arrived at the upper bound in Equation (2.27). This technique is generally applicable making it unnecessary to give separate proofs of the upper bounds in this section.

The following results given in Natvig (1982a) are firmly based on Theorem 2.15. The first theorem is also given in independent work by Block and Savits (1982) and Butler (1982).

**Theorem 2.28:** Let $(C, \phi)$ be an MMS where $X_1, \ldots, X_n$ are associated. Furthermore, for $j \in \{1, \ldots, M\}$ let $\boldsymbol{y}^j_k = (y^j_{1k}, \ldots, y^j_{nk})$, $k = 1, \ldots, n^j_\phi$ be its minimal path vectors to level $j$ and $\boldsymbol{z}^j_k = (z^j_{1k}, \ldots, z^j_{nk})$, $k = 1, \ldots, m^j_\phi$ be its minimal cut vectors to level $j$ and

$$C^j_\phi(\boldsymbol{y}^j_k), k = 1, \ldots, n^j_\phi \text{ and } D^j_\phi(\boldsymbol{z}^j_k), k = 1, \ldots, m^j_\phi$$

the corresponding minimal path and cut sets to level $j$. Let

$$\ell^{*j}_\phi = \prod_{k=1}^{m^j_\phi} P(\cup_{i \in D^j_\phi(\boldsymbol{z}^j_k)}(X_i > z^j_{ik})) \quad \bar{\ell}^{*j}_\phi = \prod_{k=1}^{n^j_\phi} P(\cup_{i \in C^j_\phi(\boldsymbol{y}^j_k)}(X_i < y^j_{ik})).$$

Then

$$\ell^{*j}_\phi \le p^j_\phi \le 1 - \bar{\ell}^{*j}_\phi. \tag{2.29}$$

Furthermore, let

$$\ell^{**j}_\phi(\boldsymbol{P}_\phi) = \prod_{k=1}^{m^j_\phi} \coprod_{i \in D^j_\phi(\boldsymbol{z}^j_k)} p_i^{z^j_{ik}+1} \qquad \bar{\ell}^{**j}_\phi(\boldsymbol{Q}_\phi) = \prod_{k=1}^{n^j_\phi} \coprod_{i \in C^j_\phi(\boldsymbol{y}^j_k)} q_i^{y^j_{ik}}$$

If $X_1, \ldots, X_n$ are independent, then

$$\ell^{*j}_\phi = \ell^{**j}_\phi(\boldsymbol{P}_\phi) \le p^j_\phi \le 1 - \bar{\ell}^{**j}_\phi(\boldsymbol{Q}_\phi) = 1 - \bar{\ell}^{*j}_\phi. \tag{2.30}$$

**Proof:** The minimal cut parallel structures $\boldsymbol{J}_{\boldsymbol{z}_k^j}(\boldsymbol{X}), k = 1, \ldots, m_\phi^j$ to level $j$ are obviously nondecreasing functions of $X_1, \ldots, X_n$. Hence, by Property $P_3$ of associated random variables they are associated. By applying (ii) of Theorem 2.15 and Theorem 3.1, page 32 of Barlow and Proschan (1975a), we get

$$p_\phi^j = P[\prod_{k=1}^{m_\phi^j} \boldsymbol{J}_{\boldsymbol{z}_k^j}(\boldsymbol{X}) = 1] \geq \prod_{k=1}^{m_\phi^j} P(\boldsymbol{J}_{\boldsymbol{z}_k^j}(\boldsymbol{X}) = 1).$$

Hence, the left inequality of Equation (2.29) is proved. Equation (2.30) follows immediately from Equation (2.29) since now $X_1, \ldots, X_n$ are independent.

**Theorem 2.29:** Let $(C, \phi)$ be an MMS where $X_1, \ldots, X_n$ are independent, and let $\ell_\phi^{**j}(\boldsymbol{P}_\phi)$ and $1 - \bar{\ell}_\phi^{**j}(\boldsymbol{Q}_\phi)$ be defined as in Theorem 2.28. Then

(i) $\ell_\phi^{**j}(\boldsymbol{P}_\phi)$ is a nondecreasing and $1 - \bar{\ell}_\phi^{**j}(\boldsymbol{Q}_\phi)$ a nonincreasing function in each argument.

(ii) $\ell_\phi^{**j}(\boldsymbol{P}_\phi) <^1 p_\phi^j <^2 1 - \bar{\ell}_\phi^{**j}(\boldsymbol{Q}_\phi)$ for $0 < r_i^j < 1, i = 1, \ldots, n$, $j \in \{0, M\}$ if at least two minimal path sets to level $j$ overlap and at least two minimal cut sets to level $j$ overlap.

The inequality 1 is replaced by equality when there is only one minimal cut set to level $j$ and there is equality in 2 when there is only one minimal path set to level $j$.

**Proof:** The proof of (i) is trivial. To prove (ii), assume the two minimal cut sets to level $j$ $D_\phi^j(z_1^j)$ and $D_\phi^j(z_2^j)$ overlap. Introduce the minimal cut parallel structures $\boldsymbol{J}_{\boldsymbol{z}_k^j}(\boldsymbol{X}), k = 1, \ldots, m_\phi^j$ to level $j$. Then, since $0 < r_i^j < 1, i = 1, \ldots, n,\ j \in \{0, M\}$ neither of the following two random variables $\boldsymbol{J}_{\boldsymbol{z}_1^j}(\boldsymbol{X})$ and $\prod_{k=2}^{m_\phi^j} \boldsymbol{J}_{\boldsymbol{z}_k^j}(\boldsymbol{X})$ is identically equal to 0 or 1. Since $D_\phi^j(z_1^j)$ and $D_\phi^j(z_2^j)$ overlap, then it also follows that the two random variables are dependent. Furthermore, they are associated by properties $P_5$ and $P_3$ of associated random variables. By applying Exercise 6, page 31 in Barlow and Proschan (1975a) we have

$$\text{Cov}\left[\boldsymbol{J}_{\boldsymbol{z}_1^j}(\boldsymbol{X}), \prod_{k=2}^{m_\phi^j} \boldsymbol{J}_{\boldsymbol{z}_k^j}(\boldsymbol{X})\right] > 0,$$

which is equivalent to

$$p_\phi^j = E\left[\prod_{k=1}^{m_\phi^j} \boldsymbol{J}_{\boldsymbol{z}_k^j}(\boldsymbol{X})\right] > E[\boldsymbol{J}_{\boldsymbol{z}_1^j}(\boldsymbol{X})]E\left[\prod_{k=2}^{m_\phi^j} \boldsymbol{J}_{\boldsymbol{z}_k^j}(\boldsymbol{X})\right].$$

Finally by applying Theorem 3.1, page 32 of Barlow and Proschan (1975a) on the right-hand side, inequality 1 follows. The equality in 1, when there is only one minimal cut set to level $j$, is straightforward from the proof above.

By combining Equations (2.28) and (2.29) we arrive at the following corollary.

**Corollary 2.30:** Make the same assumptions as in the first part of Theorem 2.28 and define for $j = 1, \ldots, M$

$$L_\phi^j = \max[\ell_\phi'^j(\boldsymbol{P}_\phi), \ell_\phi^{*j}] \qquad \bar{L}_\phi^j = \max[\bar{\ell}_\phi'^j(\boldsymbol{Q}_\phi), \bar{\ell}_\phi^{*j}].$$

Then

$$L_\phi^j \leq p_\phi^j \leq 1 - \bar{L}_\phi^j. \tag{2.31}$$

An objection to these bounds is that $\ell_\phi^{*j}$ and $\bar{\ell}_\phi^{*j}$ seem very complex. This is dealt with in the next corollary, the price paid being stronger assumptions.

**Corollary 2.31:** Make the same assumptions as in the second part of Theorem 2.28 and define for $j = 1, \ldots, M$

$$L_\phi^{*j}(\boldsymbol{P}_\phi) = \max[\ell_\phi'^j(\boldsymbol{P}_\phi), \ell_\phi^{**j}(\boldsymbol{P}_\phi)]$$

$$\bar{L}_\phi^{*j}(\boldsymbol{Q}_\phi) = \max[\bar{\ell}_\phi'^j(\boldsymbol{Q}_\phi), \bar{\ell}_\phi^{**j}(\boldsymbol{Q}_\phi)].$$

Then

$$L_\phi^j = L_\phi^{*j}(\boldsymbol{P}_\phi) \leq p_\phi^j \leq 1 - \bar{L}_\phi^{*j}(\boldsymbol{Q}_\phi) = 1 - \bar{L}_\phi^j. \tag{2.32}$$

Due to a counterexample in Butler (1982) the bounds on $p_\phi^j$ are not certain to be nonincreasing in $j$. Corollary 3.1 of the latter paper hence inspires the next two corollaries, proofs being straightforward.

**Corollary 2.32:** Make the same assumptions as in the first part of Theorem 2.28 and define for $j = 1, \ldots, M$

$$B_\phi^j = \max_{j \leq k \leq M}[L_\phi^k] \qquad \bar{B}_\phi^j = \max_{1 \leq k \leq j}[\bar{L}_\phi^k].$$

Then

$$L_\phi^j \le B_\phi^j \le p_\phi^j \le 1 - \bar{B}_\phi^j \le 1 - \bar{L}_\phi^j. \tag{2.33}$$

**Corollary 2.33:** Make the same assumptions as in the second part of Theorem 2.28 and define for $j = 1, \ldots, M$

$$B_\phi^{*j}(\boldsymbol{P}_\phi) = \max_{j \le k \le M} [L_\phi^{*k}(\boldsymbol{P}_\phi)] \qquad \bar{B}_\phi^{*j}(\boldsymbol{Q}_\phi) = \max_{1 \le k \le j} [\bar{L}_\phi^{*k}(\boldsymbol{Q}_\phi)].$$

Then

$$\begin{aligned} L_\phi^{*j}(\boldsymbol{P}_\phi) \le B_\phi^j = B_\phi^{*j}(\boldsymbol{P}_\phi) \le p_\phi^j \le 1 - \bar{B}_\phi^{*j}(\boldsymbol{Q}_\phi) \\ = 1 - \bar{B}_\phi^j \le 1 - \bar{L}_\phi^{*j}(\boldsymbol{Q}_\phi). \end{aligned} \tag{2.34}$$

We conclude this section by generalizing Theorem 3.2 and Corollary 3.3, page 33 of Barlow and Proschan (1975a) where some simple dynamic models are considered. Let $\{X_i(t), t \ge 0\}$ denote the stochastic process representing the state of the $i$th component as a function of time $t, i = 1, \ldots, n$. Assume the components are in the perfect functioning state $M$ at $t = 0$ and are not repaired. Introduce the random variables

$$T_i^j = \inf\{t : X_i(t) \le j\} \quad i = 1, \ldots, n, j = 0, \ldots, M - 1,$$

representing the lifelength in the states $\{j + 1, \ldots, M\}$ of the $i$th component.

**Theorem 2.34:** If $T_1^j, \ldots, T_n^j$ are associated random variables, then

$$P[\cap_{i=1}^n (T_i^j > t_i)] \ge \prod_{i=1}^n P(T_i^j > t_i) \tag{2.35}$$

$$P[\cap_{i=1}^n (T_i^j \le t_i)] \ge \prod_{i=1}^n P(T_i^j \le t_i). \tag{2.36}$$

**Proof:**

$$P[\cap_{i=1}^n (T_i^j > t_i)] = P[\cap_{i=1}^n (X_i(t_i) \ge j + 1)] = P\left[\prod_{i=1}^n I_{j+1}(X_i(t_i)) = 1\right].$$

Since $I_{j+1}(X_i(t_i))$ is nondecreasing in $T_i^j, i = 1, \ldots, n$ it follows from property $P_3$ of associated random variables that these random variables are associated. Equation (2.35) now follows from Theorem 3.1, page 32 in Barlow and Proschan (1975a). Equation (2.36) is proved similarly.

**Corollary 2.35:** If $T_1^j, \ldots, T_n^j$ are associated random variables, then

$$P[\min_{1\leq i\leq n} T_i^j > t] \geq \prod_{i=1}^{n} P(T_i^j > t) \tag{2.37}$$

$$P[\max_{1\leq i\leq n} T_i^j > t] \leq \coprod_{i=1}^{n} P(T_i^j > t). \tag{2.38}$$

The proof follows from Theorem 2.34.

## 2.5 Stochastic performance of binary type multistate strongly coherent systems

In this section we mainly demonstrate that all results on stochastic performance obtained by Barlow and Wu (1978) extend either to a binary type multistate monotone system (BTMMS) or to a binary type multistate strongly coherent system (BTMSCS). We shall, however, also try to indicate that most of the theory for a binary coherent system (BCS) can be extended to a BTMSCS.

We start by generalizing Theorem 2.1 of Barlow and Wu (1978).

**Theorem 2.36:** Consider a BTMMS having binary monotone structure functions $\phi_1, \ldots, \phi_M$. Let $p_{\phi_j}$ be the reliability of $\phi_j$, i.e.

$$p_{\phi_j} = E\phi_j(\boldsymbol{I}_j(\boldsymbol{X})) \quad j = 1, \ldots, M. \tag{2.39}$$

Then

$$r_\phi^0 = 1 - p_{\phi_1}$$

$$r_\phi^j = p_{\phi_j} - p_{\phi_{j+1}} \quad j = 1, \ldots, M-1$$

$$r_\phi^M = p_{\phi_M}$$

$$p_\phi^j = p_{\phi_j} \quad j = 1, \ldots, M.$$

Furthermore, the performance function of the system, $h_\phi$, is given by $h_\phi = \sum_{j=1}^{M} p_{\phi_j}$.

**Proof:** The results follow immediately from Theorem 2.8 and Equation (2.15).

In Barlow and Wu (1978) $X_1, \ldots, X_n$ are assumed to be independent. Note that this is not assumed in the theorem above. Note also that in order to compute exact expressions for $r_\phi^0, \ldots, r_\phi^M$ and $h_\phi$, or to give upper and lower bounds for these quantities, we can just apply binary theory on $p_{\phi_1}, \ldots, p_{\phi_M}$.

If the components are independent, or some simpler assumptions on the dependence are made, the binary version of Theorem 2.22 improved by Satyanarayana and Chang (1983) and the binary version of Theorem 2.23 improved by Satyanarayana and Prabhakar (1978) and Satyanarayana (1982), can be applied to compute exact values of $p_{\phi_1}, \ldots, p_{\phi_M}$. If not, upper and lower bounds for $p_{\phi_1}, \ldots, p_{\phi_M}$ can be found from Natvig (1980) by choosing suitable modular decompositions for each $\phi_j$, $j = 1, \ldots, M$. Note especially that by assuming $X_1, \ldots, X_n$ to be associated random variables, it follows by Property $P_3$ of associated random variables that $I_j(X_1), \ldots, I_j(X_n)$ are associated for fixed $j \in \{1, \ldots, M\}$. This is often needed when applying Natvig (1980).

When components are repaired, the following definition seems natural.

**Definition 2.37:** Consider a BTMMS having binary monotone structure functions $\phi_1, \ldots, \phi_M$. Furthermore, consider a time interval $I = [t_A, t_B]$ and let $\tau(I) = \tau \cap I$, where $\tau$ is an index set contained in $[0, \infty)$. Let $j \in \{1, \ldots, M\}$. The availability, $p_{\phi_j}^{(I)}$ and the unavailability, $q_{\phi_j}^{(I)}$ to level $j$ in the time interval $I$ for this system are given by

$$p_{\phi_j}^{(I)} = P[\phi_j(I_j(\boldsymbol{X}(s))) = 1 \text{ for all } s \in \tau(I)]$$

$$q_{\phi_j}^{(I)} = P[\phi_j(I_j(\boldsymbol{X}(s))) = 0 \text{ for all } s \in \tau(I)].$$

By applying the theory of Natvig (1980) upper and lower bounds on $p_{\phi_j}^{(I)}$ and $q_{\phi_j}^{(I)}$ can be obtained from component availabilities and unavailabilities in the case of maintained, interdependent components.

We now return to Barlow and Wu (1978) and generalize their Proposition 2.3, which is again a generalization of the Moore–Shannon Theorem. To do this we need some notation. If $X_1, \ldots, X_n$ are independent, Equation (2.39) is written in the form

$$p_{\phi_j} = p_{\phi_j}(\boldsymbol{p}^j) \quad j = 1, \ldots, M, \tag{2.40}$$

where

$$\boldsymbol{p}^j = (p_1^j, \ldots, p_n^j). \tag{2.41}$$

If, in particular, $p_i^j = p_0^j, i = 1, \ldots, n$, we write

$$p_{\phi_j} = p_{\phi_j}(p_0^j) \quad j = 1, \ldots, M. \tag{2.42}$$

**Theorem 2.38:** Consider a BTMSCS where $X_1, \dots, X_n$ are independent and $n \geq 2$. Let

$$(r_i^0, r_i^1, \dots, r_i^M) = (\alpha_0, \alpha_1, \dots, \alpha_M) = \boldsymbol{\alpha} \quad i = 1, \dots, n.$$

Assume $p_{\phi_j}(p_0^j) = p_0^j$ for some $0 < p_0^j < 1$, $j = 1, \dots, M$. Then

$$\sum_{k=j}^{M} \alpha_k \leq p_0^j \quad j = 1, \dots, M \Rightarrow p_{\phi_j}(\sum_{k=j}^{M} \alpha_k) \leq \sum_{k=j}^{M} \alpha_k \quad j = 1, \dots, M \tag{2.43}$$

$$\sum_{k=j}^{M} \alpha_k \geq p_0^j \quad j = 1, \dots, M \Rightarrow p_{\phi_j}(\sum_{k=j}^{M} \alpha_k) \geq \sum_{k=j}^{M} \alpha_k \quad j = 1, \dots, M. \tag{2.44}$$

**Proof:** The result follows immediately from the Moore–Shannon Theorem, Theorem 5.4, page 46 in Barlow and Proschan (1975a).

The theorem allows us to compare the performance distribution of an arbitrary BTMSCS with identical components to the common performance distribution of its components. Note for $j = 1, \dots, M$ that if $\phi_j$ has no path sets or cut sets of size 1, then, from the Moore–Shannon Theorem, there exists $0 < p_0^j < 1$ such that $p_{\phi_j}(p_0^j) = p_0^j$.

We conclude this section by looking briefly into the measure of component importance suggested by Barlow and Wu (1978), extended to a BTMSCS by the following definition.

**Definition 2.39:** Consider a BTMSCS. A vector $\boldsymbol{x}$ is a critical vector for component $i$ and level $j$, $j = 0 \dots, M$ iff $x_i = j \Leftrightarrow \phi(\boldsymbol{x}) = j$. The probability importance of component $i$ with respect to system state $j$, $I^{i,j}$, is defined by $I^{i,j} = P[X_i = j \Leftrightarrow \phi(\boldsymbol{X}) = j]$.

For the case where $X_1, \dots, X_n$ are independent, Barlow and Wu (1978) give results on $I^{i,j}$ for their multistate coherent system. These results can readily be extended to a BTMSCS. However, we do not feel that Definition 2.39 is the most reasonable one. Let $\{0, 1, \dots, j-1\}$ correspond to the failure state when a binary approach is applied. Then the following definition reduces to the binary one given by Birnbaum (1969).

**Definition 2.40:** Consider a BTMSCS. A vector $\boldsymbol{x}$ is a critical vector for component $i$ and level $j$, $j = 0 \dots, M$ iff $x_i \geq j \Leftrightarrow \phi(\boldsymbol{x}) \geq j$. The probability importance of component $i$ with respect to system state $j$, $I_B^{i,j}$, is defined by $I_B^{i,j} = P[X_i \geq j \Leftrightarrow \phi(\boldsymbol{X}) \geq j]$.

The following result is well known in the binary case and is proved along the same lines.

**Theorem 2.41:** Consider a BTMSCS where $X_1, \ldots, X_n$ are independent. Then for $i = 1, \ldots, n$, $j = 0. \ldots, M$

$$I_B^{i,j} = p_{\phi_j}(1_i, \boldsymbol{p}^j) - p_{\phi_j}(0_i, \boldsymbol{p}^j).$$

**Proof:**

$$\begin{aligned} I_B^{i,j} &= P[\phi_j(1_i, \boldsymbol{I}_j(\boldsymbol{X})) - \phi_j(0_i, \boldsymbol{I}_j(\boldsymbol{X})) = 1] \\ &= E[\phi_j(1_i, \boldsymbol{I}_j(\boldsymbol{X})) - \phi_j(0_i, \boldsymbol{I}_j(\boldsymbol{X}))] \\ &= E[\phi_j(1_i, \boldsymbol{I}_j(\boldsymbol{X}))] - E[\phi_j(0_i, \boldsymbol{I}_j(\boldsymbol{X}))], \end{aligned}$$

leading to the result.

## 2.6 Exercises

2.1 Verify Equation (2.3).

2.2 Show that the $k$-out-of-$n$ system having structure function $\phi(\boldsymbol{x}) = x_{(n-k+1)}$ is a special case of an MMS.

2.3 Show that the structure functions given by Equations (1.1)–(1.6) are examples of an MMS.

2.4 Show that the dual system of an MMS is an MMS.

2.5 Show that if an MMS $(C, \phi)$ has a modular decomposition $\{(A_k, \chi_k)\}_{k=1}^r$ with organizing structure function $\psi$, $\phi^D$ has a modular decomposition $\{(A_k, \chi_k^D)\}_{k=1}^r$ with organizing structure function $\psi^D$.

2.6 Show that MSCS, MCS and MWCS are generalizations of the multistate coherent system suggested by El-Neweihi *et al.* (1978) given in Definition 1.10.

2.7 Show that the MMS of Table 2.1 is an MSCS, but not an MCS suggested by El-Neweihi *et al.* (1978).

2.8 Show that the MMS of Table 2.2 is an MWCS, but neither an MSCS nor an MCS.

2.9 Show that the MMS represented by the simple network in Figure 1.1, with structure function tabulated in Table 1.1, is an MSCS, but not a BTMMS.

2.10 Show that the MMS of Table 2.4 is an MCS suggested by El-Neweihi *et al.* (1978), but not a BTMMS.

2.11 Show that the minimal path and cut vectors for the simple network system of Figure 1.1 are given respectively in Tables 2.5 and 2.6.

2.12 Show that the minimal path and cut vectors for the structure function $\phi_1$ given by Equation (1.3) are given respectively in Tables 2.7 and 2.8.

2.13 Show that the minimal path and cut vectors for the structure function $\phi_2$ given by Equation (1.4) are given respectively in Tables 2.9 and 2.10.

2.14 Show that the minimal path and cut vectors for the module given by Equation (2.4) are given respectively in Tables 2.11 and 2.12.

2.15 Show that the minimal path and cut vectors for the organizing structure function given by Equation (2.5) are given respectively in Tables 2.13 and 2.14.

2.16 Prove Theorem 2.18.

2.17 Prove Equation (2.23).

2.18 Verify Equation (2.26).

2.19 Prove Equation (2.36).

2.20 Prove Equation (2.38).

# 3

# Bounds for system availabilities and unavailabilities

Agrawal and Barlow (1984) demonstrate that computational complexity makes it impossible to arrive at exact reliabilities associated with a large binary coherent system even if the system can be represented by a graph and its components are unrealistically assumed to be independent and not maintained. Therefore, bounds seem to be a necessity. In this chapter upper and lower bounds as given in Funnemark and Natvig (1985), Natvig (1986) and Natvig (1993) for the availabilities and unavailabilities, to any level, in a fixed time interval are arrived at for an MMS based on corresponding information on the multistate components. These are assumed to be maintained and interdependent. In the latter paper, sufficient conditions are given for some of these bounds to be strict, and also exactly correct, contributing to the understanding of the nature of the bounds and to their applicability. In fact, exactly correct expressions are obtainable just for trivial systems. Such bounds are of great interest when trying to predict the performance process of the system. The bounds given generalize the existing corresponding bounds established by Natvig (1980) for a BMS in traditional binary theory, again generalizing and extending work by Esary and Proschan (1970), Bodin (1970) and Barlow

*Multistate Systems Reliability Theory with Applications* Bent Natvig

and Proschan (1975a). The bounds also generalize and extend the ones given in Theorems 2.27–2.29 and Corollaries 2.30–2.33 for an MMS at a fixed point of time. The bounds are, in Section 3.8, applied to the offshore electrical power generation system presented in Section 1.2 and then in Chapter 4 to an offshore gas pipeline network.

## 3.1 Performance processes of the components and the system

We start by giving a series of basic definitions. Let $\tau$ be an index set contained in $[0, \infty)$.

**Definition 3.1:** The performance process of the $i$th component, $i = 1, \ldots, n$ is a stochastic process $\{X_i(t), t \in \tau\}$, where for each fixed $t \in \tau$, $X_i(t)$ is a random variable which takes values in $S_i$. $X_i(t)$ denotes the state of the $i$th component at time $t$. The joint performance process for the components $\{\boldsymbol{X}(t), t \in \tau\} = \{(X_1(t), \ldots, X_n(t)), t \in \tau\}$ is the corresponding vector stochastic process. The performance process of an MMS with structure function $\phi$ is a stochastic process $\{\phi(\boldsymbol{X}(t)), t \in \tau\}$, where for each fixed $t \in \tau, \phi(\boldsymbol{X}(t))$ is a random variable which takes values in $S$. $\phi(\boldsymbol{X}(t))$ denotes the system state at time $t$.

We assume that the sample functions of the performance process of a component are continuous from the right on $\tau$. It then follows that the sample functions of $\{\phi(\boldsymbol{X}(t)), t \in \tau\}$ are also continuous from the right on $\tau$. Now consider a time interval $I = [t_A, t_B] \subset [0, \infty)$ and let $\tau(I) = \tau \cap I$.

**Definition 3.2:** The marginal performance processes $\{X_i(t), t \in \tau\}$, $i = 1, \ldots, n$ are independent in the time interval $I$ iff, for any integer $m$ and $\{t_1, \ldots, t_m\} \subset \tau(I)$ the random vectors $\{X_1(t_1), \ldots, X_1(t_m)\}, \ldots, \{X_n(t_1), \ldots, X_n(t_m)\}$ are independent.

**Definition 3.3:** A modular decomposition $\{(A_k, \chi_k)\}_{k=1}^r$ of an MMS $(C, \phi)$ consists of totally independent modules in the time interval $I$ iff, for any integer $m$ and $\{t_1, \ldots, t_m\} \subset \tau(I)$ the random vectors $\{\boldsymbol{X}^{A_1}(t_1), \ldots, \boldsymbol{X}^{A_1}(t_m)\}, \ldots, \{\boldsymbol{X}^{A_k}(t_1), \ldots, \boldsymbol{X}^{A_k}(t_m)\}$ are independent.

**Definition 3.4:** The joint performance process for the components $\{\boldsymbol{X}(t), t \in \tau\}$ is associated in the time interval $I$ iff, for any integer $m$

and $\{t_1, \ldots, t_m\} \subset \tau(I)$ the random variables in the array

$$\begin{array}{c} X_1(t_1) \ldots X_1(t_m) \\ \vdots \\ X_n(t_1) \ldots X_n(t_m) \end{array}$$

are associated.

This definition obviously applies to a marginal performance process too.

**Definition 3.5:** Let $i = 1, \ldots, n$, $j = 1, \ldots, M$. The availability, $p_i^{j(I)}$, and the unavailability, $q_i^{j(I)}$, to level $j$ in the time interval $I$ of the $i$th component are given by

$$p_i^{j(I)} = P[X_i(s) \geq j \ \ \forall s \in \tau(I)] \qquad q_i^{j(I)} = P[X_i(s) < j \ \ \forall s \in \tau(I)].$$

The availability, $p_\phi^{j(I)}$, and the unavailability , $q_\phi^{j(I)}$, to level $j$ in the time interval $I$ for an MMS with structure function $\phi$ are given by

$$p_\phi^{j(I)} = P[\phi(\boldsymbol{X}(s)) \geq j \ \ \forall s \in \tau(I)]$$

$$q_\phi^{j(I)} = P[\phi(\boldsymbol{X}(s)) < j \ \ \forall s \in \tau(I)].$$

Finally, introduce the $n \times M$ component availability and unavailability matrices

$$\boldsymbol{P}_\phi^{(I)} = \{p_i^{j(I)}\}_{\substack{i=1,\ldots,n \\ j=1,\ldots,M}} \qquad \boldsymbol{Q}_\phi^{(I)} = \{q_i^{j(I)}\}_{\substack{i=1,\ldots,n \\ j=1,\ldots,M}}.$$

Note that

$$p_\phi^{j(I)} + q_\phi^{j(I)} \leq 1, \tag{3.1}$$

with equality for the case $\tau(I) = [t, t]$. From Definition 2.2, remembering the dual level $j^D = M - j + 1$ for $(C, \phi^D)$ introduced in Theorem 2.18, we obviously have the following duality relation

$$p_{\phi^D}^{j^D(I)} = q_\phi^{j(I)}. \tag{3.2}$$

Note that $(\phi^D)^D = \phi$ and $(j^D)^D = j$. Hence, Equation (3.2) implies

$$p_\phi^{j(I)} = q_{\phi^D}^{j^D(I)}. \tag{3.3}$$

When we write down duality relations later in this chapter, we present only one of the two versions, the other one being straightforward. Let

$$p_i^{Dj^D(I)} = P[X_i^D(s) \geq j^D \;\; \forall s \in \tau(I)]$$
$$q_i^{Dj^D(I)} = P[X_i^D(s) < j^D \;\; \forall s \in \tau(I)].$$

We also introduce the dual matrices by the convention

$$\boldsymbol{P}_{\phi^D}^{D(I)} = \{p_i^{Dj^D(I)}\}_{\substack{i=1,\ldots,n \\ j=1,\ldots,M}} \qquad \boldsymbol{Q}_{\phi^D}^{D(I)} = \{q_i^{Dj^D(I)}\}_{\substack{i=1,\ldots,n \\ j=1,\ldots,M}}, \tag{3.4}$$

and hence, parallel to Equation (3.2)

$$\boldsymbol{P}_{\phi^D}^{D(I)} = \boldsymbol{Q}_{\phi}^{(I)}. \tag{3.5}$$

In Natvig (1993) it is stated that for the case where the marginal performance processes of the components are independent in $I$, no additional assumption that each of these is associated in $I$ is needed to establish any strict or non-strict bounds. Neither Esary and Proschan (1970) and Natvig (1980) treating the binary case nor Funnemark and Natvig (1985) and Natvig (1986) treating the multistate case were aware of this. Accordingly, this was not taken into account in the offshore electrical power generation case study considered in Natvig *et al.* (1986).

In Section 3.2 basic bounds on the availability, $p_\phi^{j(I)}$ and the unavailability, $q_\phi^{j(I)}$ to level $j$ in the time interval $I$ are arrived at, whereas improved bounds are given in Section 3.3 by using modular decompositions. Such bounds as the latter are in fact the only ones of practical interest for large systems, as has also been pointed out by Butler (1982). The reason is that the bounds in Section 3.2 are based on all minimal path and cut vectors of the system. To arrive at these for a system of a large number of components seems impossible even for present-day computers. However, the number of components of each module, and the number of modules, may be chosen to be moderate, making it possible to arrive at the minimal path and cut vectors both of the organizing structure and of each module. The strategy is then to arrive at bounds for the availabilities and unavailabilities for the modules and then insert these into the bounds for the availabilities and unavailabilities for the organizing structure. This finally leads to improved bounds for the availabilities and unavailabilities for the system. Furthermore, in Section 3.4 the special case $I = [t, t]$ is treated, giving improved bounds using modular decompositions. Due to the mentioned inaccuracy in Theorem 3.1 in Butler (1982), inherited by Funnemark and Natvig (1985), some necessary corrections on these improved bounds have been made. These corrections are mainly along the lines in these papers, but partly inspired by Gåsemyr (2010). In Section 3.5 strict and exactly correct bounds are given. The explicit bounds are based on the availabilities and unavailabilities of the components. These are determined in Section 3.6 for the case where the performance processes of the components are Markovian. In Section 3.7 the theory is applied to the simple network system given in Figure 1.1, and the chapter is concluded in Section 3.8 by applying the theory to the offshore electrical power generation system presented in Section 1.2.

## 3.2 Basic bounds in a time interval

We start by generalizing the Min–Max bounds given for a fixed point of time in Theorem 2.27 to a time interval $I$.

**Theorem 3.6:** Let $(C, \phi)$ be an MMS. Furthermore, for $j \in \{1, \ldots, M\}$ let $\boldsymbol{y}_k^j = (y_{1k}^j, \ldots, y_{nk}^j)$, $k = 1, \ldots, n_\phi^j$ be its minimal path vectors to level $j$ and $\boldsymbol{z}_k^j = (z_{1k}^j, \ldots, z_{nk}^j)$, $k = 1, \ldots, m_\phi^j$ be its minimal cut vectors to level $j$ and

$$C_\phi^j(\boldsymbol{y}_k^j), k = 1, \ldots, n_\phi^j \text{ and } D_\phi^j(\boldsymbol{z}_k^j), k = 1, \ldots, m_\phi^j$$

the corresponding minimal path and cut sets to level $j$. Let

$$\ell_\phi^{\prime\prime j(I)} = \max_{1 \le k \le n_\phi^j} P[\cap_{i \in C_\phi^j(\boldsymbol{y}_k^j)} (X_i(s) \ge y_{ik}^j) \ \forall s \in \tau(I)]$$

$$u_\phi^{\prime\prime j(I)} = \min_{1 \le k \le m_\phi^j} P[\cup_{i \in D_\phi^j(\boldsymbol{z}_k^j)} (X_i(s) > z_{ik}^j) \ \forall s \in \tau(I)]$$

$$\bar{\ell}_\phi^{\prime\prime j(I)} = \max_{1 \le k \le m_\phi^j} P[\cap_{i \in D_\phi^j(\boldsymbol{z}_k^j)} (X_i(s) \le z_{ik}^j) \ \forall s \in \tau(I)]$$

$$\bar{u}_\phi^{\prime\prime j(I)} = \min_{1 \le k \le n_\phi^j} P[\cup_{i \in C_\phi^j(\boldsymbol{y}_k^j)} (X_i(s) < y_{ik}^j) \ \forall s \in \tau(I)].$$

Then

$$\ell_\phi^{\prime\prime j(I)} \le p_\phi^{j(I)} \le u_\phi^{\prime\prime j(I)} \tag{3.6}$$

$$\bar{\ell}_\phi^{\prime\prime j(I)} \le q_\phi^{j(I)} \le \bar{u}_\phi^{\prime\prime j(I)}. \tag{3.7}$$

Furthermore, $\ell_\phi^{\prime\prime j(I)}$ and $u_\phi^{\prime\prime j(I)}$ are nonincreasing and $\bar{\ell}_\phi^{\prime\prime j(I)}$ and $\bar{u}_\phi^{\prime\prime j(I)}$ non-decreasing in $j$ as is true respectively for $p_\phi^{j(I)}$ and $q_\phi^{j(I)}$. Let

$$\ell_\phi^{\prime j(I)}(\boldsymbol{P}_\phi^{(I)}) = \max_{1 \leq k \leq n_\phi^j} \prod_{i \in C_\phi^j(\boldsymbol{y}_k^j)} p_i^{y_{ik}^j(I)}$$

$$\bar{\ell}_\phi^{\prime j(I)}(\boldsymbol{Q}_\phi^{(I)}) = \max_{1 \leq k \leq m_\phi^j} \prod_{i \in D_\phi^j(\boldsymbol{z}_k^j)} q_i^{z_{ik}^j+1(I)}.$$

If the joint performance process of the system's components is associated in $I$, or the marginal performance processes of the components are independent in $I$, then

$$\ell_\phi^{\prime j(I)}(\boldsymbol{P}_\phi^{(I)}) \leq p_\phi^{j(I)} \leq \inf_{t \in \tau(I)} [1 - \bar{\ell}_\phi^{\prime j([t,t])}(\boldsymbol{Q}_\phi^{([t,t])})] \leq 1 - \bar{\ell}_\phi^{\prime j(I)}(\boldsymbol{Q}_\phi^{(I)}) \tag{3.8}$$

$$\bar{\ell}_\phi^{\prime j(I)}(\boldsymbol{Q}_\phi^{(I)}) \leq q_\phi^{j(I)} \leq \inf_{t \in \tau(I)} [1 - \ell_\phi^{\prime j([t,t])}(\boldsymbol{P}_\phi^{([t,t])})] \leq 1 - \ell_\phi^{\prime j(I)}(\boldsymbol{P}_\phi^{(I)}). \tag{3.9}$$

Actually, when the marginal performance processes of the components are independent in $I$, the lower bounds of Equations (3.6) and (3.7) reduce to the corresponding ones of Equations (3.8) and (3.9). Furthermore, $\ell_\phi^{\prime j(I)}(\boldsymbol{P}_\phi^{(I)})$ and $\bar{\ell}_\phi^{\prime j(I)}(\boldsymbol{Q}_\phi^{(I)})$ are respectively nonincreasing and nondecreasing in $j$.

**Proof:** Following the proof of Theorem 2.27 we easily get, for $k = 1, \ldots, n_\phi^j$ and $r = 1, \ldots, m_\phi^j$

$$P[J_{\boldsymbol{y}_k^j}(\boldsymbol{X}(s)) = 1 \ \ \forall s \in \tau(I)] \leq p_\phi^{j(I)} \leq P[J_{\boldsymbol{z}_r^j}(\boldsymbol{X}(s)) = 1 \ \ \forall s \in \tau(I)].$$

Equation (3.6) now follows by maximizing over $k$ and minimizing over $r$. Remembering Theorem 2.18 and Equation (3.2), applying Equation (3.6) on the dual structure and dual level, gives Equation (3.7). Since for all $k \in \{1, \ldots, n_\phi^{j+1}\}$ there exists $l \in \{1, \ldots, n_\phi^j\}$ such that $\boldsymbol{y}_k^{j+1} \geq \boldsymbol{y}_l^j$, it follows that $\ell_\phi^{\prime\prime j(I)}$ and $\ell_\phi^{\prime j(I)}(\boldsymbol{P}_\phi^{(I)})$ are nonincreasing and $\bar{u}_\phi^{\prime\prime j(I)}$ nondecreasing in $j$. The corresponding properties of the three other bounds follow similarly.

Let, as in Esary and Proschan (1970), $S$ be a countable subset of $\tau(I)$ that is dense in $\tau(I)$. Since the sample functions of $\{X_i(t), t \in \tau\}, i =$

$1, \ldots, n$ are continuous from the right on $\tau$, then

$$\ell_{\phi}^{\prime\prime j(I)} = \max_{1 \leq k \leq n_{\phi}^{j}} P[\cap_{i \in C_{\phi}^{j}(\boldsymbol{y}_{k}^{j})}(X_i(s) \geq y_{ik}^{j}) \ \ \forall s \in S].$$

Let $S_m = \{s_1, \ldots, s_m\}$ $m = 1, 2, \ldots$ be subsets of $S$ such that $S_m \uparrow S$ as $m \to \infty$. By monotone convergence

$$\max_{1 \leq k \leq n_{\phi}^{j}} P[\cap_{i \in C_{\phi}^{j}(\boldsymbol{y}_{k}^{j})}(X_i(s) \geq y_{ik}^{j}) \ \ \forall s \in S_m] \downarrow \ell_{\phi}^{\prime\prime j(I)},$$

as $m \to \infty$. Remembering the indicators given by Equation (2.1) we have

$$\begin{aligned}
&\max_{1 \leq k \leq n_{\phi}^{j}} P[\cap_{i \in C_{\phi}^{j}(\boldsymbol{y}_{k}^{j})}(X_i(s) \geq y_{ik}^{j}) \ \ \forall s \in S_m] \\
&= \max_{1 \leq k \leq n_{\phi}^{j}} P[\prod_{i \in C_{\phi}^{j}(\boldsymbol{y}_{k}^{j})} \prod_{l=1}^{m} I_{y_{ik}^{j}}(X_i(s_l)) = 1] \\
&\geq \max_{1 \leq k \leq n_{\phi}^{j}} \prod_{i \in C_{\phi}^{j}(\boldsymbol{y}_{k}^{j})} P[X_i(s) \geq y_{ik}^{j} \ \ \forall s \in S_m].
\end{aligned}$$

This inequality follows by applying Theorem 3.1, page 32 of Barlow and Proschan (1975a) since the assumption that the joint performance process of the system's components is associated in $I$, implies that the random variables $\prod_{l=1}^{m} I_{y_{ik}^{j}}(X_i(s_l))$, $i = 1, \ldots, n$ are associated by Property $P_3$ of associated random variables. Finally, by monotone convergence

$$\max_{1 \leq k \leq n_{\phi}^{j}} \prod_{i \in C_{\phi}^{j}(\boldsymbol{y}_{k}^{j})} P[X_i(s) \geq y_{ik}^{j} \ \ \forall s \in S_m] \downarrow \ell_{\phi}^{\prime j(I)}(\boldsymbol{P}_{\phi}^{(I)}),$$

as $m \to \infty$ and the lower bound of Equation (3.8) follows from the corresponding one of Equation (3.6). The lower bound of Equation (3.9) follows by applying the lower bound of Equation (3.8) on the dual structure and dual level, as shown for a fixed point of time after the proof of Theorem 2.27.

To prove the upper bounds of Equation (3.8) we apply the lower bound of Equation (3.9) to get

$$\begin{aligned}
p_{\phi}^{j(I)} &\leq \inf_{t \in \tau(I)} p_{\phi}^{j([t,t])} = \inf_{t \in \tau(I)} [1 - q_{\phi}^{j([t,t])}] \leq \inf_{t \in \tau(I)} [1 - \bar{\ell}_{\phi}^{\prime j([t,t])}(\boldsymbol{Q}_{\phi}^{([t,t])})] \\
&\leq 1 - \bar{\ell}_{\phi}^{\prime j(I)}(\boldsymbol{Q}_{\phi}^{(I)}).
\end{aligned}$$

The upper bounds of Equation (3.9) are proved similarly. When the marginal performance processes of the components are independent in $I$, a proper argument can be used to verify the rather obvious fact that $\min_{s\in\tau(I)} X_1(s), \min_{s\in\tau(I)} X_2(s), \ldots, \min_{s\in\tau(I)} X_n(s)$ are independent random variables. Hence,

$$\ell_\phi^{\prime\prime j(I)} = \max_{1\le k\le n_\phi^j} P[\cap_{i\in C_\phi^j(\boldsymbol{y}_k^j)}(\min_{s\in\tau(I)} X_i(s) \ge y_{ik}^j)] = \ell_\phi^{\prime j(I)}(\boldsymbol{P}_\phi^{(I)}).$$

Accordingly, under this condition the lower bound of Equation (3.6) reduces to the corresponding one of Equation (3.8). A similar argument using $\max_{s\in\tau(I)}$ instead of $\min_{s\in\tau(I)}$ shows that the lower bound of Equation (3.7) reduces to the corresponding one of Equation (3.9). Having established the lower bounds also under this condition, the corresponding upper bounds follow as above.

The bounds in Equations (3.6) and (3.7) are very general, but seem of little practical value due to their complexity.

We next generalize the bounds given for a fixed point of time in Theorem 2.28 to a time interval $I$.

**Theorem 3.7:** Let $(C, \phi)$ be an MMS. Furthermore, for $j \in \{1, \ldots, M\}$ let $\boldsymbol{y}_k^j = (y_{1k}^j, \ldots, y_{nk}^j)$, $k = 1, \ldots, n_\phi^j$ be its minimal path vectors to level $j$ and $\boldsymbol{z}_k^j = (z_{1k}^j, \ldots, z_{nk}^j)$, $k = 1, \ldots, m_\phi^j$ be its minimal cut vectors to level $j$ and

$$C_\phi^j(\boldsymbol{y}_k^j), k = 1, \ldots, n_\phi^j \quad \text{and} \quad D_\phi^j(\boldsymbol{z}_k^j), k = 1, \ldots, m_\phi^j$$

the corresponding minimal path and cut sets to level $j$. Let

$$\ell_\phi^{*j(I)} = \prod_{k=1}^{m_\phi^j} P(\cup_{i\in D_\phi^j(\boldsymbol{z}_k^j)}(X_i(s) > z_{ik}^j) \ \forall s \in \tau(I))$$

$$\bar{\ell}_\phi^{*j(I)} = \prod_{k=1}^{n_\phi^j} P(\cup_{i\in C_\phi^j(\boldsymbol{y}_k^j)}(X_i(s) < y_{ik}^j) \ \forall s \in \tau(I)).$$

If the joint performance process of the system's components is associated in $I$, then

$$\ell_\phi^{*j(I)} \le p_\phi^{j(I)} \le \inf_{t\in\tau(I)} [1 - \bar{\ell}_\phi^{*j([t,t])}] \le 1 - \bar{\ell}_\phi^{*j(I)} \tag{3.10}$$

$$\bar{\ell}_\phi^{*j(I)} \le q_\phi^{j(I)} \le \inf_{t\in\tau(I)} [1 - \ell_\phi^{*j([t,t])}] \le 1 - \ell_\phi^{*j(I)}. \tag{3.11}$$

Furthermore, let

$$\ell_\phi^{**j(I)}(\boldsymbol{P}_\phi^{(I)}) = \prod_{k=1}^{m_\phi^j} \coprod_{i \in D_\phi^j(\boldsymbol{z}_k^j)} p_i^{z_{ik}^j+1(I)}$$

$$\bar{\ell}_\phi^{**j(I)}(\boldsymbol{Q}_\phi^{(I)}) = \prod_{k=1}^{n_\phi^j} \coprod_{i \in C_\phi^j(\boldsymbol{y}_k^j)} q_i^{y_{ik}^j(I)}.$$

If the marginal performance processes of the components are independent in $I$, then

$$\begin{aligned} \ell_\phi^{**j(I)}(\boldsymbol{P}_\phi^{(I)}) \le p_\phi^{j(I)} &\le \inf_{t \in \tau(I)} [1 - \bar{\ell}_\phi^{**j([t,t])}(\boldsymbol{Q}_\phi^{([t,t])})] \\ &\le 1 - \bar{\ell}_\phi^{**j(I)}(\boldsymbol{Q}_\phi^{(I)}) \end{aligned} \tag{3.12}$$

$$\begin{aligned} \bar{\ell}_\phi^{**j(I)}(\boldsymbol{Q}_\phi^{(I)}) \le q_\phi^{j(I)} &\le \inf_{t \in \tau(I)} [1 - \ell_\phi^{**j([t,t])}(\boldsymbol{P}_\phi^{([t,t])})] \\ &\le 1 - \ell_\phi^{**j(I)}(\boldsymbol{P}_\phi^{(I)}). \end{aligned} \tag{3.13}$$

Actually, when the marginal performance processes of the components are independent and each of them is associated in $I$, we have

$$\ell_\phi^{**j(I)}(\boldsymbol{P}_\phi^{(I)}) \le \ell_\phi^{*j(I)} \tag{3.14}$$

$$\bar{\ell}_\phi^{**j(I)}(\boldsymbol{Q}_\phi^{(I)}) \le \bar{\ell}_\phi^{*j(I)}. \tag{3.15}$$

**Proof:** Again let $S$ be a countable subset of $\tau(I)$ that is dense in $\tau(I)$. Since the sample functions of $\{X_i(t), t \in \tau\}, i = 1, \ldots, n$ are continuous from the right on $\tau$, then by applying (ii) of Theorem 2.15

$$p_\phi^{j(I)} = P\left[\prod_{k=1}^{m_\phi^j} \boldsymbol{J}_{\boldsymbol{z}_k^j}(\boldsymbol{X}(s)) = 1 \ \ \forall s \in S\right].$$

Let $S_m = \{s_1, \ldots, s_m\}$ $m = 1, 2, \ldots$ be subsets of $S$ such that $S_m \uparrow S$ as $m \to \infty$. By monotone convergence

$$P\left[\prod_{k=1}^{m_\phi^j} \boldsymbol{J}_{\boldsymbol{z}_k^j}(\boldsymbol{X}(s)) = 1 \ \ \forall s \in S_m\right] \downarrow p_\phi^{j(I)},$$

as $m \to \infty$. We have

$$P\left[\prod_{k=1}^{m_\phi^j} J_{z_k^j}(X(s)) = 1 \ \ \forall s \in S_m\right] = P\left[\prod_{k=1}^{m_\phi^j}\prod_{l=1}^{m} J_{z_k^j}(X(s_l)) = 1\right]$$

$$\geq \prod_{k=1}^{m_\phi^j} P[J_{z_k^j}(X(s)) = 1 \ \ \forall s \in S_m].$$

This inequality follows by applying Theorem 3.1, page 32 of Barlow and Proschan (1975a) since the assumption that the joint performance process of the system's components is associated in $I$ implies that the random variables $\prod_{l=1}^{m} J_{z_k^j}(X(s_l)), k = 1, \ldots, m_\phi^j$ are associated by Property $P_3$ of associated random variables. Finally, by monotone convergence

$$\prod_{k=1}^{m_\phi^j} P[J_{z_k^j}(X(s)) = 1 \ \ \forall s \in S_m] \downarrow \ell_\phi^{*j(I)},$$

as $m \to \infty$. Hence, the left inequality of Equation (3.10) is proved.

Next we prove the left inequality of Equation (3.12). From (ii) of Theorem 2.15 we have

$$\begin{aligned} p_\phi^{j(I)} &= P[\forall k \in \{1, \ldots, m_\phi^j\} \ \forall s \in \tau(I) \ \exists i \in D_\phi^j(z_k^j) \text{ such that} \\ &\quad X_i(s) \geq z_{ik}^j + 1] \\ &\geq P[\forall k \in \{1, \ldots, m_\phi^j\} \ \exists i \in D_\phi^j(z_k^j) \text{ such that} \\ &\quad X_i(s) \geq z_{ik}^j + 1 \ \forall s \in \tau(I)]. \end{aligned}$$

Introduce, for $k \in \{1, \ldots, m_\phi^j\}$, the following generalization to an interval $I$ of the minimal cut parallel structure $J_{z_k^j}(x)$ to level $j$.

$$J_{z_k^j}(x^{\min(I)}) = \begin{cases} 1 & \text{if } \exists i \in D_\phi^j(z_k^j) \text{ such that } \min_{s \in \tau(I)} x_i(s) \geq z_{ik}^j + 1 \\ 0 & \text{otherwise} \end{cases}. \tag{3.16}$$

Then the inequality above can be written

$$p_\phi^{j(I)} \geq P\left[\prod_{k=1}^{m_\phi^j} J_{z_k^j}(X^{\min(I)}) = 1\right]. \tag{3.17}$$

Since the marginal performance processes of the components are independent in $I$, as in the proof of Theorem 3.6 $\min_{s\in\tau(I)} X_1(s), \min_{s\in\tau(I)} X_2(s), \ldots, \min_{s\in\tau(I)} X_n(s)$ are independent random variables. Furthermore, $\boldsymbol{J}_{\boldsymbol{z}_k^j}(\boldsymbol{X}^{\min(I)}), k \in \{1, \ldots, m_\phi^j\}$ are nondecreasing functions of $\min_{s\in\tau(I)} X_1(s), \min_{s\in\tau(I)} X_2(s), \ldots, \min_{s\in\tau(I)} X_n(s)$ and hence they are associated by Properties $P_5$ and $P_3$ of associated random variables. By applying Theorem 3.1, page 32 of Barlow and Proschan (1975a) we finally get

$$p_\phi^{j(I)} \geq \prod_{k=1}^{m_\phi^j} P[\boldsymbol{J}_{\boldsymbol{z}_k^j}(\boldsymbol{X}^{\min(I)}) = 1]$$

$$= \prod_{k=1}^{m_\phi^j} \coprod_{i\in D_\phi^j(\boldsymbol{z}_k^j)} p_i^{z_{ik}^j+1(I)} = \ell_\phi^{**j(I)}(\boldsymbol{P}_\phi^{(I)}).$$

Hence, the left inequality of Equation (3.12) is proved. The rest of the inequalities in Equations (3.10)–(3.13) are proved as in the proof of Equations (3.8) and (3.9). When the marginal performance processes of the components are independent and each of them is associated in $I$, it follows from Property $P_4$ of associated random variables and Definitions 3.2 and 3.4 that the joint performance process of the system's components is associated in $I$. Hence, the lower bounds of Equations (3.10) and (3.11) hold. We now get

$$\begin{aligned}
\ell_\phi^{*j(I)} &= \prod_{k=1}^{m_\phi^j} P[\forall s \in \tau(I)\ \exists i \in D_\phi^j(z_k^j) \text{ such that } X_i(s) \geq z_{ik}^j + 1] \\
&\geq \prod_{k=1}^{m_\phi^j} P[\exists i \in D_\phi^j(z_k^j) \text{ such that } X_i(s) \geq z_{ik}^j + 1\ \forall s \in \tau(I)] \\
&= \ell_\phi^{**j(I)}(\boldsymbol{P}_\phi^{(I)}),
\end{aligned}$$

having applied the fact that the marginal performance processes of the components are independent in $I$. Hence, the bound of Equation (3.14) is proved. The bound of Equation (3.15) follows by applying Equation (3.14) on the dual structure and dual level.

By combining Equations (3.8) and (3.10) respectively Equations (3.9) and (3.11) we arrive at the following corollary generalizing Corollary 2.30 to a time interval $I$.

**Corollary 3.8:** Make the same assumptions as in the first part of Theorem 3.7 and define, for $j = 1, \ldots, M$

$$L_\phi^{j(I)} = \max[\ell_\phi^{\prime j(I)}(\boldsymbol{P}_\phi^{(I)}), \ell_\phi^{*j(I)}] \quad \bar{L}_\phi^{j(I)} = \max[\bar{\ell}_\phi^{\prime j(I)}(\boldsymbol{Q}_\phi^{(I)}), \bar{\ell}_\phi^{*j(I)}].$$

Then

$$L_\phi^{j(I)} \le p_\phi^{j(I)} \le \inf_{t \in \tau(I)} [1 - \bar{L}_\phi^{j([t,t])}] \le 1 - \bar{L}_\phi^{j(I)} \tag{3.18}$$

$$\bar{L}_\phi^{j(I)} \le q_\phi^{j(I)} \le \inf_{t \in \tau(I)} [1 - L_\phi^{j([t,t])}] \le 1 - L_\phi^{j(I)}. \tag{3.19}$$

An objection to these bounds is that $\ell_\phi^{*j(I)}$ and $\bar{\ell}_\phi^{*j(I)}$ seem very complex. This is dealt with in the next corollary, generalizing Corollary 2.31 to a time interval $I$.

**Corollary 3.9:** Make the same assumptions as in the second part of Theorem 3.7 and define, for $j = 1, \ldots, M$

$$L_\phi^{*j(I)}(\boldsymbol{P}_\phi^{(I)}) = \max[\ell_\phi^{\prime j(I)}(\boldsymbol{P}_\phi^{(I)}), \ell_\phi^{**j(I)}(\boldsymbol{P}_\phi^{(I)})]$$

$$\bar{L}_\phi^{*j(I)}(\boldsymbol{Q}_\phi^{(I)}) = \max[\bar{\ell}_\phi^{\prime j(I)}(\boldsymbol{Q}_\phi^{(I)}), \bar{\ell}_\phi^{**j(I)}(\boldsymbol{Q}_\phi^{(I)})].$$

Then

$$L_\phi^{*j(I)}(\boldsymbol{P}_\phi^{(I)}) \le p_\phi^{j(I)} \le \inf_{t \in \tau(I)} [1 - \bar{L}_\phi^{*j([t,t])}(\boldsymbol{Q}_\phi^{([t,t])})] \le 1 - \bar{L}_\phi^{*j(I)}(\boldsymbol{Q}_\phi^{(I)}) \tag{3.20}$$

$$\bar{L}_\phi^{*j(I)}(\boldsymbol{Q}_\phi^{(I)}) \le q_\phi^{j(I)} \le \inf_{t \in \tau(I)} [1 - L_\phi^{*j([t,t])}(\boldsymbol{P}_\phi^{([t,t])})] \le 1 - L_\phi^{*j(I)}(\boldsymbol{P}_\phi^{(I)}). \tag{3.21}$$

Due to the counterexample in Butler (1982) when $I = [t, t]$ the bounds on $p_\phi^{j(I)}$ and $q_\phi^{j(I)}$ are not certain to be respectively nonincreasing and nondecreasing in $j$. This leads to the next two corollaries, generalizing Corollaries 2.32 and 2.33 to a time interval $I$.

**Corollary 3.10:** Make the same assumptions as in the first part of Theorem 3.7 and define, for $j = 1, \ldots, M$

$$B_\phi^{j(I)} = \max_{j \le k \le M} [L_\phi^{k(I)}]$$

$$\bar{B}_\phi^{j(I)} = \max_{1 \le k \le j} [\bar{L}_\phi^{k(I)}].$$

Then

$$L_\phi^{j(I)} \leq B_\phi^{j(I)} \leq p_\phi^{j(I)} \leq \inf_{t\in\tau(I)} [1 - \bar{B}_\phi^{j([t,t])}] \leq 1 - \bar{B}_\phi^{j(I)} \leq 1 - \bar{L}_\phi^{j(I)} \tag{3.22}$$

$$\bar{L}_\phi^{j(I)} \leq \bar{B}_\phi^{j(I)} \leq q_\phi^{j(I)} \leq \inf_{t\in\tau(I)} [1 - B_\phi^{j([t,t])}] \leq 1 - B_\phi^{j(I)} \leq 1 - L_\phi^{j(I)}. \tag{3.23}$$

**Corollary 3.11:** Make the same assumptions as in the second part of Theorem 3.7 and define, for $j = 1, \ldots, M$

$$B_\phi^{*j(I)}(\boldsymbol{P}_\phi^{(I)}) = \max_{j\leq k\leq M} [L_\phi^{*k(I)}(\boldsymbol{P}_\phi^{(I)})]$$

$$\bar{B}_\phi^{*j(I)}(\boldsymbol{Q}_\phi^{(I)}) = \max_{1\leq k\leq j} [\bar{L}_\phi^{*k(I)}(\boldsymbol{Q}_\phi^{(I)})].$$

Then

$$L_\phi^{*j(I)}(\boldsymbol{P}_\phi^{(I)}) \leq B_\phi^{*j(I)}(\boldsymbol{P}_\phi^{(I)}) \leq p_\phi^{j(I)} \leq \inf_{t\in\tau(I)} [1 - \bar{B}_\phi^{*j([t,t])}(\boldsymbol{Q}_\phi^{([t,t])})]$$

$$\leq 1 - \bar{B}_\phi^{*j(I)}(\boldsymbol{Q}_\phi^{(I)}) \leq 1 - \bar{L}_\phi^{*j(I)}(\boldsymbol{Q}_\phi^{(I)}) \tag{3.24}$$

$$\bar{L}_\phi^{*j(I)}(\boldsymbol{Q}_\phi^{(I)}) \leq \bar{B}_\phi^{*j(I)}(\boldsymbol{Q}_\phi^{(I)}) \leq q_\phi^{j(I)} \leq \inf_{t\in\tau(I)} [1 - B_\phi^{*j([t,t])}(\boldsymbol{P}_\phi^{([t,t])})]$$

$$\leq 1 - B_\phi^{*j(I)}(\boldsymbol{P}_\phi^{(I)}) \leq 1 - L_\phi^{*j(I)}(\boldsymbol{P}_\phi^{(I)}). \tag{3.25}$$

As for the corresponding bounds for a fixed point of time given in Section 2.4, it is easy to check that all bounds given in this section lie inside the interval [0, 1]. On the other hand, the upper bounds given, except in Equations (3.6) and (3.7) of Theorem 3.6, are poor if $p_\phi^{j(I)} + q_\phi^{j(I)}$ is not close to 1. We have not, however, been able to arrive at better bounds.

## 3.3 Improved bounds in a time interval using modular decompositions

We start by listing some duality relations which come out from the proofs in the last section

$$\ell_{\phi^D}''^{j^D(I)} = \bar{\ell}_\phi''^{j(I)} \tag{3.26}$$

$$u''^{j^D(I)}_{\phi^D} = \bar{u}''^{j(I)}_{\phi} \tag{3.27}$$

$$\ell'^{j^D(I)}_{\phi^D}(\boldsymbol{P}^{D(I)}_{\phi^D}) = \bar{\ell}'^{j(I)}_{\phi}(\boldsymbol{Q}^{(I)}_{\phi}) \tag{3.28}$$

$$L^{j^D(I)}_{\phi^D} = \bar{L}^{j(I)}_{\phi} \tag{3.29}$$

$$L^{*j^D(I)}_{\phi^D}(\boldsymbol{P}^{D(I)}_{\phi^D}) = \bar{L}^{*j(I)}_{\phi}(\boldsymbol{Q}^{(I)}_{\phi}) \tag{3.30}$$

$$B^{j^D(I)}_{\phi^D} = \bar{B}^{j(I)}_{\phi} \tag{3.31}$$

$$B^{*j^D(I)}_{\phi^D}(\boldsymbol{P}^{D(I)}_{\phi^D}) = \bar{B}^{*j(I)}_{\phi}(\boldsymbol{Q}^{(I)}_{\phi}). \tag{3.32}$$

We now consider an MMS $(C, \phi)$ with modular decomposition given by Definition 1.6. Obviously, the results of the previous section can be applied on $\psi$ and $\chi_k$, $k = 1, \ldots, r$ as well. Introduce the following $r \times M$ module availability and unavailability matrices

$$\boldsymbol{P}^{(I)}_{\psi} = \{p^{j(I)}_{\chi_k}\}_{\substack{k=1,\ldots,r\\ j=1,\ldots,M}} \qquad \boldsymbol{Q}^{(I)}_{\psi} = \{q^{j(I)}_{\chi_k}\}_{\substack{k=1,\ldots,r\\ j=1,\ldots,M}}, \tag{3.33}$$

and correspondingly define the following $r \times M$ matrices

$$\boldsymbol{\ell}''^{(I)}_{\psi}, \bar{\boldsymbol{\ell}}''^{(I)}_{\psi}, \boldsymbol{\ell}'^{(I)}_{\psi}(\boldsymbol{P}^{(I)}_{\phi}), \bar{\boldsymbol{\ell}}'^{(I)}_{\psi}(\boldsymbol{Q}^{(I)}_{\phi}), \boldsymbol{\ell}^{*(I)}_{\psi}, \bar{\boldsymbol{\ell}}^{*(I)}_{\psi}, \boldsymbol{L}^{(I)}_{\psi}, \bar{\boldsymbol{L}}^{(I)}_{\psi},$$
$$\boldsymbol{L}^{*(I)}_{\psi}(\boldsymbol{P}^{(I)}_{\phi}), \bar{\boldsymbol{L}}^{*(I)}_{\psi}(\boldsymbol{Q}^{(I)}_{\phi}), \boldsymbol{B}^{(I)}_{\psi}, \bar{\boldsymbol{B}}^{(I)}_{\psi}, \boldsymbol{B}^{*(I)}_{\psi}(\boldsymbol{P}^{(I)}_{\phi}), \bar{\boldsymbol{B}}^{*(I)}_{\psi}(\boldsymbol{Q}^{(I)}_{\phi}).$$

The two theorems to follow are generalizations of Theorems 2.6 and 2.7 in Natvig (1980). In the first one we find lower bounds that are improvements of the lower bounds in Equations (3.6) and (3.7) of Theorem 3.6. This is by no means proved with regard to the upper bounds.

**Theorem 3.12:** Let $(C, \phi)$ be an MMS with modular decomposition given by Definition 1.6 consisting of totally independent modules in the time interval $I$. Then for $j = 1, \ldots, M$

$$\begin{aligned}
\ell''^{j(I)}_{\phi} &\leq^6 B^{*j(I)}_{\psi}(\boldsymbol{\ell}''^{(I)}_{\psi}) \leq^5 B^{*j(I)}_{\psi}(\boldsymbol{P}^{(I)}_{\psi}) \leq^4 p^{j(I)}_{\phi} \\
&\leq^3 \inf_{t \in \tau(I)} [1 - \bar{B}^{*j([t,t])}_{\psi}(\boldsymbol{Q}^{([t,t])}_{\psi})] \leq^2 1 - \bar{B}^{*j(I)}_{\psi}(\boldsymbol{Q}^{(I)}_{\psi}) \\
&\leq^1 1 - \bar{B}^{*j(I)}_{\psi}(\bar{\boldsymbol{\ell}}''^{(I)}_{\psi})
\end{aligned} \tag{3.34}$$

$$u''^{j(I)}_{\phi} \leq^8 \inf_{t \in \tau(I)} [1 - \bar{\ell}'^{j([t,t])}_{\psi}(\bar{\boldsymbol{\ell}}''^{([t,t])}_{\psi})] \leq^7 1 - \bar{\ell}'^{j(I)}_{\psi}(\bar{\boldsymbol{\ell}}''^{(I)}_{\psi}) \tag{3.35}$$

$$\bar{\ell}_\phi^{\prime\prime j(I)} \leq^{14} \bar{B}_\psi^{*j(I)}(\bar{\boldsymbol{\ell}}_\psi^{\prime\prime(I)}) \leq^{13} \bar{B}_\psi^{*j(I)}(\boldsymbol{Q}_\psi^{(I)}) \leq^{12} q_\phi^{j(I)}$$

$$\leq^{11} \inf_{t\in\tau(I)} [1 - B_\psi^{*j([t,t])}(\boldsymbol{P}_\psi^{([t,t])})] \leq^{10} 1 - B_\psi^{*j(I)}(\boldsymbol{P}_\psi^{(I)})$$

$$\leq^{9} 1 - B_\psi^{*j(I)}(\boldsymbol{\ell}_\psi^{\prime\prime(I)}) \tag{3.36}$$

$$\bar{u}_\phi^{\prime\prime j(I)} \leq^{16} \inf_{t\in\tau(I)} [1 - \ell_\psi^{\prime j([t,t])}(\boldsymbol{\ell}_\psi^{\prime\prime([t,t])})] \leq^{15} 1 - \ell_\psi^{\prime j(I)}(\boldsymbol{\ell}_\psi^{\prime\prime(I)}). \tag{3.37}$$

**Proof:** From the assumptions of the theorem it follows that the marginal performance processes of the modules of $(C, \phi)$ are independent in $I$. Hence, Corollary 3.11 can be applied by considering the modules as components and the inequalities 2, 3, 4, 10, 11 and 12 follow. Furthermore, Equations (3.6) and (3.7) of Theorem 3.6, can be applied on all modules. Hence, the inequalities 5, 9 and 1, 13 follow since $B_\psi^{*j(I)}(\cdot)$ and $\bar{B}_\psi^{*j(I)}(\cdot)$ are nondecreasing functions in each matrix element. In addition, the inequalities 6, 15 and 16 follow by applying respectively 14, 7 and 8 on the dual structure and dual level, remembering Equations (3.26), (3.32), (3.28) and (3.27). The inequalities 14, 7 and 8 are proved in Section A.1 of Appendix A.

Note, as revealed in Natvig (1993) we do not need the assumption that the marginal performance process of each module is associated in $I$ to apply Corollary 3.11 as claimed in Funnemark and Natvig (1985).

In the next theorem we find bounds that are improvements of the bounds in Equations (3.8) and (3.9) of Theorem 3.6. Actually, we also prove the bounds under somewhat different assumptions.

**Theorem 3.13:** Let $(C, \phi)$ be an MMS with modular decomposition given by Definition 1.6. Assume the marginal performance processes of the modules to be independent in the time interval $I$ and furthermore the joint performance process for the components of each module to be associated in $I$. Then for $j = 1, \ldots, M$

$$\ell_\phi^{\prime j(I)}(\boldsymbol{P}_\phi^{(I)}) \leq^{7} B_\psi^{*j(I)}(\boldsymbol{B}_\psi^{(I)}) \leq^{6} B_\psi^{*j(I)}(\boldsymbol{P}_\psi^{(I)}) \leq^{5} p_\phi^{j(I)}$$

$$\leq^{4} \inf_{t\in\tau(I)} [1 - \bar{B}_\psi^{*j([t,t])}(\boldsymbol{Q}_\psi^{([t,t])})] \leq^{3} 1 - \bar{B}_\psi^{*j(I)}(\boldsymbol{Q}_\psi^{(I)})$$

$$\leq^{2} 1 - \bar{B}_\psi^{*j(I)}(\bar{\boldsymbol{B}}_\psi^{(I)}) \leq^{1} 1 - \bar{\ell}_\phi^{\prime j(I)}(\boldsymbol{Q}_\phi^{(I)}) \tag{3.38}$$

$$\bar{\ell}_\phi^{\prime j(I)}(\boldsymbol{Q}_\phi^{(I)}) \leq^{14} \bar{B}_\psi^{*j(I)}(\bar{\boldsymbol{B}}_\psi^{(I)}) \leq^{13} \bar{B}_\psi^{*j(I)}(\boldsymbol{Q}_\psi^{(I)}) \leq^{12} q_\phi^{j(I)}$$

$$\leq^{11} \inf_{t\in\tau(I)} [1 - B_{\psi}^{*j([t,t])}(\boldsymbol{P}_{\psi}^{([t,t])})] \leq^{10} 1 - B_{\psi}^{*j(I)}(\boldsymbol{P}_{\psi}^{(I)})$$

$$\leq^{9} 1 - B_{\psi}^{*j(I)}(\boldsymbol{B}_{\psi}^{(I)}) \leq^{8} 1 - \ell_{\phi}^{\prime j(I)}(\boldsymbol{P}_{\phi}^{(I)}). \tag{3.39}$$

**Proof:** Since we have assumed that the marginal performance processes of the modules of $(C, \phi)$ are independent in $I$, Corollary 3.11 can again be applied by considering the modules as components and the inequalities 3, 4, 5, 10, 11 and 12 follow. Furthermore, Corollary 3.10 can be applied on all modules. Hence, the inequalities 6, 9, 2 and 13 follow since $B_{\psi}^{*j(I)}(\cdot)$ and $\bar{B}_{\psi}^{*j(I)}(\cdot)$ are nondecreasing functions in each matrix element. Furthermore, the inequalities 7 and 8 are equivalent. The same is true for 1 and 14. Since 7 follows by applying 14 on the dual structure and dual level, remembering Equations (3.28), (3.31) and (3.32), what remains to prove is inequality 14. This is done in Section A.2 of Appendix A.

The following new bounds are improvements of the bounds in Corollary 3.11.

**Theorem 3.14:** Let $(C, \phi)$ be an MMS with modular decomposition given by Definition 1.6. Assume the marginal performance processes of the components to be independent in the time interval $I$. Then, for $j = 1, \ldots, M$

$$B_{\psi}^{*j(I)}(\boldsymbol{B}_{\psi}^{*(I)}(\boldsymbol{P}_{\phi}^{(I)})) \leq^{6} B_{\psi}^{*j(I)}(\boldsymbol{P}_{\psi}^{(I)}) \leq^{5} p_{\phi}^{j(I)}$$

$$\leq^{4} \inf_{t\in\tau(I)} [1 - \bar{B}_{\psi}^{*j([t,t])}(\boldsymbol{Q}_{\psi}^{([t,t])})] \leq^{3} 1 - \bar{B}_{\psi}^{*j(I)}(\boldsymbol{Q}_{\psi}^{(I)})$$

$$\leq^{2} 1 - \bar{B}_{\psi}^{*j(I)}(\bar{\boldsymbol{B}}_{\psi}^{*(I)}(\boldsymbol{Q}_{\phi}^{(I)})). \tag{3.40}$$

$$\bar{B}_{\psi}^{*j(I)}(\bar{\boldsymbol{B}}_{\psi}^{*(I)}(\boldsymbol{Q}_{\phi}^{(I)})) \leq^{13} \bar{B}_{\psi}^{*j(I)}(\boldsymbol{Q}_{\psi}^{(I)}) \leq^{12} q_{\phi}^{j(I)}$$

$$\leq^{11} \inf_{t\in\tau(I)} [1 - B_{\psi}^{*j([t,t])}(\boldsymbol{P}_{\psi}^{([t,t])})] \leq^{10} 1 - B_{\psi}^{*j(I)}(\boldsymbol{P}_{\psi}^{(I)})$$

$$\leq^{9} 1 - B_{\psi}^{*j(I)}(\boldsymbol{B}_{\psi}^{*(I)}(\boldsymbol{P}_{\phi}^{(I)})). \tag{3.41}$$

Remembering Theorem 2.17, if the minimal path vectors and minimal cut vectors to level $j$ of $(C, \phi)$ are respectively exactly given by

$$\boldsymbol{y}_{\phi_{\ell}^{j}m}^{j} \quad m = 1, \ldots, n_{\phi_{\ell}^{j}}^{j}, \quad \ell = 1, \ldots, n_{\psi}^{j},$$

and

$$\boldsymbol{z}_{\phi_{\ell}^{j}m}^{j} \quad m = 1, \ldots, m_{\phi_{\ell}^{j}}^{j}, \quad \ell = 1, \ldots, m_{\psi}^{j},$$

then

$$B_{\phi}^{*j(I)}(\boldsymbol{P}_{\phi}^{(I)}) \leq^{7} B_{\psi}^{*j(I)}(\boldsymbol{B}_{\psi}^{*(I)}(\boldsymbol{P}_{\phi}^{(I)}))$$

$$1 - \bar{B}_{\psi}^{*j(I)}(\bar{\boldsymbol{B}}_{\psi}^{*(I)}(\boldsymbol{Q}_{\phi}^{(I)})) \leq^{1} 1 - \bar{B}_{\phi}^{*j(I)}(\boldsymbol{Q}_{\phi}^{(I)}). \tag{3.42}$$

$$\bar{B}_{\phi}^{*j(I)}(\boldsymbol{Q}_{\phi}^{(I)}) \leq^{14} \bar{B}_{\psi}^{*j(I)}(\bar{\boldsymbol{B}}_{\psi}^{*(I)}(\boldsymbol{Q}_{\phi}^{(I)}))$$

$$1 - B_{\psi}^{*j(I)}(\boldsymbol{B}_{\psi}^{*(I)}(\boldsymbol{P}_{\phi}^{(I)})) \leq^{8} 1 - B_{\phi}^{*j(I)}(\boldsymbol{P}_{\phi}^{(I)}). \tag{3.43}$$

**Proof:** Since we have assumed that the marginal performance processes of the components of $(C, \phi)$ are independent in $I$, the same is true for the modules. Hence, Corollary 3.11 can again be applied by considering the modules as components and the inequalities 3, 4, 5, 10, 11 and 12 follow. Furthermore, Corollary 3.11 can be applied on all modules. Hence, the inequalities 6, 9, 2 and 13 follow since $B_{\psi}^{*j(I)}(\cdot)$ and $\bar{B}_{\psi}^{*j(I)}(\cdot)$ are nondecreasing functions in each matrix element. Furthermore, the inequalities 7 and 8 are equivalent. The same is true for 1 and 14. Since 14 follows by applying 7 on the dual structure and dual level, remembering Equation (3.32), what remains to prove is inequality 7. For $I = [t, t]$, and since all components are independent at $t$, it follows from Equation (2.30) of Theorem 2.28 for $j = 1, \ldots, M$ that

$$\ell_{\phi}^{*j} = \ell_{\phi}^{**j}(\boldsymbol{P}_{\phi}).$$

This relation can obviously also be applied on each module. If the minimal cut vectors to level $j$ of $(C, \phi)$ are exactly

$$z^{j}_{\phi^{j}_{\ell} m} \quad m = 1, \ldots, m^{j}_{\phi^{j}_{\ell}}, \quad \ell = 1, \ldots, m^{j}_{\psi},$$

it hence follows from Equation (A.7) that

$$\ell_{\phi}^{**j}(\boldsymbol{P}_{\phi}) \leq \ell_{\psi}^{**j}(\boldsymbol{\ell}_{\psi}^{**}(\boldsymbol{P}_{\phi})). \tag{3.44}$$

However, applying an argument given in Gåsemyr (2010), since this is a purely algebraic relation in $\boldsymbol{P}_{\phi}$, and $\boldsymbol{P}_{\phi}$ and $\boldsymbol{P}_{\phi}^{(I)}$ share the same properties, $\boldsymbol{P}_{\phi}$ can be replaced by $\boldsymbol{P}_{\phi}^{(I)}$ in Equation (3.44). Furthermore, since the marginal performance processes of the components are independent in $I$, it follows from Theorem 3.6 that

$$\bar{\ell}_{\phi}^{\prime\prime j(I)} = \bar{\ell}_{\phi}^{\prime j(I)}(\boldsymbol{Q}_{\phi}^{(I)}).$$

This relation can also be applied on each module. Inserting these equalities into Equations (A.2) and (A.3) and applying the resulting equality on the dual structure and dual level, remembering Equation (3.28), we get

$$\ell_{\phi}^{\prime j(I)}(\boldsymbol{P}_{\phi}^{(I)}) = \ell_{\psi}^{\prime j(I)}(\boldsymbol{\ell}_{\psi}^{\prime (I)}(\boldsymbol{P}_{\phi}^{(I)})). \tag{3.45}$$

Hence, from the extended Equation (3.44) and Equation (3.45), we get

$$\begin{aligned}
L_{\phi}^{*j(I)}(\boldsymbol{P}_{\phi}^{(I)}) &= \max[\ell_{\phi}^{\prime j(I)}(\boldsymbol{P}_{\phi}^{(I)}), \ell_{\phi}^{**j(I)}(\boldsymbol{P}_{\phi}^{(I)})] \\
&\leq \max[\ell_{\psi}^{\prime j(I)}(\boldsymbol{\ell}_{\psi}^{\prime(I)}(\boldsymbol{P}_{\phi}^{(I)})), \ell_{\psi}^{**j(I)}(\boldsymbol{\ell}_{\psi}^{**(I)}(\boldsymbol{P}_{\phi}^{(I)}))] \\
&\leq \max[\ell_{\psi}^{\prime j(I)}(\boldsymbol{L}_{\psi}^{*(I)}(\boldsymbol{P}_{\phi}^{(I)})), \ell_{\psi}^{**j(I)}(\boldsymbol{L}_{\psi}^{*(I)}(\boldsymbol{P}_{\phi}^{(I)}))] \\
&\leq \max[\ell_{\psi}^{\prime j(I)}(\boldsymbol{B}_{\psi}^{*(I)}(\boldsymbol{P}_{\phi}^{(I)})), \ell_{\psi}^{**j(I)}(\boldsymbol{B}_{\psi}^{*(I)}(\boldsymbol{P}_{\phi}^{(I)}))] \\
&= L_{\psi}^{*j(I)}(\boldsymbol{B}_{\psi}^{*(I)}(\boldsymbol{P}_{\phi}^{(I)})).
\end{aligned}$$

Hence,

$$B_{\phi}^{*j(I)}(\boldsymbol{P}_{\phi}^{(I)}) \leq B_{\psi}^{*j(I)}(\boldsymbol{B}_{\psi}^{*(I)}(\boldsymbol{P}_{\phi}^{(I)})),$$

and inequality 7 is proved.

**Theorem 3.15:** Let $(C, \phi)$ be an MMS with modular decomposition given by Definition 1.6. Furthermore, for $k = 1, \ldots, \ell$, $1 \leq \ell \leq r$ assume that $(A_k, \chi_k)$ has a modular decomposition $\{(B_{km}, \Omega_{km})\}_{m=1}^{s_k}$ with organizing structure function $\sigma_k$. Assume the marginal performance processes of the components are independent in the time interval $I$. Let

$$\boldsymbol{\Omega} = (\Omega_{11}, \ldots, \Omega_{1s_1}, \ldots, \Omega_{\ell 1}, \ldots, \Omega_{\ell s_\ell}, \chi_{\ell+1}, \ldots, \chi_r)$$

and introduce

$$\theta(\boldsymbol{\Omega}) = \psi[\sigma_1(\Omega_{11}, \ldots, \Omega_{1s_1}), \ldots, \sigma_\ell(\Omega_{\ell 1}, \ldots, \Omega_{\ell s_\ell}), \chi_{\ell+1}, \ldots, \chi_r].$$

Remembering Theorem 2.17, assume that the minimal path vectors and minimal cut vectors to level $j$ of $\theta(\boldsymbol{\Omega})$ are respectively exactly given by

$$\boldsymbol{y}_{\theta_\ell^j m}^j \quad m = 1, \ldots, n_{\theta_\ell^j}^j, \quad \ell = 1, \ldots, n_\psi^j,$$

and

$$\boldsymbol{z}_{\theta_\ell^j m}^j \quad m = 1, \ldots, m_{\theta_\ell^j}^j, \quad \ell = 1, \ldots, m_\psi^j.$$

Also write the matrix given by Equation (3.33) in terms of two matrices by

$$\boldsymbol{P}_{\psi}^{(I)} = (\boldsymbol{P}_{\psi(1)}^{(I)}, \boldsymbol{P}_{\psi(2)}^{(I)}),$$

where $\boldsymbol{P}_{\psi^{(1)}}^{(I)}$ is an $\ell \times M$ matrix and $\boldsymbol{P}_{\psi^{(2)}}^{(I)}$ is an $(r-\ell) \times M$ matrix. With obvious extensions of this notation we have, for $j = 1, \ldots, M$

$$\begin{aligned}
B_\theta^{*j(I)}(\boldsymbol{P}_{\sigma_1}^{(I)}, \ldots, \boldsymbol{P}_{\sigma_\ell}^{(I)}, \boldsymbol{B}_{\psi^{(2)}}^{*(I)}(\boldsymbol{P}_\phi^{(I)})) &\leq^6 B_\psi^{*j(I)}(\boldsymbol{B}_\psi^{*(I)}(\boldsymbol{P}_\theta^{(I)})) \\
&\leq^5 B_\psi^{*j(I)}(\boldsymbol{P}_{\psi^{(1)}}^{(I)}, \boldsymbol{B}_{\psi^{(2)}}^{*(I)}(\boldsymbol{P}_\phi^{(I)})) \leq^4 p_\phi^{j(I)} \\
&\leq^3 1 - \bar{B}_\psi^{*j(I)}(\boldsymbol{Q}_{\psi^{(1)}}^{(I)}, \bar{\boldsymbol{B}}_{\psi^{(2)}}^{*(I)}(\boldsymbol{Q}_\phi^{(I)})) \leq^2 1 - \bar{B}_\psi^{*j(I)}(\bar{\boldsymbol{B}}_\psi^{*(I)}(\boldsymbol{Q}_\theta^{(I)})) \\
&\leq^1 1 - \bar{B}_\theta^{*j(I)}(\boldsymbol{Q}_{\sigma_1}^{(I)}, \ldots, \boldsymbol{Q}_{\sigma_\ell}^{(I)}, \bar{\boldsymbol{B}}_{\psi^{(2)}}^{*(I)}(\boldsymbol{Q}_\phi^{(I)})).
\end{aligned} \tag{3.46}$$

**Proof:** By applying the inequalities 7 and 6 of Theorem 3.14 on the structure $\theta$, we get

$$\begin{aligned}
B_\theta^{*j(I)}(\boldsymbol{P}_{\sigma_1}^{(I)}, \ldots, \boldsymbol{P}_{\sigma_\ell}^{(I)}, \boldsymbol{P}_{\psi^{(2)}}^{(I)}) &\leq B_\psi^{*j(I)}(\boldsymbol{B}_\psi^{*(I)}(\boldsymbol{P}_\theta^{(I)})) \\
&\leq B_\psi^{*j(I)}(\boldsymbol{P}_{\psi^{(1)}}^{(I)}, \boldsymbol{P}_{\psi^{(2)}}^{(I)}).
\end{aligned}$$

By replacing $\boldsymbol{P}_{\psi^{(2)}}^{(I)}$ by $\boldsymbol{B}_{\psi^{(2)}}^{*(I)}(\boldsymbol{P}_\phi^{(I)})$ the inequalities 6 and 5 follow. Since $B_\psi^{*j(I)}(\cdot)$ is nondecreasing in each argument, we have, from Equation (3.24) of Corollary 3.11 applied on the modules of $\psi^{(2)}$ and inequality 5 of Theorem 3.14

$$\begin{aligned}
B_\psi^{*j(I)}(\boldsymbol{P}_{\psi^{(1)}}^{(I)}, \boldsymbol{B}_{\psi^{(2)}}^{*(I)}(\boldsymbol{P}_\phi^{(I)})) &\leq B_\psi^{*j(I)}(\boldsymbol{P}_{\psi^{(1)}}^{(I)}, \boldsymbol{P}_{\psi^{(2)}}^{(I)}) = B_\psi^{*j(I)}(\boldsymbol{P}_\psi^{(I)}) \\
&\leq p_\phi^{j(I)}.
\end{aligned}$$

Hence, inequality 4 follows. Finally, by applying the inequalities 6, 5 and 4 on the dual structure and dual level, in the latter case applying Equation (3.1), the inequalities 1, 2 and 3 respectively follow.

## 3.4 Improved bounds at a fixed point of time using modular decompositions

In this section we specialize $I = [t, t]$ and obtain improved bounds for system availabilities at time $t$ using modular decompositions. Hence, $I$ is excluded from the notation. The bounds obtained will automatically give bounds for the unavailabilities at time $t$, since, as stated in Section 3.1

$$q_\phi^j = 1 - p_\phi^j. \tag{3.47}$$

Note also that if we assume that all components, and hence the system, are in the perfect functioning state at $t = 0$, but are not maintained, $p_\phi^j$ is just the system reliability to level $j$ at time $t$.

Let $\mathbf{1}^{a\times b}$ be an $a \times b$ matrix with all elements being 1. We then obviously have

$$\boldsymbol{Q}_\phi = \mathbf{1}^{n\times M} - \boldsymbol{P}_\phi \qquad \boldsymbol{Q}_\psi = \mathbf{1}^{r\times M} - \boldsymbol{P}_\psi.$$

The first two theorems are generalizations of Theorems 3.1 and 3.2 of Natvig (1980).

**Theorem 3.16:** Let $(C, \phi)$ be an MMS with modular decomposition given by Definition 1.6 consisting of totally independent modules at time $t$. Then, for $j = 1, \ldots, M$

$$\ell_\phi''^{j} \leq^{10} B_\psi^{*j}(\boldsymbol{\ell}_\psi'') \left\{ \begin{array}{l} \leq^{8} B_\psi^{*j}(\boldsymbol{P}_\psi) \leq^{6} \\ \leq^{9} p_\psi^{j}(\boldsymbol{\ell}_\psi'') \leq^{7} \end{array} \right\} p_\phi^j$$

$$\left\{ \begin{array}{l} \leq^{4} 1 - \bar{B}_\psi^{*j}(\mathbf{1}^{r\times M} - \boldsymbol{P}_\psi) \leq^{2} \\ \leq^{5} p_\psi^{j}(\mathbf{1}^{r\times M} - \bar{\boldsymbol{\ell}}_\psi'') \leq^{3} \end{array} \right\} 1 - \bar{B}_\psi^{*j}(\bar{\boldsymbol{\ell}}_\psi'') \leq^{1} 1 - \bar{\ell}_\phi''^{j}. \tag{3.48}$$

**Proof:** The inequalities 1, 2, 4, 6, 8 and 10 follow immediately from the inequalities 14, 13, 3, 4, 5 and 6 of Theorem 3.12 by specializing for the case treated in this section. Since the modules are totally independent at time $t$, extending a part of the proof of Theorem 3.1 in Butler (1982), we have

$$p_\phi^j = p_\psi^j(\boldsymbol{P}_\psi). \tag{3.49}$$

From (i) of Theorem 2.21 it follows that $p_\psi^j(\cdot)$ is nondecreasing in each matrix element. Hence, the inequalities 7 and 5 follow from Equation (2.27) of Theorem 2.27 and the inequalities 9 and 3 from Corollary 2.33, noting that for these bounds the matrix arguments $\boldsymbol{\ell}_\psi''$ and $\mathbf{1}^{r\times M} - \bar{\boldsymbol{\ell}}_\psi''$ have the same properties as $\boldsymbol{P}_\psi$.

Note that due to Equation (3.47) we have no objections against the upper bounds arrived at in the two previous sections, when applied here. However, any upper bound arrived at is now equivalent to an established lower bound and nothing is gained. Note also that Theorem 3.12 can be considered a generalization of the main part of Theorem 3.16. Correspondingly, Theorem 3.14 can be considered an extension of the main parts of the following Corollary 3.18.

**Theorem 3.17:** Make the same assumptions as in Theorem 3.16. Assume furthermore that for each module the states of the components at time $t$ are associated random variables. Then, for $j = 1, \ldots, M$

$$B_{\psi}^{*j}(\boldsymbol{B}_{\psi}) \left\{ \begin{array}{c} \leq^8 B_{\psi}^{*j}(\boldsymbol{P}_{\psi}) \leq^6 \\ \leq^9 p_{\psi}^{j}(\boldsymbol{B}_{\psi}) \leq^7 \end{array} \right\} p_{\phi}^{j}$$

$$\left\{ \begin{array}{c} \leq^4 1 - \bar{B}_{\psi}^{*j}(\mathbf{1}^{r\times M} - \boldsymbol{P}_{\psi}) \leq^2 \\ \leq^5 p_{\psi}^{j}(\mathbf{1}^{r\times M} - \bar{\boldsymbol{B}}_{\psi}) \leq^3 \end{array} \right\} 1 - \bar{B}_{\psi}^{*j}(\bar{\boldsymbol{B}}_{\psi}). \tag{3.50}$$

Remembering Theorem 2.17, if the minimal path and cut vectors to level $j$ of $(C, \phi)$ are exactly given by

$$\boldsymbol{y}^{j}_{\phi_\ell^j m} \quad m = 1, \ldots, n^{j}_{\phi_\ell^j}, \quad \ell = 1, \ldots, n^{j}_{\psi},$$

and

$$\boldsymbol{z}^{j}_{\phi_\ell^j m} \quad m = 1, \ldots, m^{j}_{\phi_\ell^j}, \quad \ell = 1, \ldots, m^{j}_{\psi},$$

then

$$B_{\phi}^{j} \leq^{10} B_{\psi}^{*j}(\boldsymbol{B}_{\psi}) \tag{3.51}$$

$$1 - \bar{B}_{\psi}^{*j}(\bar{\boldsymbol{B}}_{\psi}) \leq^{1} 1 - \bar{B}_{\phi}^{j}. \tag{3.52}$$

**Proof:** The inequalities 2, 4, 6 and 8 follow immediately from the inequalities 2, 4, 5 and 6 of Theorem 3.13 by specializing for the case treated in this section. Remembering Equation (3.49), the inequalities 7 and 5 follow from Corollary 2.32 and the inequalities 9 and 3 from Corollary 2.33. Since 1 follows by applying 10 on the dual structure and dual level, remembering Equations (3.31) and (3.32), what remains to prove is inequality 10. This is done in Section A.3 of Appendix A.

**Corollary 3.18:** Let $(C, \phi)$ be an MMS with modular decomposition given by Definition 1.6 having independent components at time $t$. Then, for $j = 1, \ldots, M$

$$B_{\psi}^{*j}(\boldsymbol{B}_{\psi}^{*}(\boldsymbol{P}_{\phi})) \left\{ \begin{array}{c} \leq^8 B_{\psi}^{*j}(\boldsymbol{P}_{\psi}) \leq^6 \\ \leq^9 p_{\psi}^{j}(\boldsymbol{B}_{\psi}^{*}(\boldsymbol{P}_{\phi})) \leq^7 \end{array} \right\} p_{\phi}^{j}$$

$$\left\{ \begin{array}{c} \leq^4 1 - \bar{B}_{\psi}^{*j}(\mathbf{1}^{r\times M} - \boldsymbol{P}_{\psi}) \leq^2 \\ \leq^5 p_{\psi}^{j}(\mathbf{1}^{r\times M} - \bar{\boldsymbol{B}}_{\psi}^{*}(\mathbf{1}^{n\times M} - \boldsymbol{P}_{\phi})) \leq^3 \end{array} \right\} 1 - \bar{B}_{\psi}^{*j}(\bar{\boldsymbol{B}}_{\psi}^{*}(\mathbf{1}^{n\times M} - \boldsymbol{P}_{\phi})) \tag{3.53}$$

Remembering Theorem 2.17, if the minimal path and cut vectors to level $j$ of $(C, \phi)$ are exactly given by

$$\mathbf{y}^j_{\phi^j_\ell m} \quad m = 1, \ldots, n^j_{\phi^j_\ell}, \quad \ell = 1, \ldots, n^j_\psi,$$

and

$$\mathbf{z}^j_{\phi^j_\ell m} \quad m = 1, \ldots, m^j_{\phi^j_\ell}, \quad \ell = 1, \ldots, m^j_\psi,$$

then

$$B^{*j}_\phi(\boldsymbol{P}_\phi) \leq^{10} B^{*j}_\psi(\boldsymbol{B}^*_\psi(\boldsymbol{P}_\phi)) \tag{3.54}$$

$$1 - \bar{B}^{*j}_\psi(\bar{\boldsymbol{B}}^*_\psi(\mathbf{1}^{n\times M} - \boldsymbol{P}_\phi)) \leq^1 1 - \bar{B}^{*j}_\phi(\mathbf{1}^{n\times M} - \boldsymbol{P}_\phi). \tag{3.55}$$

**Proof:** Since $I = [t, t]$ and all components are independent at $t$, it follows from Equation (2.30) of Theorem 2.28 for $j = 1, \ldots, M$

$$\ell^{*j}_\phi = \ell^{**j}_\phi(\boldsymbol{P}_\phi) \qquad \bar{\ell}^{*j}_\phi = \bar{\ell}^{**j}_\phi(\mathbf{1}^{n\times M} - \boldsymbol{P}_\phi)$$

and hence

$$B^j_\phi = B^{*j}_\phi(\boldsymbol{P}_\phi) \qquad \bar{B}^j_\phi = \bar{B}^{*j}_\phi(\mathbf{1}^{n\times M} - \boldsymbol{P}_\phi).$$

These latter relations can obviously also be applied on each module. Hence Corollary 3.18 follows from Theorem 3.17.

Note that even for the special case of the latter corollary our results improve the inequalities of Theorems 3.1 and 3.2 of Butler (1982). The reason is that this paper does not utilize the bounds of Equation (2.28) of Theorem 2.27 at all when improving bounds by using modular decompositions.

We end this section by generalizing Theorems 3.3 and 3.4 in Natvig (1980). The three theorems to follow essentially tell us that it is advantageous to decompose modules with unknown availabilities and do nothing with the remaining ones.

**Theorem 3.19:** Make the same assumptions as in Theorem 3.16. Furthermore, for $k = 1, \ldots, \ell$, $1 \leq \ell \leq r$ assume that $(A_k, \chi_k)$ has a modular decomposition $\{(B_{km}, \Omega_{km})\}_{m=1}^{s_k}$ consisting of totally independent modules at time $t$ and with organizing structure function $\sigma_k$. Let

$$\boldsymbol{\Omega} = (\Omega_{11}, \ldots, \Omega_{1s_1}, \ldots, \Omega_{\ell 1}, \ldots, \Omega_{\ell s_\ell}, \chi_{\ell+1}, \ldots, \chi_r)$$

and introduce

$$\theta(\boldsymbol{\Omega}) = \psi[\sigma_1(\Omega_{11}, \ldots, \Omega_{1s_1}), \ldots, \sigma_\ell(\Omega_{\ell 1}, \ldots, \Omega_{\ell s_\ell}), \chi_{\ell+1}, \ldots, \chi_r].$$

Also write the matrix given by Equation (3.33) in terms of two matrices by

$$\boldsymbol{P}_{\psi} = (\boldsymbol{P}_{\psi(1)}, \boldsymbol{P}_{\psi(2)}),$$

where $\boldsymbol{P}_{\psi(1)}$ is an $\ell \times M$ matrix and $\boldsymbol{P}_{\psi(2)}$ is an $(r-\ell) \times M$ matrix. Remembering Theorem 2.17, if the minimal path and cut vectors to level $j$ of $\theta(\boldsymbol{\Omega})$ are exactly given by

$$\boldsymbol{y}^j_{\theta^j_\ell m} \quad m = 1, \ldots, n^j_{\theta^j_\ell}, \quad \ell = 1, \ldots, n^j_{\psi},$$

and

$$\boldsymbol{z}^j_{\theta^j_\ell m} \quad m = 1, \ldots, m^j_{\theta^j_\ell}, \quad \ell = 1, \ldots, m^j_{\psi},$$

then, with obvious extensions of the notation, we have for $j = 1, \ldots, M$

$$\begin{aligned}
&B^{*j}_{\theta}(\boldsymbol{P}_{\sigma_1}, \ldots, \boldsymbol{P}_{\sigma_\ell}, \boldsymbol{\ell}''_{\psi(2)}) \leq^{10} B^{*j}_{\psi}(\boldsymbol{P}_{\psi(1)}, \boldsymbol{\ell}''_{\psi(2)}) \leq^{9} p^j_{\phi} \\
&\leq^{8} 1 - \bar{B}^{*j}_{\psi}(\boldsymbol{1}^{\ell \times M} - \boldsymbol{P}_{\psi(1)}, \bar{\boldsymbol{\ell}}''_{\psi(2)}) \\
&\leq^{7} 1 - \bar{B}^{*j}_{\theta}(\boldsymbol{1}^{s_1 \times M} - \boldsymbol{P}_{\sigma_1}, \ldots, \boldsymbol{1}^{s_\ell \times M} - \boldsymbol{P}_{\sigma_\ell}, \bar{\boldsymbol{\ell}}''_{\psi(2)}).
\end{aligned} \tag{3.56}$$

Without the latter assumptions we have

$$\begin{aligned}
&p^j_{\psi}(\boldsymbol{\ell}''_{\psi(1)}, \boldsymbol{P}_{\psi(2)}) \leq^{6} p^j_{\psi}(\boldsymbol{B}^*_{\psi(1)}(\boldsymbol{\ell}''_{\sigma_1}, \ldots, \boldsymbol{\ell}''_{\sigma_\ell}), \boldsymbol{P}_{\psi(2)}) \\
&\leq^{5} p^j_{\theta}(\boldsymbol{\ell}''_{\sigma_1}, \ldots, \boldsymbol{\ell}''_{\sigma_\ell}, \boldsymbol{P}_{\psi(2)}) \leq^{4} p^j_{\phi} \\
&\leq^{3} p^j_{\theta}(\boldsymbol{1}^{s_1 \times M} - \bar{\boldsymbol{\ell}}''_{\sigma_1}, \ldots, \boldsymbol{1}^{s_\ell \times M} - \bar{\boldsymbol{\ell}}''_{\sigma_\ell}, \boldsymbol{P}_{\psi(2)}) \\
&\leq^{2} p^j_{\psi}(\boldsymbol{1}^{\ell \times M} - \bar{\boldsymbol{B}}^*_{\psi(1)}(\bar{\boldsymbol{\ell}}''_{\sigma_1}, \ldots, \bar{\boldsymbol{\ell}}''_{\sigma_\ell}), \boldsymbol{P}_{\psi(2)}) \leq^{1} p^j_{\psi}(\boldsymbol{1}^{\ell \times M} - \bar{\boldsymbol{\ell}}''_{\psi(1)}, \boldsymbol{P}_{\psi(2)}).
\end{aligned} \tag{3.57}$$

**Proof:** By applying the inequalities 10 and 8 of Corollary 3.18, we get

$$B^{*j}_{\theta}(\boldsymbol{P}_{\sigma_1}, \ldots, \boldsymbol{P}_{\sigma_\ell}, \boldsymbol{P}_{\psi(2)}) \leq B^{*j}_{\psi}(\boldsymbol{P}_{\psi(1)}, \boldsymbol{P}_{\psi(2)}).$$

By replacing $\boldsymbol{P}_{\psi(2)}$ by $\boldsymbol{\ell}''_{\psi(2)}$ the inequality 10 follows. From the inequalities 10 and 9 of Theorem 3.16, we have, for $k = 1, \ldots, \ell$

$$\ell''^j_{\chi_k} \leq B^{*j}_{\sigma_k}(\boldsymbol{\ell}''_{\sigma_k}) \leq p^j_{\sigma_k}(\boldsymbol{\ell}''_{\sigma_k}).$$

Since $p^j_{\psi}(\cdot)$ is nondecreasing in each argument, we then have

$$\begin{aligned}
&p^j_{\psi}(\boldsymbol{\ell}''_{\psi(1)}, \boldsymbol{P}_{\psi(2)}) \leq p^j_{\psi}(\boldsymbol{B}^*_{\psi(1)}(\boldsymbol{\ell}''_{\sigma_1}, \ldots, \boldsymbol{\ell}''_{\sigma_\ell}), \boldsymbol{P}_{\psi(2)}) \\
&\leq p^j_{\psi}(\boldsymbol{P}_{\psi(1)}(\boldsymbol{\ell}''_{\sigma_1}, \ldots, \boldsymbol{\ell}''_{\sigma_\ell}), \boldsymbol{P}_{\psi(2)}) = p^j_{\theta}(\boldsymbol{\ell}''_{\sigma_1}, \ldots, \boldsymbol{\ell}''_{\sigma_\ell}, \boldsymbol{P}_{\psi(2)}),
\end{aligned}$$

and inequalities 6 and 5 follow. Since $B_{\psi}^{*j}(\cdot)$ is nondecreasing in each argument, we have, from Equation (2.27) of Theorem 2.27 and inequality 6 of Theorem 3.16

$$B_{\psi}^{*j}(\boldsymbol{P}_{\psi(1)}, \boldsymbol{\ell}''_{\psi(2)}) \leq B_{\psi}^{*j}(\boldsymbol{P}_{\psi(1)}, \boldsymbol{P}_{\psi(2)}) = B_{\psi}^{*j}(\boldsymbol{P}_{\psi}) \leq p_{\phi}^{j}.$$

Hence, inequality 9 follows. Similarly,

$$p_{\theta}^{j}(\boldsymbol{\ell}''_{\sigma_1}, \ldots, \boldsymbol{\ell}''_{\sigma_\ell}, \boldsymbol{P}_{\psi(2)}) \leq p_{\theta}^{j}(\boldsymbol{P}_{\sigma_1}, \ldots, \boldsymbol{P}_{\sigma_\ell}, \boldsymbol{P}_{\psi(2)}) = p_{\psi}^{j}(\boldsymbol{P}_{\psi}) = p_{\phi}^{j},$$

having applied Equation (2.27) of Theorem 2.27 and Equation (3.49), and inequality 4 follows. Finally, by applying the inequalities 10 and 9 on the dual structure and dual level, the inequalities 7 and 8 respectively follow. The inequalities 1, 2 and 3 are proved completely parallel to 6, 5 and 4.

**Theorem 3.20:** Make the same assumptions as in Theorem 3.19. Furthermore, assume that for each of the modules $(B_{km}, \Omega_{km}), k = 1, \ldots, \ell, m = 1, \ldots, s_k$ and $(A_k, \chi_k), k = \ell + 1, \ldots, r$ the states of the components at time $t$ are associated random variables. Remembering Theorem 2.17, if the minimal path and cut vectors to level $j$ of $\theta(\boldsymbol{\Omega})$ are exactly given by

$$\boldsymbol{y}^{j}_{\theta_\ell^j m} \quad m = 1, \ldots, n^{j}_{\theta_\ell^j}, \quad \ell = 1, \ldots, n^{j}_{\psi},$$

and

$$\boldsymbol{z}^{j}_{\theta_\ell^j m} \quad m = 1, \ldots, m^{j}_{\theta_\ell^j}, \quad \ell = 1, \ldots, m^{j}_{\psi},$$

then, for $j = 1, \ldots, M$

$$\begin{aligned} B_{\theta}^{*j}(\boldsymbol{P}_{\sigma_1}, \ldots, \boldsymbol{P}_{\sigma_\ell}, \boldsymbol{B}_{\psi(2)}) &\leq^{10} B_{\psi}^{*j}(\boldsymbol{P}_{\psi(1)}, \boldsymbol{B}_{\psi(2)}) \leq^{9} p_{\phi}^{j} \\ &\leq^{8} 1 - \bar{B}_{\psi}^{*j}(\boldsymbol{1}^{\ell \times M} - \boldsymbol{P}_{\psi(1)}, \bar{\boldsymbol{B}}_{\psi(2)})) \\ &\leq^{7} 1 - \bar{B}_{\theta}^{*j}(\boldsymbol{1}^{s_1 \times M} - \boldsymbol{P}_{\sigma_1}, \ldots, \boldsymbol{1}^{s_\ell \times M} - \boldsymbol{P}_{\sigma_\ell}, \bar{\boldsymbol{B}}_{\psi(2)}). \end{aligned} \tag{3.58}$$

If the minimal path and cut vectors to level $j$ of $\sigma_k, k = 1, \ldots, \ell$ are exactly given by

$$\boldsymbol{y}^{j}_{\sigma_{k_\ell^j} m} \quad m = 1, \ldots, n^{j}_{\sigma_{k_\ell^j}}, \quad \ell = 1, \ldots, n^{j}_{\psi},$$

and

$$\boldsymbol{z}^{j}_{\sigma_{k_\ell^j} m} \quad m = 1, \ldots, m^{j}_{\sigma_{k_\ell^j}}, \quad \ell = 1, \ldots, m^{j}_{\psi},$$

then, for $j = 1, \ldots, M$

$$p_{\psi}^{j}(\boldsymbol{B}_{\psi(1)}, \boldsymbol{P}_{\psi(2)}) \leq^{6} p_{\psi}^{j}(\boldsymbol{B}^{*}_{\psi(1)}(\boldsymbol{B}_{\sigma_1}, \ldots, \boldsymbol{B}_{\sigma_\ell}), \boldsymbol{P}_{\psi(2)})$$

$$\leq^5 p_\theta^j(\boldsymbol{B}_{\sigma_1}, \ldots, \boldsymbol{B}_{\sigma_\ell}, \boldsymbol{P}_{\psi(2)}) \leq^4 p_\phi^j$$
$$\leq^3 p_\theta^j(\mathbf{1}^{s_1 \times M} - \bar{\boldsymbol{B}}_{\sigma_1}, \ldots, \mathbf{1}^{s_\ell \times M} - \bar{\boldsymbol{B}}_{\sigma_\ell}, \boldsymbol{P}_{\psi(2)})$$
$$\leq^2 p_\psi^j(\mathbf{1}^{\ell \times M} - \bar{\boldsymbol{B}}^*_{\psi(1)}(\bar{\boldsymbol{B}}_{\sigma_1}, \ldots, \bar{\boldsymbol{B}}_{\sigma_\ell}), \boldsymbol{P}_{\psi(2)})$$
$$\leq^1 p_\psi^j(\mathbf{1}^{\ell \times M} - \bar{\boldsymbol{B}}_{\psi(1)}, \boldsymbol{P}_{\psi(2)}). \tag{3.59}$$

**Proof:** The proof is very similar to the one for Theorem 3.19 applying Theorem 3.17 instead of Theorem 3.16 on the systems $\sigma_k, k = 1, \ldots, \ell$. We also apply Corollary 2.32 instead of Theorem 2.27.

**Corollary 3.21:** Make the same assumptions as in Theorem 3.19. Furthermore, assume that all components are independent at time $t$. Remembering Theorem 2.17, if the minimal path and cut vectors to level $j$ of $\theta(\boldsymbol{\Omega})$ are exactly given by

$$\boldsymbol{y}^j_{\theta^j_\ell m} \quad m = 1, \ldots, n^j_{\theta^j_\ell}, \quad \ell = 1, \ldots, n^j_\psi,$$

and

$$\boldsymbol{z}^j_{\theta^j_\ell m} \quad m = 1, \ldots, m^j_{\theta^j_\ell}, \quad \ell = 1, \ldots, m^j_\psi,$$

then, for $j = 1, \ldots, M$

$$B_\theta^{*j}(\boldsymbol{P}_{\sigma_1}, \ldots, \boldsymbol{P}_{\sigma_\ell}, \boldsymbol{B}^*_{\psi(2)}(\boldsymbol{P}_\phi)) \leq^{12} B_\psi^{*j}(\boldsymbol{B}^*_\psi(\boldsymbol{P}_\theta))$$
$$\leq^{11} B_\psi^{*j}(\boldsymbol{P}_{\psi(1)}, \boldsymbol{B}^*_{\psi(2)}(\boldsymbol{P}_\phi)) \leq^{10} p_\phi^j$$
$$\leq^9 1 - \bar{B}_\psi^{*j}(\mathbf{1}^{\ell \times M} - \boldsymbol{P}_{\psi(1)}, \bar{\boldsymbol{B}}^*_{\psi(2)}(\mathbf{1}^{n \times M} - \boldsymbol{P}_\phi))$$
$$\leq^8 1 - \bar{B}_\psi^{*j}(\bar{\boldsymbol{B}}^*_\psi(\mathbf{1}^{(\sum_{k=1}^{\ell} s_k + r - \ell) \times M} - \boldsymbol{P}_\theta)$$
$$\leq^7 1 - \bar{B}_\theta^{*j}(\mathbf{1}^{s_1 \times M} - \boldsymbol{P}_{\sigma_1}, \ldots, \mathbf{1}^{s_\ell \times M} - \boldsymbol{P}_{\sigma_\ell}, \bar{\boldsymbol{B}}^*_{\psi(2)}(\mathbf{1}^{n \times M} - \boldsymbol{P}_\phi)). \tag{3.60}$$

If the minimal path and cut vectors to level $j$ of $\sigma_k, k = 1, \ldots, \ell$ are exactly given by

$$\boldsymbol{y}^j_{\sigma_{k\ell}^j m} \quad m = 1, \ldots, n^j_{\sigma_{k\ell}^j}, \quad \ell = 1, \ldots, n^j_\psi,$$

and

$$\boldsymbol{z}^j_{\sigma_{k\ell}^j m} \quad m = 1, \ldots, m^j_{\sigma_{k\ell}^j}, \quad \ell = 1, \ldots, m^j_\psi,$$

then, for $j = 1, \ldots, M$

$$p_\psi^j(\boldsymbol{B}^*_{\psi(1)}(\boldsymbol{P}_\phi), \boldsymbol{P}_{\psi(2)}) \leq^6 p_\psi^j(\boldsymbol{B}^*_{\psi(1)}(\boldsymbol{B}^*_{\sigma_1}(\boldsymbol{P}_\phi), \ldots, \boldsymbol{B}^*_{\sigma_\ell}(\boldsymbol{P}_\phi)), \boldsymbol{P}_{\psi(2)})$$

$$\le^5 p_\theta^j(\boldsymbol{B}_{\sigma_1}^*(\boldsymbol{P}_\phi), \ldots, \boldsymbol{B}_{\sigma_\ell}^*(\boldsymbol{P}_\phi), \boldsymbol{P}_{\psi(2)}) \le^4 p_\phi^j$$
$$\le^3 p_\theta^j(\mathbf{1}^{s_1\times M} - \bar{\boldsymbol{B}}_{\sigma_1}^*(\mathbf{1}^{n\times M} - \boldsymbol{P}_\phi), \ldots, \mathbf{1}^{s_\ell\times M} - \bar{\boldsymbol{B}}_{\sigma_\ell}^*(\mathbf{1}^{n\times M} - \boldsymbol{P}_\phi), \boldsymbol{P}_{\psi(2)})$$
$$\le^2 p_\psi^j(\mathbf{1}^{\ell\times M} - \bar{\boldsymbol{B}}_{\psi(1)}^*(\bar{\boldsymbol{B}}_{\sigma_1}^*(\mathbf{1}^{n\times M} - \boldsymbol{P}_\phi), \ldots, \bar{\boldsymbol{B}}_{\sigma_\ell}^*(\mathbf{1}^{n\times M} - \boldsymbol{P}_\phi)), \boldsymbol{P}_{\psi(2)})$$
$$\le^1 p_\psi^j(\mathbf{1}^{\ell\times M} - \bar{\boldsymbol{B}}_{\psi(1)}^*(\mathbf{1}^{n\times M} - \boldsymbol{P}_\phi), \boldsymbol{P}_{\psi(2)}). \tag{3.61}$$

**Proof:** Equation (3.60) follows from Theorem 3.15 by specializing $I = [t, t]$. In the same way as Corollary 3.18 follows from Theorem 3.17, Equation (3.61) follows from Theorem 3.20.

Let us try to summarize the implications of Theorems 3.19, 3.20 and Corollary 3.21. Consider an MMS $(C, \phi)$ with modular decomposition consisting of totally independent modules at time $t$ and with organizing structure function $\psi$ having known availability functions $p_\psi^j$, $j = 1, \ldots, M$, which are considered functions of the availabilities of the modules as given in Equation (3.49). Furthermore, assume that each module with unknown availabilities at time $t$ has a modular decomposition consisting of modules having either the specific properties mentioned in Theorems 3.19, 3.20 or Corollary 3.21 and in addition having unknown availabilities at $t$. Finally, assume that the organizing structure function $\theta$ of the refined modular decomposition of $(C, \phi)$ has known availability functions $p_\theta^j$, $j = 1, \ldots, M$. Then the inequalities 1, 2, 3, 4, 5 and 6 tell us that the bounds based on the refined modular decomposition are better than the bounds based on the original one.

The remaining inequalities tell us in a way just the opposite. Now the $p_\psi^j$s and the $p_\theta^j$s are supposed to be unknown. However, this time each module with known availabilities at $t$ is decomposed into modules with known availabilities at $t$. Now the bounds based on the refined modular decomposition are worse. If, in this latter case, the $p_\psi^j$s and the $p_\theta^j$s are known, for instance in the situation of Theorem 3.19, it readily follows that

$$p_\theta^j(\boldsymbol{P}_{\sigma_1}, \ldots, \boldsymbol{P}_{\sigma_\ell}, \boldsymbol{\ell}''_{\psi(2)}) = p_\psi^j(\boldsymbol{P}_{\psi(1)}, \boldsymbol{\ell}''_{\psi(2)}) \le p_\phi^j$$
$$\le p_\psi^j(\boldsymbol{P}_{\psi(1)}, \mathbf{1}^{(r-\ell)\times M} - \bar{\boldsymbol{\ell}}''_{\psi(2)}) = p_\theta^j(\boldsymbol{P}_{\sigma_1}, \ldots, \boldsymbol{P}_{\sigma_\ell}, \mathbf{1}^{(r-\ell)\times M} - \bar{\boldsymbol{\ell}}''_{\psi(2)}).$$

Parallel relations are proved for the situations in Theorem 3.20 and Corollary 3.21. Hence, now the bounds based on the refined modular decomposition and the bounds based on the original one are equally good.

## 3.5 Strict and exactly correct bounds

The strict Min–Max bounds from Natvig (1993) given in the following theorem for a time interval $I$ were new at the time even in the special case of binary monotone systems considered at a fixed point of time. The corresponding non-strict bounds are given in Theorem 3.6.

**Theorem 3.22:** Let $(C, \phi)$ be an MMS where the marginal performance processes of the components are independent in $I$ and assume that there are at least two minimal path sets to level $j$ and at least two minimal cut sets to level $j$ for $j = 1, \ldots, M$. Finally, assume that $p_i^{j(I)}$ is strictly decreasing and $q_i^{j(I)}$ strictly increasing in $j$ and that $p_i^{M(I)} > 0$, $q_i^{1(I)} > 0$, $i = 1, \ldots, n$. Then

$$\ell_\phi^{\prime j(I)}(\boldsymbol{P}_\phi^{(I)}) <^1 p_\phi^{j(I)} <^2 1 - \bar{\ell}_\phi^{\prime j(I)}(\boldsymbol{Q}_\phi^{(I)}) \tag{3.62}$$

$$\bar{\ell}_\phi^{\prime j(I)}(\boldsymbol{Q}_\phi^{(I)}) <^3 q_\phi^{j(I)} <^4 1 - \ell_\phi^{\prime j(I)}(\boldsymbol{P}_\phi^{j(I)}). \tag{3.63}$$

The inequality 1 is replaced by equality when there is only one minimal path set to level $j$ and there is equality in 3 when there is only one minimal cut set to level $j$.

**Proof:** We will prove inequality 1. Inequality 3 follows by applying inequality 1 on the dual structure and dual level. Inequalities 2 and 4 follow immediately from Equation (3.1) and the inequalities 3 and 1 respectively. From (i) of Theorem 2.15 we get

$$\begin{aligned} p_\phi^{j(I)} &= P\left[\coprod_{k=1}^{n_\phi^j} \boldsymbol{J}_{\boldsymbol{y}_k^j}(\boldsymbol{X}(s)) = 1 \ \ \forall s \in \tau(I)\right] \\ &= P[\forall s \in \tau(I)\ \exists k \in \{1, \ldots, n_\phi^j\} \text{ such that } \boldsymbol{J}_{\boldsymbol{y}_k^j}(\boldsymbol{X}(s)) = 1] \\ &\geq P[\exists k \in \{1, \ldots, n_\phi^j\} \text{ such that } \boldsymbol{J}_{\boldsymbol{y}_k^j}(\boldsymbol{X}(s)) = 1\ \forall s \in \tau(I)]. \end{aligned}$$

From the definition of $\ell_\phi^{\prime j(I)}(\boldsymbol{P}_\phi^{(I)})$ in Theorem 3.6 it follows that for some $r \in \{1, \ldots, n_\phi^j\}$

$$\ell_\phi^{\prime j(I)}(\boldsymbol{P}_\phi^{(I)}) = \prod_{i \in C_\phi^j(\boldsymbol{y}_r^j)} p_i^{y_{ir}^j(I)}. \tag{3.64}$$

Since there are at least two minimal path sets to level $j$, we can choose $k \in \{1, \ldots, n_\phi^j\} - \{r\}$. It immediately follows that

$$\begin{aligned} p_\phi^{j(I)} &\geq P[\{\boldsymbol{J}_{\boldsymbol{y}_r^j}(\boldsymbol{X}(s)) = 1\ \forall s \in \tau(I)\} \cup \{\boldsymbol{J}_{\boldsymbol{y}_k^j}(\boldsymbol{X}(s)) = 1\ \forall s \in \tau(I)\}] \\ &= P[\boldsymbol{J}_{\boldsymbol{y}_r^j}(\boldsymbol{X}(s)) = 1\ \forall s \in \tau(I)] + P[\boldsymbol{J}_{\boldsymbol{y}_k^j}(\boldsymbol{X}(s)) = 1\ \forall s \in \tau(I)] \\ &- P[\boldsymbol{J}_{\boldsymbol{y}_r^j}(\boldsymbol{X}(s)) \cdot \boldsymbol{J}_{\boldsymbol{y}_k^j}(\boldsymbol{X}(s)) = 1\ \forall s \in \tau(I)]. \end{aligned} \tag{3.65}$$

By the independence of the marginal performance processes of the components in $I$, we have, as in the proof of Theorem 3.6

$$\begin{aligned} P[\boldsymbol{J}_{\boldsymbol{y}_r^j}(\boldsymbol{X}(s)) = 1\ \forall s \in \tau(I)] &= P[\cap_{i \in C_\phi^j(\boldsymbol{y}_r^j)} (\min_{s \in \tau(I)} X_i(s) \geq y_{ir}^j)] \\ &= \prod_{i \in C_\phi^j(\boldsymbol{y}_r^j)} p_i^{y_{ir}^j(I)}. \end{aligned}$$

The same argument works for the two other terms. It then follows from Equations (3.64) and (3.65) that

$$p_\phi^{j(I)} \geq \ell_\phi^{\prime j(I)}(\boldsymbol{P}_\phi^{(I)}) + \prod_{i \in C_\phi^j(\boldsymbol{y}_k^j)} p_i^{y_{ik}^j(I)} - \prod_{i \in C_\phi^j(\boldsymbol{y}_r^j) \cup C_\phi^j(\boldsymbol{y}_k^j)} p_i^{\max(y_{ir}^j, y_{ik}^j)(I)}.$$

Since now $p_i^{j(I)}$ is strictly decreasing in $j$ and $0 < p_i^{j(I)} < 1$, $i = 1, \ldots, n$, $j = 1, \ldots, M$, inequality 1 follows. The equality in 1, when there is only one minimal path set to level $j$, is straightforward from the proof above.

We next generalize the strict bounds given in Theorem 2.29 for a fixed point of time to a time interval $I$. The corresponding non-strict bounds are given in Theorem 3.7.

**Theorem 3.23:** Let $(C, \phi)$ be an MMS where the marginal performance processes of the components are independent in $I$ and assume that at least two minimal path sets to level $j$ overlap and at least two minimal cut sets to level $j$ overlap for $j = 1, \ldots, M$. Finally, assume that $p_i^{M(I)} > 0, q_i^{1(I)} > 0, i = 1, \ldots, n$. Then

$$\ell_\phi^{**j(I)}(\boldsymbol{P}_\phi^{(I)}) <^1 p_\phi^{j(I)} <^2 1 - \bar{\ell}_\phi^{**j(I)}(\boldsymbol{Q}_\phi^{(I)}) \tag{3.66}$$

$$\bar{\ell}_\phi^{**j(I)}(\boldsymbol{Q}_\phi^{(I)}) <^3 q_\phi^{j(I)} <^4 1 - \ell_\phi^{**j(I)}(\boldsymbol{P}_\phi^{j(I)}). \tag{3.67}$$

**Proof:** Again we only have to prove inequality 1. Remember the indicator functions $J_{z_k^j}(x^{\min(I)}),\ k \in \{1, \dots, m_\phi^j\}$ introduced in Equation (3.16). Since the marginal performance processes of the components are independent in $I$, we have

$$P[J_{z_k^j}(X^{\min(I)}) = 1] = \coprod_{i \in D_\phi^j(z_k^j)} p_i^{z_{ik}^j+1(I)}. \tag{3.68}$$

Furthermore, since $0 < p_i^{j(I)} < 1, i = 1, \dots n,\ j = 1, \dots, M$, it follows that the random variables $J_{z_k^j}(X^{\min(I)}),\ k = 1, \dots, m_\phi^j$ are not identically equal to 0 or 1. Assume the two minimal cut sets to level $j$ $D_\phi^j(z_1^j)$ and $D_\phi^j(z_2^j)$ overlap. Since neither of the two random variables $J_{z_1^j}(X^{\min(I)})$ and $\prod_{k=2}^{m_\phi^j} J_{z_k^j}(X^{\min(I)})$ is identically equal to 0 or 1, it follows that they are dependent. Because $J_{z_k^j}(X^{\min(I)}), k = 1, \dots, m_\phi^j$ were shown to be associated random variables, by applying Exercise 6, page 31 in Barlow and Proschan (1975a) we have

$$\mathrm{Cov}[J_{z_1^j}(X^{\min(I)}), \prod_{k=2}^{m_\phi^j} J_{z_k^j}(X^{\min(I)})] > 0.$$

We then get, from Equation (3.17)

$$\begin{aligned} p_\phi^{j(I)} &\geq E \prod_{k=1}^{m_\phi^j} J_{z_k^j}(X^{\min(I)}) \\ &> E J_{z_1^j}(X^{\min(I)}) E \prod_{k=2}^{m_\phi^j} J_{z_k^j}(X^{\min(I)}). \end{aligned}$$

By finally applying Theorem 3.1, page 32 of Barlow and Proschan (1975a) on the right-hand side we get, from Equation (3.68)

$$\begin{aligned} p_\phi^{j(I)} &> \prod_{k=1}^{m_\phi^j} E J_{z_k^j}(X^{\min(I)}) = \prod_{k=1}^{m_\phi^j} \coprod_{i \in D_\phi^j(z_k^j)} p_i^{z_{ik}^j+1(I)} \\ &= \ell_\phi^{**j(I)}(\boldsymbol{P}_\phi^{(I)}), \end{aligned}$$

and the inequality 1 is proved.

Note that for a proper interval $I$, inequality in 1 is not replaced by equality when there is only one minimal cut set to level $j$ and there is not equality in 3 when there is only one minimal path set to level $j$. We now combine Theorems 3.22 and 3.23. Corresponding non-strict bounds are given in Corollary 3.11.

**Theorem 3.24:** Make the assumptions of Theorems 3.22 and 3.23. Then for any integer $m$ and $T_m = \{t_1, \ldots, t_m\} \in \tau(I)$ and $j = 1, \ldots, M$

$$B_\phi^{*j(I)}(\boldsymbol{P}_\phi^{(I)}) <^1 p_\phi^{j(I)} <^2 \min_{t \in T_m}[1 - \bar{B}_\phi^{*j([t,t])}(\boldsymbol{Q}_\phi^{([t,t])})]$$

$$\leq^3 1 - \bar{B}_\phi^{*j(I)}(\boldsymbol{Q}_\phi^{(I)}) \tag{3.69}$$

$$\bar{B}_\phi^{*j(I)}(\boldsymbol{Q}_\phi^{(I)}) <^4 q_\phi^{j(I)} <^5 \min_{t \in T_m}[1 - B_\phi^{*j([t,t])}(\boldsymbol{P}_\phi^{([t,t])})]$$

$$\leq^6 1 - B_\phi^{*j(I)}(\boldsymbol{P}_\phi^{(I)}). \tag{3.70}$$

The inequality 1 is replaced by equality when there is only one minimal path set to level $j$ and there is equality in 4 when there is only one minimal cut set to level $j$.

**Proof:** From Theorems 3.22 and 3.23 we have

$$B_\phi^{*j(I)}(\boldsymbol{P}_\phi^{(I)}) < \max_{j \leq k \leq M} p_\phi^{k(I)} = p_\phi^{j(I)}.$$

Hence, inequality 1 is proved. Furthermore,

$$\begin{aligned} p_\phi^{j(I)} &= \min_{1 \leq k \leq j} p_\phi^{k(I)} \leq \min_{1 \leq k \leq j} \min_{t \in T_m} p_\phi^{k([t,t])} \\ &< \min_{t \in T_m} \min_{1 \leq k \leq j} \{1 - \max[\bar{\ell}_\phi'^{k([t,t])}(\boldsymbol{Q}_\phi^{([t,t])}), \bar{\ell}_\phi^{**k([t,t])}(\boldsymbol{Q}_\phi^{([t,t])})]\} \\ &= \min_{t \in T_m}[1 - \bar{B}_\phi^{*j([t,t])}(\boldsymbol{Q}_\phi^{([t,t])})] \leq 1 - \bar{B}_\phi^{*j(I)}(\boldsymbol{Q}_\phi^{(I)}). \end{aligned}$$

The strict inequality follows by applying Theorems 3.22 and 3.23 for a fixed $t \in T_m$. The last inequality is established by inspecting $\bar{B}_\phi^{*j(I)}(\boldsymbol{Q}_\phi^{(I)})$. Hence, inequalities 2 and 3 are proved. Inequalities 4, 5 and 6 are proved by applying respectively inequalities 1, 2 and 3 on the dual structure and the dual level.

It is worth noting that if we replace $\min_{t \in T_m}$ by $\inf_{t \in \tau(I)}$, the inequalities 2 and 5 are no longer strict.

We next specialize $I = [t, t]$ and obtain improved strict bounds for system availabilities at time $t$ by using modular decompositions.

Corresponding non-strict bounds are given in Equation (3.53) of Corollary 3.18. Since $p_i^j + q_i^j = 1$ in this case, $p_i^j$ being strictly decreasing in $j$ is equivalent to $q_i^j$ being strictly increasing in $j$.

**Theorem 3.25:** Let $(C, \phi)$ be an MMS with modular decomposition given by Definition 1.6 having independent components at time $t$. Assume there exists a module $\chi_m$ with set of states $S$ and that every level of this module is relevant to the same level of the system. Assume also that both for the organizing structure function and for the structure functions of the modules at least two minimal path sets to level $j$ overlap and at least two minimal cut sets to level $j$ overlap. Finally, assume that $0 < r_i^j < 1$, $i = 1, \ldots, n$, $0 < r_{\chi k}^j < 1$, $k = 1, \ldots, r$, $j = 0, \ldots, M$. Then, for $j = 1, \ldots, M$

$$B_\psi^{*j}(\boldsymbol{B}_\psi^*(\boldsymbol{P}_\phi)) \left\{ \begin{array}{c} <^7 B_\psi^{*j}(\boldsymbol{P}_\psi) <^5 \\ <^8 p_\psi^j(\boldsymbol{B}_\psi^*(\boldsymbol{P}_\phi)) <^6 \end{array} \right\} p_\phi^j$$
$$\left\{ \begin{array}{c} <^3 1 - \bar{B}_\psi^{*j}(\boldsymbol{1}^{r\times M} - \boldsymbol{P}_\psi) <^1 \\ <^4 p_\psi^j(\boldsymbol{1}^{r\times M} - \bar{\boldsymbol{B}}_\psi^*(\boldsymbol{1}^{n\times M} - \boldsymbol{P}_\phi)) <^2 \end{array} \right\} 1 - \bar{B}_\psi^{*j}(\bar{\boldsymbol{B}}_\psi^*(\boldsymbol{1}^{n\times M} - \boldsymbol{P}_\phi)). \tag{3.71}$$

**Proof:** Applying Theorem 3.24 on each module we get, for $k = 1, \ldots, r$

$$B_{\chi k}^{*j}(\boldsymbol{P}_\phi) < p_{\chi k}^j. \tag{3.72}$$

Since $B_\psi^{*j}(\cdot)$ is strictly increasing in each matrix element, inequality 7 follows. The availabilities of each module are assumed to be strictly decreasing in $j$. Hence, using Equation (3.49), inequalities 8 and 5 follow by applying Theorem 3.24 on the organizing structure function $\psi$. Finally, from Theorem 2.20 $p_\psi^j(\boldsymbol{P}_\psi)$ is strictly increasing in $p_{\chi m}^j$. Hence, inequality 6 follows from Equations (3.72) and (3.49). Inequalities 1, 2, 3 and 4 follow by respectively applying inequalities 7, 8, 5 and 6 on the dual structure and dual level.

We conclude this section by briefly commenting on some of the sufficient conditions given to establish the strict bounds. First of all we have assumed that the marginal performance processes of the components are independent in $I$ or at time $t$. This simply reflects the fact that when dependencies between components come into play in reliability, anything can happen, even that bounds for availabilities and unavailabilities are just the exactly correct values. The assumption that $p_i^{M(I)} > 0$, $q_i^{1(I)} > 0$, $i = 1, \ldots, n$ is just a technical one to avoid uninteresting special cases. The assumptions on the minimal path and cut sets are really crucial in order to arrive at the strict bounds. Moreover, knowing when the bounds just give the exactly correct values is obviously important when applying the theory, since the upper bounds can be rather bad for long time intervals.

## 3.6 Availabilities and unavailabilities of the components

As mentioned at the start of this chapter, and seen all through it so far, bounds for the availabilities and unavailabilities in a time interval $I$ for an MMS are based on corresponding information on the multistate components. Remember Definition 3.5 denoting the availability and unavailability to level $j$ in the time interval $I$ for the $i$th component of an MMS by $p_i^{j(I)}$ and $q_i^{j(I)}$ respectively, $i = 1, \ldots, n$, $j = 1, \ldots, M$. In this section we establish these quantities in the case where $\tau = [0, \infty)$ and the performance processes of the components are Markovian. Introduce

$$p_i^{(k,\ell)}(t_1, t_2) = P(X_i(t_2) = \ell | X_i(t_1) = k).$$

Furthermore, denote the corresponding transition probabilities when $E$ is a set of absorbing states by $p_i^{(k,\ell)E}(t_1, t_2)$. Finally, assume that at time $t = 0$ all components are in the perfect functioning state $M$; i.e. $\boldsymbol{X}(0) = \boldsymbol{M}$. Then remembering Equation (2.7) we get, for $I = [t_1, t_2]$

$$p_i^{j(I)} = \sum_{k \in S_{i,j}^1} p_i^{(M,k)}(0, t_1)[1 - \sum_{\ell \in S_{i,j}^0} p_i^{(k,\ell)S_{i,j}^0}(t_1, t_2)] \tag{3.73}$$

$$q_i^{j(I)} = \sum_{k \in S_{i,j}^0} p_i^{(M,k)}(0, t_1)[1 - \sum_{\ell \in S_{i,j}^1} p_i^{(k,\ell)S_{i,j}^1}(t_1, t_2)]. \tag{3.74}$$

Note that we get $q_i^{j(I)}$ from $p_i^{j(I)}$ by interchanging the 'dual' sets $S_{i,j}^1$ and $S_{i,j}^0$.

Now let

$$\lambda_i^{(k,\ell)}(s) = \lim_{h \to 0} p_i^{(k,\ell)}(s, s + h)/h \qquad k \neq \ell,$$

be the transition intensities of $\{X_i(t), t \in [0, \infty)\}$. For simplicity we assume that the performance processes of the components are time-homogeneous, i.e.

$$p_i^{(k,\ell)}(t_1, t_2) = p_i^{(k,\ell)}(t_2 - t_1)$$
$$\lambda_i^{(k,\ell)}(s) = \lambda_i^{(k,\ell)} \qquad \text{for all } s \in [0, \infty),\ k \neq \ell.$$

Consequently, all that is needed to arrive at expressions for $p_i^{j(I)}$ and $q_i^{j(I)}$, and hence bounds for $p_\phi^{j(I)}$ and $q_\phi^{j(I)}$, are these time-independent transition intensities.

Now, for $i = 1, \ldots, n$ introduce the matrices of component transition probabilities

$$\boldsymbol{P}_i(t) = \{p_i^{(k,\ell)}(t)\}_{k\in S_i, \ell\in S_i}.$$

Since the set $S_i$ is finite, the performance processes of the components are conservative, implying that the corresponding intensity matrices are of the form $(i = 1, \ldots, n)$

$$\boldsymbol{A}_i = \begin{bmatrix} -\sum\limits_{k\in S_i-\{0\}} \lambda^{(0,k)} & \cdots\cdots\cdots & \lambda^{(0,M)} \\ \vdots & & \vdots \\ \vdots & & \vdots \\ \lambda^{(M,0)} & \cdots\cdots\cdots & -\sum\limits_{k\in S_i-\{M\}} \lambda^{(M,k)} \end{bmatrix}$$

Denoting $|S_i|$ the cardinality of the set $S_i$, $\boldsymbol{A}_i$ is a $|S_i| \times |S_i|$ matrix.

By applying standard theory for finite-state continuous-time Markov processes (see Karlin and Taylor, 1975), we have

$$\boldsymbol{P}_i(t) = \exp(\boldsymbol{A}_i(t)) = \mathbf{I} + \sum_{n=1}^{\infty} \frac{\boldsymbol{A}_i^n t^n}{n!}, \tag{3.75}$$

where $\mathbf{I}$ is the identity matrix and the initial condition is $\boldsymbol{P}_i(0) = \mathbf{I}$. Introduce the Lagrange interpolation coefficients for $k = 1, \ldots, |S_i|$

$$\boldsymbol{L}_k(\boldsymbol{A}_i) = \prod_{j=1,\ j\neq k}^{|S_i|} \frac{\boldsymbol{A}_i - \gamma_j \mathbf{I}}{\gamma_k - \gamma_j}, \tag{3.76}$$

where $\gamma_1, \ldots, \gamma_{|S_i|}$ are the eigenvalues of $\boldsymbol{A}_i$. From Apostol (1969) we then get

$$\boldsymbol{P}_i(t) = \sum_{k=1}^{|S_i|} \exp(\gamma_k) \boldsymbol{L}_k(\boldsymbol{A}_i). \tag{3.77}$$

Hence, having found a way of determining the component transition probability matrices, $p_i^{j(I)}$ and $q_i^{j(I)}$ are established by Equations (3.73) and (3.74).

Now consider first the simple special case with a binary $i$th component, i.e. $S_i = \{0, M\}$, where $M = 3$, with intensity matrix given by

$$\boldsymbol{A}_i = \begin{bmatrix} -\mu & \mu \\ \lambda & -\lambda \end{bmatrix}. \tag{3.78}$$

By solving the equation

$$\det\{\boldsymbol{A}_i - \gamma \mathbf{I}\} = 0,$$

we easily find

$$\gamma_1 = 0, \quad \gamma_2 = -(\lambda + \mu).$$

From Equations (3.76) and (3.77) we get

$$p_i^{(3,3)}(t) = \mu/(\lambda + \mu) + [\lambda/(\lambda + \mu)] \exp -(\lambda + \mu)t \tag{3.79}$$

$$p_i^{(3,0)}(t) = \lambda/(\lambda + \mu) - [\lambda/(\lambda + \mu)] \exp -(\lambda + \mu)t. \tag{3.80}$$

Then, from Equations (3.73) and (3.74), we finally get

$$\begin{aligned} p_i^{1(I)} &= p_i^{2(I)} = p_i^{3(I)} = p_i^{(3,3)}(t_1) \exp(-\lambda(t_2 - t_1)) \\ q_i^{1(I)} &= q_i^{2(I)} = q_i^{3(I)} = p_i^{(3,0)}(t_1) \exp(-\mu(t_2 - t_1)). \end{aligned} \tag{3.81}$$

Now assume that the $i$th component has the states $S_i = \{0, 1, 2\}$ with intensity matrix given by

$$\boldsymbol{A}_i = \begin{bmatrix} -(\lambda^{(0,1)} + \lambda^{(0,2)}) & \lambda^{(0,1)} & \lambda^{(0,2)} \\ \lambda^{(1,0)} & -(\lambda^{(1,0)} + \lambda^{(1,2)}) & \lambda^{(1,2)} \\ \lambda^{(2,0)} & \lambda^{(2,1)} & -(\lambda^{(2,0)} + \lambda^{(2,1)}) \end{bmatrix}. \tag{3.82}$$

By solving the equation

$$\det\{\boldsymbol{A}_i - \gamma \mathbf{I}\} = 0,$$

we find, after some algebra

$$\gamma_1 = 0, \quad \gamma_2 = -(B + \sqrt{B^2 - 4C})/2, \quad \gamma_3 = -(B - \sqrt{B^2 - 4C})/2,$$

where

$$B = \lambda^{(2,0)} + \lambda^{(2,1)} + \lambda^{(1,2)} + \lambda^{(0,2)} + \lambda^{(1,0)} + \lambda^{(0,1)}$$

$$C = \lambda^{(2,0)}\lambda^{(1,0)} + \lambda^{(1,2)}\lambda^{(2,0)} + \lambda^{(1,0)}\lambda^{(2,1)} + \lambda^{(0,2)}\lambda^{(2,1)} + \lambda^{(0,2)}\lambda^{(1,0)} + \lambda^{(0,2)}\lambda^{(1,2)} + \lambda^{(2,0)}\lambda^{(0,1)} + \lambda^{(1,2)}\lambda^{(0,1)} + \lambda^{(2,1)}\lambda^{(0,1)}.$$

Hence, the choice of just three possible states of the components has the advantage of leading to a second-order equation for these eigenvalues.

After more algebra we get, from Equations (3.76) and (3.77)

$$\begin{aligned}
p_i^{(2,0)}(t) &= [\lambda^{(2,0)}(\lambda^{(1,2)} + \lambda^{(1,0)}) + \lambda^{(1,0)}\lambda^{(2,1)}]/(\gamma_2\gamma_3) \\
&+ [\lambda^{(1,0)}\lambda^{(2,1)} - \lambda^{(2,0)}(\lambda^{(0,1)} + \lambda^{(0,2)} + \lambda^{(2,0)} + \lambda^{(2,1)} + \gamma_3)] \\
&\exp(\gamma_2 t)/(\gamma_2(\gamma_2 - \gamma_3)) \\
&+ [\lambda^{(1,0)}\lambda^{(2,1)} - \lambda^{(2,0)}(\lambda^{(0,1)} + \lambda^{(0,2)} + \lambda^{(2,0)} + \lambda^{(2,1)} + \gamma_2)] \\
&\exp(\gamma_3 t)/(\gamma_3(\gamma_3 - \gamma_2)) \\
p_i^{(2,1)}(t) &= [\lambda^{(2,1)}\lambda^{(0,2)} + \lambda^{(2,1)}\lambda^{(0,1)} + \lambda^{(2,0)}\lambda^{(0,1)}]/(\gamma_2\gamma_3) \\
&+ [\lambda^{(2,1)}(\lambda^{(0,2)} + \lambda^{(0,1)} + \gamma_2) + \lambda^{(2,0)}\lambda^{(0,1)}]\exp(\gamma_2 t)/(\gamma_2(\gamma_2 - \gamma_3)) \\
&+ [\lambda^{(2,1)}(\lambda^{(0,2)} + \lambda^{(0,1)} + \gamma_3) + \lambda^{(2,0)}\lambda^{(0,1)}]\exp(\gamma_3 t)/(\gamma_3(\gamma_3 - \gamma_2)) \\
p_i^{(2,2)}(t) &= [(\lambda^{(1,0)} + \lambda^{(1,2)})\lambda^{(0,2)} + \lambda^{(0,1)}\lambda^{(1,2)}](\gamma_2\gamma_3) \\
&+ [\lambda^{(0,2)}\lambda^{(2,0)} + \lambda^{(1,2)}\lambda^{(2,1)} + (\lambda^{(2,0)} + \lambda^{(2,1)})^2 + \gamma_3(\lambda^{(2,0)} + \lambda^{(2,1)})] \\
&\exp(\gamma_2 t)/(\gamma_2(\gamma_2 - \gamma_3)) \\
&+ [\lambda^{(0,2)}\lambda^{(2,0)} + \lambda^{(1,2)}\lambda^{(2,1)} + (\lambda^{(2,0)} + \lambda^{(2,1)})^2 + \gamma_2(\lambda^{(2,0)} + \lambda^{(2,1)})] \\
&\exp(\gamma_3 t)/(\gamma_3(\gamma_3 - \gamma_2)) \\
p_i^{(1,0)}(t) &= [\lambda^{(2,0)}(\lambda^{(1,2)} + \lambda^{(1,0)}) + \lambda^{(1,0)}\lambda^{(2,1)}]/(\gamma_2\gamma_3) \\
&+ [\lambda^{(1,2)}\lambda^{(2,0)} - \lambda^{(1,0)}(\lambda^{(0,1)} + \lambda^{(0,2)} + \lambda^{(1,0)} + \lambda^{(1,2)} + \gamma_3)] \\
&\exp(\gamma_2 t)/(\gamma_2(\gamma_2 - \gamma_3)) \\
&+ [\lambda^{(1,2)}\lambda^{(2,0)} - \lambda^{(1,0)}(\lambda^{(0,1)} + \lambda^{(0,2)} + \lambda^{(1,0)} + \lambda^{(1,2)} + \gamma_2)] \\
&\exp(\gamma_3 t)/(\gamma_3(\gamma_3 - \gamma_2)) \\
p_i^{(1,2)}(t) &= [(\lambda^{(1,0)} + \lambda^{(1,2)})\lambda^{(0,2)} + \lambda^{(0,1)}\lambda^{(1,2)}]/(\gamma_2\gamma_3) \\
&+ [\lambda^{(1,0)}\lambda^{(0,2)} - \lambda^{(1,2)}(\lambda^{(1,0)} + \lambda^{(1,2)} + \lambda^{(2,0)} + \lambda^{(2,1)} + \gamma_3)]
\end{aligned}$$

$$\exp(\gamma_2 t)/(\gamma_2(\gamma_2-\gamma_3))$$
$$+[\lambda^{(1,0)}\lambda^{(0,2)}-\lambda^{(1,2)}(\lambda^{(1,0)}+\lambda^{(1,2)}+\lambda^{(2,0)}+\lambda^{(2,1)}+\gamma_2)]$$
$$\exp(\gamma_3 t)/(\gamma_3(\gamma_3-\gamma_2))$$
$$p_i^{(0,2)}(t)=[(\lambda^{(1,0)}+\lambda^{(1,2)})\lambda^{(0,2)}+\lambda^{(0,1)}\lambda^{(1,2)}]/(\gamma_2\gamma_3)$$
$$+[\lambda^{(0,2)}(\lambda^{(1,0)}+\lambda^{(1,2)}+\gamma_2)+\lambda^{(0,1)}\lambda^{(1,2)}]\exp(\gamma_2 t)/(\gamma_2(\gamma_2-\gamma_3))$$
$$+[\lambda^{(0,2)}(\lambda^{(1,0)}+\lambda^{(1,2)}+\gamma_3)+\lambda^{(0,1)}\lambda^{(1,2)}]\exp(\gamma_3 t)/(\gamma_3(\gamma_3-\gamma_2)). \tag{3.83}$$

By specializing $\lambda^{(0,1)}=\lambda^{(0,2)}=0$ in $p_i^{(2,0)}(t)$ and $p_i^{(1,0)}(t)$, we get, respectively, $p_i^{(2,0)\{0\}}(t)$ and $p_i^{(1,0)\{0\}}(t)$. Similarly, by specializing $\lambda^{(2,1)}=\lambda^{(2,0)}=0$ in $p_i^{(1,2)}(t)$ and $p_i^{(0,2)}(t)$, we get, respectively, $p_i^{(1,2)\{2\}}(t)$ and $p_i^{(0,2)\{2\}}(t)$.

From Equations (3.73) and (3.74) we see that for this special case we have calculated all that is needed

$$p_i^{2(I)}=p_i^{(2,2)}(t_1)\exp(-(\lambda^{(2,0)}+\lambda^{(2,1)})(t_2-t_1))$$
$$p_i^{1(I)}=p_i^{(2,2)}(t_1)[1-p_i^{(2,0)\{0\}}(t_2-t_1)]+p_i^{(2,1)}(t_1)[1-p_i^{(1,0)\{0\}}(t_2-t_1)]$$
$$q_i^{2(I)}=p_i^{(2,1)}(t_1)[1-p_i^{(1,2)\{2\}}(t_2-t_1)]+p_i^{(2,0)}(t_1)[1-p_i^{(0,2)\{2\}}(t_2-t_1)]$$
$$q_i^{1(I)}=p_i^{(2,0)}(t_1)\exp(-(\lambda^{(0,1)}+\lambda^{(0,2)})(t_2-t_1)). \tag{3.84}$$

## 3.7 The simple network system revisited

We now apply the theory to the simple network system given in Figure 1.1. We still consider the time interval $I=[t_1,t_2]$ and assume that the marginal performance processes of the two components are independent in $[0,t_2]$ and also that the two branches of each component fail and are repaired/replaced independently of each other. All branches have the same instantaneous failure rate $\lambda$ and repair/replacement rate $\mu$. Hence, for $i=1,2$ we have

$$\lambda^{(0,1)}=2\mu,\quad \lambda^{(1,3)}=\mu,\quad \lambda^{(3,1)}=2\lambda,$$
$$\lambda^{(1,0)}=\lambda,\lambda^{(0,3)}=\lambda^{(3,0)}=0,$$

and hence we have

$$\boldsymbol{A}_i=\begin{bmatrix}-2\mu & 2\mu & 0\\ \lambda & -(\lambda+\mu) & \mu\\ 0 & 2\lambda & -2\lambda\end{bmatrix}.$$

This is a special case of the intensity matrix given by Equation (3.82). Hence, by specializing, the desired component availabilities and unavailabilities are obtained from Equations (3.83) and (3.84) noting that component state 3 in our simple network system corresponds to state 2 in these equations. Explicit results are given in Funnemark and Natvig (1985).

Since the bounds in Equations (3.10) and (3.11) of Theorem 3.7 cannot be established as functions of component availabilities, these bounds are dropped along with the ones which are based on those given in Corollaries 3.8 and 3.10. Because the marginal performance processes of the two components are assumed to be independent in the time interval $[0, t_2]$, Theorem 3.6, Equations (3.12) and (3.13) of Theorem 3.7, Corollaries 3.9 and 3.11 and Theorems 3.22–3.24 can be applied. In the following we concentrate on bounds on the availabilities. Since, according to Theorem 3.6 under the mentioned condition, the lower bound of Equation (3.6) reduces to the corresponding one of Equation (3.8), we treat only the latter one. It should be mentioned that, since from Table 2.5 there is only one minimal path set both to levels 1 and 3, it follows from Theorem 3.22 that for these levels the lower bound of Equation (3.8) is exactly correct. Applying Tables 2.5 and 2.6, noting that $p_i^{2(I)} = p_i^{3(I)}$ the lower bounds are given by

$$\begin{aligned}
\ell_\phi^{\prime 1(I)}(\boldsymbol{P}_\phi^{(I)}) &= \ell_\phi^{**1(I)}(\boldsymbol{P}_\phi^{(I)}) = L_\phi^{*1(I)}(\boldsymbol{P}_\phi^{(I)}) \\
&= B_\phi^{*1(I)}(\boldsymbol{P}_\phi^{(I)}) = (p_i^{1(I)})^2 = p_\phi^{1(I)} \\
\ell_\phi^{\prime 2(I)}(\boldsymbol{P}_\phi^{(I)}) &= p_i^{1(I)} p_i^{3(I)} \leq p_i^{1(I)} p_i^{3(I)} \max[1, p_i^{1(I)}(2 - p_i^{3(I)})] \\
&= L_\phi^{*2(I)}(\boldsymbol{P}_\phi^{(I)}) = B_\phi^{*2(I)}(\boldsymbol{P}_\phi^{(I)}) \leq p_\phi^{2(I)} \\
\ell_\phi^{\prime 3(I)}(\boldsymbol{P}_\phi^{(I)}) &= \ell_\phi^{**3(I)}(\boldsymbol{P}_\phi^{(I)}) = L_\phi^{*3(I)}(\boldsymbol{P}_\phi^{(I)}) \\
&= B_\phi^{*3(I)}(\boldsymbol{P}_\phi^{(I)}) = (p_i^{3(I)})^2 = p_\phi^{3(I)}.
\end{aligned} \tag{3.85}$$

Noting that $q_i^{2(I)} = q_i^{3(I)}$ the upper bounds are given by

$$\begin{aligned}
u_\phi^{\prime\prime 1(I)} &= p_i^{1(I)} \leq 1 - q_i^{1(I)} = 1 - \bar{\ell}_\phi^{\prime 1(I)}(\boldsymbol{Q}_\phi^{(I)}) \geq (1 - q_i^{1(I)})^2 \\
&= 1 - \bar{L}_\phi^{*1(I)}(\boldsymbol{Q}_\phi^{(I)}) = 1 - \bar{B}_\phi^{*1(I)}(\boldsymbol{Q}_\phi^{(I)})
\end{aligned}$$

$$\begin{aligned}
&1 - \bar{\ell}_\phi^{\prime 2(I)}(\boldsymbol{Q}_\phi^{(I)}) = 1 - \max[q_i^{1(I)}, (q_i^{3(I)})^2] \\
&\geq 1 - \max[q_i^{1(I)}, (q_i^{1(I)} + q_i^{3(I)} - q_i^{1(I)} q_i^{3(I)})^2] = 1 - \bar{L}_\phi^{*2(I)}(\boldsymbol{Q}_\phi^{(I)}) \\
&\geq 1 - \bar{B}_\phi^{*2(I)}(\boldsymbol{Q}_\phi^{(I)}) = \min[(1 - q_i^{1(I)})^2, 1 - (q_i^{1(I)} + q_i^{3(I)} - q_i^{1(I)} q_i^{3(I)})^2]
\end{aligned}$$

$$u_{\phi}^{\prime\prime 3(I)} = p_i^{3(I)} \leq 1 - q_i^{3(I)} = 1 - \bar{\ell}_{\phi}^{\prime 3(I)}(\boldsymbol{Q}_{\phi}^{(I)}) \geq (1 - q_i^{3(I)})^2$$
$$= 1 - \bar{L}_{\phi}^{*3(I)}(\boldsymbol{Q}_{\phi}^{(I)}) = 1 - \bar{B}_{\phi}^{*3(I)}(\boldsymbol{Q}_{\phi}^{(I)}). \tag{3.86}$$

In Table 3.1 we give numerical values of the bounds for all combinations of $\lambda = 0.001, 0.01$; $\mu = 0.001, 0.01$ and $I = [100, 110], [100, 200], [1000, 1100]$. The lower bounds given are representatives of the different ones in Equation (3.85). Hence, for level 1 $p_{\phi}^{1(I)}$ is given, for level 2 $\ell_{\phi}^{\prime 2(I)}(\boldsymbol{P}_{\phi}^{(I)})$ and $L_{\phi}^{*2(I)}(\boldsymbol{P}_{\phi}^{(I)})$ are given, whereas for level 3 $p_{\phi}^{3(I)}$ is given. Similarly, the upper bounds given are representatives of the different ones in Equation (3.86). Hence, for level 1 $u_{\phi}^{\prime\prime 1(I)}$, $1 - \bar{\ell}_{\phi}^{\prime 1(I)}(\boldsymbol{Q}_{\phi}^{(I)})$ and $1 - \bar{L}_{\phi}^{*1(I)}(\boldsymbol{Q}_{\phi}^{(I)})$ are given, for level 2 $1 - \bar{\ell}_{\phi}^{\prime 2(I)}(\boldsymbol{Q}_{\phi}^{(I)})$, $1 - \bar{L}_{\phi}^{*2(I)}(\boldsymbol{Q}_{\phi}^{(I)})$ and $1 - \bar{B}_{\phi}^{*2(I)}(\boldsymbol{Q}_{\phi}^{(I)})$ are given, whereas for level 3 $u_{\phi}^{\prime\prime 3(I)}$, $1 - \bar{\ell}_{\phi}^{\prime 3(I)}(\boldsymbol{Q}_{\phi}^{(I)})$ and $1 - \bar{L}_{\phi}^{*3(I)}(\boldsymbol{Q}_{\phi}^{(I)})$ are given.

As we see from Table 3.1, the bounds for this simple network system are amazingly good. Only the upper bounds for the availability to level 2 are getting really bad; i.e. for $\lambda = \mu = 0.01$ and $I = [100, 200], [1000, 1100]$. In these cases we expect more fluctuations between the states of the two components and hence of the system during $I$. Since our bounds are based on component availabilities and unavailabilities, a lot of information is lost in these cases, and the bounds are hence poor.

Note that we always have $1 - \bar{L}_{\phi}^{*2(I)}(\boldsymbol{Q}_{\phi}^{(I)}) = 1 - \bar{B}_{\phi}^{*2(I)}(\boldsymbol{Q}_{\phi}^{(I)})$ in Table 3.1. In fact we have been unable to find values of $\lambda$, $\mu$ and $I$ such that this equality does not hold. For the special case $t_1 = t_2 = \infty$ this can be shown after some algebra. Hence, the bounds of Corollary 3.11 have not been of any help in this example.

To illustrate the application of a modular decomposition, consider the two components as modules and the two branches as components in the simple network system of Figure 1.1. The minimal path vectors and minimal cut vectors for the modules are given respectively in Tables 3.2 and 3.3. The minimal path vectors and minimal cut vectors for the organizing structure are obviously identical to the ones given respectively in Tables 2.5 and 2.6. The minimal path vectors and minimal cut vectors for the four-component system are given respectively in Tables 3.4 and 3.5.

Theorem 3.6, Equations (3.12) and (3.13) of Theorem 3.7, Corollaries 3.9 and 3.11, Theorem 3.14 and Theorems 3.22–3.24 can now be applied, concentrating on bounds on the availabilities. Since, according to Theorem 3.6, the lower bound of Equation (3.6) reduces to the corresponding one of Equation (3.8) when the marginal performance processes

Table 3.1 Bounds for the availabilities of the simple network system of Figure 1.1.

| | Level 1 | | Level 2 | | Level 3 | |
|---|---|---|---|---|---|---|
| | Lower | Upper | Lower | Upper | Lower | Upper |
| $\lambda = 0.001$ | 0.9802 | 0.9901 | 0.8025 | 0.9706 | 0.6570 | 0.8106 |
| $\mu = 0.001$ | | 0.9919 | 0.9451 | 0.9683 | | 0.8286 |
| $I = [100, 110]$ | | 0.9840 | | 0.9683 | | 0.6865 |
| $\lambda = 0.001$ | 0.9400 | 0.9695 | 0.6564 | 0.9750 | 0.4584 | 0.6770 |
| $\mu = 0.001$ | | 0.9933 | 0.8420 | 0.9732 | | 0.8420 |
| $I = [100, 200]$ | | 0.9866 | | 0.9732 | | 0.7090 |
| $\lambda = 0.001$ | 0.5862 | 0.7656 | 0.2020 | 0.6012 | 0.0696 | 0.2638 |
| $\mu = 0.001$ | | 0.8470 | 0.2685 | 0.5268 | | 0.3685 |
| $I = [1000, 1100]$ | | 0.7174 | | 0.5268 | | 0.1358 |
| $\lambda = 0.001$ | 0.9903 | 0.9952 | 0.8607 | 0.9886 | 0.7481 | 0.8649 |
| $\mu = 0.01$ | | 0.9970 | 0.9723 | 0.9880 | | 0.8932 |
| $I = [100, 110]$ | | 0.9940 | | 0.9880 | | 0.7978 |
| $\lambda = 0.001$ | 0.9667 | 0.9832 | 0.7103 | 0.9979 | 0.5219 | 0.7224 |
| $\mu = 0.01$ | | 0.9995 | 0.8923 | 0.9979 | | 0.9543 |
| $I = [100, 200]$ | | 0.9990 | | 0.9979 | | 0.9108 |
| $\lambda = 0.001$ | 0.9522 | 0.9758 | 0.6603 | 0.9954 | 0.4578 | 0.6766 |
| $\mu = 0.01$ | | 0.9989 | 0.8526 | 0.9952 | | 0.9320 |
| $I = [1000, 1100]$ | | 0.9978 | | 0.9952 | | 0.8686 |
| $\lambda = 0.01$ | 0.3429 | 0.5856 | 0.0742 | 0.2934 | 0.0161 | 0.1268 |
| $\mu = 0.001$ | | 0.6395 | 0.0814 | 0.1935 | | 0.1594 |
| $I = [100, 110]$ | | 0.4089 | | 0.1935 | | 0.0254 |
| $\lambda = 0.01$ | 0.0762 | 0.2761 | 0.0058 | 0.3394 | 0.0004 | 0.0210 |
| $\mu = 0.001$ | | 0.6989 | 0.0058 | 0.2445 | | 0.1872 |
| $I = [100, 200]$ | | 0.4884 | | 0.2445 | | 0.0350 |
| $\lambda = 0.01$ | 0.0046 | 0.0680 | 0.0001 | 0.0478 | 0.0000 | 0.0011 |
| $\mu = 0.001$ | | 0.3234 | 0.0001 | 0.0156 | | 0.0242 |
| $I = [1000, 1100]$ | | 0.1046 | | 0.0156 | | 0.0006 |
| $\lambda = 0.01$ | 0.5862 | 0.7656 | 0.2020 | 0.6012 | 0.0696 | 0.2638 |
| $\mu = 0.01$ | | 0.8470 | 0.2685 | 0.5268 | | 0.3685 |
| $I = [100, 110]$ | | 0.7174 | | 0.5268 | | 0.1358 |
| $\lambda = 0.01$ | 0.2024 | 0.4499 | 0.0196 | 0.8705 | 0.0019 | 0.0436 |
| $\mu = 0.01$ | | 0.9747 | 0.0196 | 0.8585 | | 0.6401 |
| $I = [100, 200]$ | | 0.9500 | | 0.8585 | | 0.4097 |
| $\lambda = 0.01$ | 0.1651 | 0.4063 | 0.0137 | 0.8349 | 0.0011 | 0.0338 |
| $\mu = 0.01$ | | 0.9662 | 0.0137 | 0.8182 | | 0.5937 |
| $I = [1000, 1100]$ | | 0.9335 | | 0.8182 | | 0.3525 |

Table 3.2 Minimal path vectors for the modules of the system of Figure 1.1.

| Level | Component 1 | Component 2 |
|---|---|---|
| 1 | 0 | 3 |
| 1 | 3 | 0 |
| 3 | 3 | 3 |

Table 3.3 Minimal cut vectors for the modules of the system of Figure 1.1.

| Level | Component 1 | Component 2 |
|---|---|---|
| 1 | 0 | 0 |
| 3 | 0 | 3 |
| 3 | 3 | 0 |

Table 3.4 Minimal path vectors for the four-component system of Figure 1.1.

| Level | Component 1 | Component 2 | Component 3 | Component 4 |
|---|---|---|---|---|
| 1 | 0 | 3 | 0 | 3 |
| 1 | 0 | 3 | 3 | 0 |
| 1 | 3 | 0 | 0 | 3 |
| 1 | 3 | 0 | 3 | 0 |
| 2 | 0 | 3 | 3 | 3 |
| 2 | 3 | 0 | 3 | 3 |
| 2 | 3 | 3 | 0 | 3 |
| 2 | 3 | 3 | 3 | 0 |
| 3 | 3 | 3 | 3 | 3 |

Table 3.5 Minimal cut vectors for the four-component system of Figure 1.1.

| Level | Component 1 | Component 2 | Component 3 | Component 4 |
|---|---|---|---|---|
| 1, 2 | 0 | 0 | 3 | 3 |
| 1, 2 | 3 | 3 | 0 | 0 |
| 2 | 0 | 3 | 0 | 3 |
| 2 | 0 | 3 | 3 | 0 |
| 2 | 3 | 0 | 0 | 3 |
| 2 | 3 | 0 | 3 | 0 |
| 3 | 0 | 3 | 3 | 3 |
| 3 | 3 | 0 | 3 | 3 |
| 3 | 3 | 3 | 0 | 3 |
| 3 | 3 | 3 | 3 | 0 |

of the components are independent in $I$, we treat only the latter one. As above, it should be mentioned that, since from Tables 2.5, 3.2 and 3.4 there is only one minimal path set to level 3, it follows from Theorem 3.22 that the corresponding lower bound of Equation (3.8) is exactly correct. The component availabilities and unavailabilities are now given by Equation (3.81). Applying Tables 2.5, 2.6 and 3.2–3.5, noting that $p_i^{1(I)} = p_i^{3(I)}$, the lower bounds are given by

$$\begin{aligned}
\ell_{\chi_k}^{\prime 1(I)}(\boldsymbol{P}_\phi^{(I)}) &= p_i^{1(I)} \leq \ell_{\chi_k}^{**1(I)}(\boldsymbol{P}_\phi^{(I)}) = L_{\chi_k}^{*1(I)}(\boldsymbol{P}_\phi^{(I)}) = B_{\chi_k}^{*1(I)}(\boldsymbol{P}_\phi^{(I)}) \\
&= (p_i^{1(I)})(2 - p_i^{1(I)}) \\
\ell_{\chi_k}^{\prime 3(I)}(\boldsymbol{P}_\phi^{(I)}) &= \ell_{\chi_k}^{**3(I)}(\boldsymbol{P}_\phi^{(I)}) = L_{\chi_k}^{*3(I)}(\boldsymbol{P}_\phi^{(I)}) = B_{\chi_k}^{*3(I)}(\boldsymbol{P}_\phi^{(I)}) \\
&= (p_i^{1(I)})^2 = p_{\chi_k}^{3(I)} \\
\ell_\psi^{\prime 1(I)}(\boldsymbol{B}_\psi^{*(I)}(\boldsymbol{P}_\phi^{(I)})) &= (p_i^{1(I)})^2(2 - p_i^{1(I)})^2 = \ell_\psi^{**1(I)}(\boldsymbol{B}_\psi^{*(I)}(\boldsymbol{P}_\phi^{(I)})) \\
&= L_\psi^{*1(I)}(\boldsymbol{B}_\psi^{*(I)}(\boldsymbol{P}_\phi^{(I)})) = B_\psi^{*1(I)}(\boldsymbol{B}_\psi^{*(I)}(\boldsymbol{P}_\phi^{(I)})) \\
\ell_\psi^{\prime 2(I)}(\boldsymbol{B}_\psi^{*(I)}(\boldsymbol{P}_\phi^{(I)})) &= (p_i^{1(I)})^3(2 - p_i^{1(I)}) \leq (p_i^{1(I)})^3(2 - p_i^{1(I)}) \\
&\qquad \max[1, p_i^{1(I)}(2 - p_i^{1(I)})(2 - (p_i^{1(I)})^2)] \\
&= L_\psi^{*2(I)}(\boldsymbol{B}_\psi^{*(I)}(\boldsymbol{P}\phi^{(I)})) = B_\psi^{*2(I)}(\boldsymbol{B}_\psi^{*(I)}(\boldsymbol{P}_\phi^{(I)})) \\
\ell_\psi^{\prime 3(I)}(\boldsymbol{B}_\psi^{*(I)}(\boldsymbol{P}_\phi^{(I)})) &= (p_i^{1(I)})^4 = \ell_\psi^{**3(I)}(\boldsymbol{B}_\psi^{*(I)}(\boldsymbol{P}_\phi^{(I)})) \\
&= L_\psi^{*3(I)}(\boldsymbol{B}_\psi^{*(I)}(\boldsymbol{P}_\phi^{(I)})) = B_\psi^{*3(I)}(\boldsymbol{B}_\psi^{*(I)}(\boldsymbol{P}_\phi^{(I)})) = p_\phi^{3(I)} \\
\ell_\phi^{\prime 1(I)}(\boldsymbol{P}_\phi^{(I)}) &= (p_i^{1(I)})^2 \leq \ell_\phi^{**1(I)}(\boldsymbol{P}_\phi^{(I)}) = L_\phi^{*1(I)}(\boldsymbol{P}_\phi^{(I)}) \\
&= B_\phi^{*1(I)}(\boldsymbol{P}_\phi^{(I)}) = (p_i^{1(I)})^2(2 - p_i^{1(I)})^2 \\
&= B_\psi^{*1(I)}(\boldsymbol{B}_\psi^{*(I)}(\boldsymbol{P}_\phi^{(I)})) \\
\ell_\phi^{\prime 2(I)}(\boldsymbol{P}_\phi^{(I)}) &= (p_i^{1(I)})^3 \leq (p_i^{1(I)})^3 \max[1, (p_i^{1(I)})^3(2 - p_i^{1(I)})^6] \\
&= L_\phi^{*2(I)}(\boldsymbol{P}_\phi^{(I)}) = B_\phi^{*2(I)}(\boldsymbol{P}_\phi^{(I)}) \\
\ell_\phi^{\prime 3(I)}(\boldsymbol{P}_\phi^{(I)}) &= \ell_\phi^{**3(I)}(\boldsymbol{P}_\phi^{(I)}) = L_\phi^{*3(I)}(\boldsymbol{P}_\phi^{(I)}) = B_\phi^{*3(I)}(\boldsymbol{P}_\phi^{(I)}) \\
&= (p_i^{1(I)})^4 = p_\phi^{3(I)} = B_\psi^{*3(I)}(\boldsymbol{B}_\psi^{*(I)}(\boldsymbol{P}_\phi^{(I)})). \qquad (3.87)
\end{aligned}$$

Noting that $q_i^{1(I)} = q_i^{3(I)}$ the upper bounds are given by

$$
\begin{aligned}
1 - \bar{\ell}'^{1(I)}_{\chi_k}(\boldsymbol{Q}_\phi^{(I)}) &= 1 - (q_i^{3(I)})^2 = 1 - \bar{L}^{*1(I)}_{\chi_k}(\boldsymbol{Q}_\phi^{(I)}) \\
&= 1 - \bar{B}^{*1(I)}_{\chi_k}(\boldsymbol{Q}_\phi^{(I)}) \\
u''^{3(I)}_{\chi_k} &= p_i^{1(I)} \le 1 - q_i^{1(I)} = 1 - q_i^{3(I)} = 1 - \bar{\ell}'^{3(I)}_{\chi_k}(\boldsymbol{Q}_\phi^{(I)}) \\
&\ge (1 - q_i^{3(I)})^2 = 1 - \bar{L}^{*3(I)}_{\chi_k}(\boldsymbol{Q}_\phi^{(I)}) = 1 - \bar{B}^{*3(I)}_{\chi_k}(\boldsymbol{Q}_\phi^{(I)}) \\
u''^{1(I)}_{\psi} &= p^{1(I)}_{\chi_k} \le 1 - \bar{B}^{*1(I)}_{\chi_k}(\boldsymbol{Q}_\phi^{(I)}) = 1 - (q_i^{3(I)})^2 \\
&= 1 - \bar{\ell}'^{1(I)}_{\psi}(\bar{\boldsymbol{B}}^{*(I)}_{\psi}(\boldsymbol{Q}_\phi^{(I)})) \\
&\ge (1 - (q_i^{3(I)})^2)^2 = 1 - \bar{L}^{*1(I)}_{\psi}(\bar{\boldsymbol{B}}^{*(I)}_{\psi}(\boldsymbol{Q}_\phi^{(I)})) \\
&= 1 - \bar{B}^{*1(I)}_{\psi}(\bar{\boldsymbol{B}}^{*(I)}_{\psi}(\boldsymbol{Q}_\phi^{(I)}))
\end{aligned}
$$

$$
\begin{aligned}
1 - \bar{\ell}'^{2(I)}_{\psi}(\bar{\boldsymbol{B}}^{*(I)}_{\psi}(\boldsymbol{Q}_\phi^{(I)})) &= 1 - (q_i^{3(I)} \amalg q_i^{3(I)})^2 \\
&\ge 1 - ((q_i^{3(I)})^2 \amalg q_i^{3(I)} \amalg q_i^{3(I)})^2 \\
&= 1 - \bar{L}^{*2(I)}_{\psi}(\bar{\boldsymbol{B}}^{*(I)}_{\psi}(\boldsymbol{Q}_\phi^{(I)})) \ge 1 - \max[((q_i^{3(I)})^2 \\
&\quad \amalg q_i^{3(I)} \amalg q_i^{3(I)})^2, (q_i^{3(I)})^2 \amalg (q_i^{3(I)})^2] \\
&= 1 - \bar{B}^{*2(I)}_{\psi}(\bar{\boldsymbol{B}}^{*(I)}_{\psi}(\boldsymbol{Q}_\phi^{(I)})) \\
u''^{3(I)}_{\psi} &= p^{3(I)}_{\chi_k} \le 1 - \bar{B}^{*3(I)}_{\chi_k}(\boldsymbol{Q}_\phi^{(I)}) = 1 - q_i^{3(I)} \amalg q_i^{3(I)} \\
&= 1 - \bar{\ell}'^{3(I)}_{\psi}(\bar{\boldsymbol{B}}^{*(I)}_{\psi}(\boldsymbol{Q}_\phi^{(I)})) \\
&\ge 1 - q_i^{3(I)} \amalg q_i^{3(I)} \amalg q_i^{3(I)} \amalg q_i^{3(I)} \\
&= 1 - \bar{L}^{*3(I)}_{\psi}(\bar{\boldsymbol{B}}^{*(I)}_{\psi}(\boldsymbol{Q}_\phi^{(I)})) \\
&= 1 - \bar{B}^{*3(I)}_{\psi}(\bar{\boldsymbol{B}}^{*(I)}_{\psi}(\boldsymbol{Q}_\phi^{(I)}))
\end{aligned}
$$

$$
\begin{aligned}
1 - \bar{\ell}'^{1(I)}_{\phi}(\boldsymbol{Q}_\phi^{(I)}) &= 1 - (q_i^{3(I)})^2 \ge 1 - \max[(q_i^{3(I)})^2, (q_i^{3(I)} \amalg q_i^{3(I)})^4] \\
&= 1 - \bar{L}^{*1(I)}_{\phi}(\boldsymbol{Q}_\phi^{(I)}) = 1 - \bar{B}^{*1(I)}_{\phi}(\boldsymbol{Q}_\phi^{(I)}) \\
1 - \bar{\ell}'^{2(I)}_{\phi}(\boldsymbol{Q}_\phi^{(I)}) &= 1 - (q_i^{3(I)})^2 \ge 1 - \max[(q_i^{3(I)})^2, \\
&\quad (q_i^{3(I)} \amalg q_i^{3(I)} \amalg q_i^{3(I)})^4] = 1 - \bar{L}^{*2(I)}_{\phi}(\boldsymbol{Q}_\phi^{(I)}) \\
&= 1 - \bar{B}^{*2(I)}_{\phi}(\boldsymbol{Q}_\phi^{(I)})
\end{aligned}
$$

$$\begin{aligned}1-\bar{\ell}_{\phi}^{\prime 3(I)}(\boldsymbol{Q}_{\phi}^{(I)}) &= 1-q_i^{3(I)} \geq 1-q_i^{3(I)} \amalg q_i^{3(I)} \amalg q_i^{3(I)} \amalg q_i^{3(I)}\\ &= 1-\bar{L}_{\phi}^{*3(I)}(\boldsymbol{Q}_{\phi}^{(I)}) = 1-\bar{B}_{\phi}^{*3(I)}(\boldsymbol{Q}_{\phi}^{(I)})\\ &= 1-\bar{B}_{\psi}^{*3(I)}(\bar{\boldsymbol{B}}_{\psi}^{*(I)}(\boldsymbol{Q}_{\phi}^{(I)})). \qquad (3.88)\end{aligned}$$

## 3.8 The offshore electrical power generation system revisited

We conclude this chapter by returning to the offshore electrical power generation system presented in Section 1.2, concentrating on the structure functions $\phi_1$ and $\phi_2$ given respectively by Equations (1.3) and (1.4).

We now assume that $\lambda^{(0,2)} = 0$; i.e. we always repair a completely failed component to the perfect functioning level. The resulting intensity matrix is a special case of the one given by Equation (3.82). Hence, by specializing, the desired component availabilities and unavailabilities are obtained from Equations (3.83) and (3.84) noting that the component states 2 and 4 in our offshore electrical power generation system correspond respectively to states 1 and 2 in these equations.

We then give, in Table 3.6, some numerical values for the availabilities and unavailabilities for the components $A_1$, $A_2$, $A_3$, $U$ and $L$ based on the given 'guesstimates' of the transition intensities. More and better data are

Table 3.6 Availabilities and unavailabilities for $A_1, A_2, A_3, U$ and $L$.

| Quantity | $A_1, A_2, A_3$ | $U$ | $L$ |
|---|---|---|---|
| $\lambda^{(4,0)}$ | 1.46 | 2 | 0.04 |
| $\lambda^{(4,2)}$ | 27.74 | 12 | 9.09 |
| $\lambda^{(2,0)}$ | 1.46 | 5 | 0.2 |
| $\lambda^{(0,4)}$ | 730 | 4380 | 10.95 |
| $\lambda^{(2,4)}$ | 17520 | 17520 | 21.9 |
| $p_i^{4([0.5,1])}$ | 0.000 | 0.001 | 0.007 |
| $p_i^{4([0.5,0.6])}$ | 0.054 | 0.246 | 0.282 |
| $p_i^{2([0.5,1])}$ | 0.481 | 0.367 | 0.950 |
| $p_i^{2([0.5,0.6])}$ | 0.862 | 0.818 | 0.983 |
| $q_i^{4([0.5,1])}$ | 0.000 | 0.000 | 0.000 |
| $q_i^{4([0.5,0.6])}$ | 0.000 | 0.000 | 0.036 |
| $q_i^{2([0.5,1])}$ | 0.000 | 0.000 | 0.000 |
| $q_i^{2([0.5,0.6])}$ | 0.000 | 0.000 | 0.003 |

needed to get better values. $A_1$, $A_2$ and $A_3$ are assumed to be of the same type. The time unit is years.

We assume the marginal performance processes of $A_1$, $A_2$, $A_3$, $U$ and $L$ to be independent in $[0, t_2]$. Then again Theorem 3.6, Equations (3.12) and (3.13) of Theorem 3.7, Corollaries 3.9 and 3.11 can be applied, concentrating on bounds on the availabilities. Since, according to Theorem 3.6 the lower bound of Equation (3.6) reduces to the corresponding one of Equation (3.8) when the marginal performance processes of the components are independent in $I$, we treat only the latter one. Then, by using the minimal path and cut vectors of $\phi_2$ in Tables 2.9 and 2.10, the corresponding lower bounds are given by

$$\ell'^{1(I)}_{\phi_2}(\boldsymbol{P}^{(I)}_{\phi_2}) = \max[p_1^{4(I)} p_2^{2(I)} p_4^{4(I)} p_5^{2(I)}, p_3^{2(I)} p_4^{2(I)}]$$

$$\ell'^{2(I)}_{\phi_2}(\boldsymbol{P}^{(I)}_{\phi_2}) = \max[p_3^{2(I)} p_4^{2(I)}, p_1^{4(I)} p_2^{2(I)} p_4^{4(I)} p_5^{4(I)}, p_1^{4(I)} p_2^{4(I)} p_4^{4(I)} p_5^{2(I)}]$$

$$\ell'^{3(I)}_{\phi_2}(\boldsymbol{P}^{(I)}_{\phi_2}) = \max[p_1^{4(I)} p_2^{2(I)} p_3^{2(I)} p_4^{4(I)} p_5^{2(I)}, p_3^{4(I)} p_4^{2(I)}, p_1^{4(I)} p_2^{4(I)} p_4^{4(I)} p_5^{4(I)}]$$

$$\ell'^{4(I)}_{\phi_2}(\boldsymbol{P}^{(I)}_{\phi_2}) = \max[p_3^{4(I)} p_4^{2(I)}, p_1^{4(I)} p_2^{4(I)} p_4^{4(I)} p_5^{4(I)}, p_1^{4(I)} p_2^{2(I)} p_3^{2(I)} p_4^{4(I)} p_5^{4(I)}, p_1^{4(I)} p_2^{4(I)} p_3^{2(I)} p_4^{4(I)} p_5^{2(I)}]$$

$$\ell^{**1(I)}_{\phi_2}(\boldsymbol{P}^{(I)}_{\phi_2}) = (p_1^{4(I)} \amalg p_3^{2(I)})(p_2^{2(I)} \amalg p_3^{2(I)})(p_3^{2(I)} \amalg p_4^{4(I)}) (p_3^{2(I)} \amalg p_5^{2(I)}) p_4^{2(I)}$$

$$\ell^{**2(I)}_{\phi_2}(\boldsymbol{P}^{(I)}_{\phi_2}) = (p_1^{4(I)} \amalg p_3^{2(I)})(p_2^{2(I)} \amalg p_3^{2(I)})(p_3^{2(I)} \amalg p_4^{4(I)}) (p_3^{2(I)} \amalg p_5^{2(I)}) p_4^{2(I)} (p_2^{4(I)} \amalg p_3^{2(I)} \amalg p_5^{4(I)})$$

$$\ell^{**3(I)}_{\phi_2}(\boldsymbol{P}^{(I)}_{\phi_2}) = p_4^{2(I)} (p_1^{4(I)} \amalg p_3^{4(I)})(p_2^{2(I)} \amalg p_3^{4(I)})(p_2^{4(I)} \amalg p_3^{2(I)}) (p_3^{2(I)} \amalg p_5^{4(I)})(p_3^{4(I)} \amalg p_4^{4(I)})(p_3^{4(I)} \amalg p_5^{2(I)})$$

$$\ell^{**4(I)}_{\phi_2}(\boldsymbol{P}^{(I)}_{\phi_2}) = p_4^{2(I)} (p_1^{4(I)} \amalg p_3^{4(I)})(p_2^{2(I)} \amalg p_3^{4(I)})(p_2^{4(I)} \amalg p_3^{2(I)}) (p_3^{2(I)} \amalg p_5^{4(I)})(p_3^{4(I)} \amalg p_4^{4(I)})(p_3^{4(I)} \amalg p_5^{2(I)}) (p_2^{4(I)} \amalg p_3^{4(I)} \amalg p_5^{4(I)})$$

$$B^{*j(I)}_{\phi_2}(\boldsymbol{P}^{(I)}_{\phi_2}) = L^{*j(I)}_{\phi_2}(\boldsymbol{P}^{(I)}_{\phi_2}) = \max[\ell'^{j(I)}_{\phi_2}(\boldsymbol{P}^{(I)}_{\phi_2}), \ell^{**j(I)}_{\phi_2}(\boldsymbol{P}^{(I)}_{\phi_2})], \quad j = 1, \ldots, 4. \qquad (3.89)$$

The upper bounds are given by

$$1 - \bar{\ell}'^{1(I)}_{\phi_2}(\boldsymbol{Q}^{(I)}_{\phi_2})$$
$$= 1 - \max[q_1^{4(I)} q_3^{2(I)}, q_2^{2(I)} q_3^{2(I)}, q_3^{2(I)} q_4^{4(I)}, q_3^{2(I)} q_5^{2(I)}, q_4^{2(I)}]$$
$$\geq 1 - \max[q_1^{4(I)} q_3^{2(I)}, q_2^{2(I)} q_3^{2(I)}, q_3^{2(I)} q_4^{4(I)}, q_3^{2(I)} q_5^{2(I)}, q_4^{2(I)},$$
$$(q_1^{4(I)} \amalg q_2^{2(I)} \amalg q_4^{4(I)} \amalg q_5^{2(I)})(q_3^{2(I)} \amalg q_4^{2(I)})] = 1 - \bar{L}^{*1(I)}_{\phi_2}(\boldsymbol{Q}^{(I)}_{\phi_2})$$
$$= 1 - \bar{B}^{*1(I)}_{\phi_2}(\boldsymbol{Q}^{(I)}_{\phi_2})$$

$$1 - \bar{\ell}'^{2(I)}_{\phi_2}(\boldsymbol{Q}^{(I)}_{\phi_2})$$
$$= 1 - \max[q_1^{4(I)} q_3^{2(I)}, q_2^{2(I)} q_3^{2(I)}, q_3^{2(I)} q_4^{4(I)}, q_3^{2(I)} q_5^{2(I)}, q_4^{2(I)},$$
$$q_2^{4(I)} q_3^{2(I)} q_5^{4(I)}] \geq 1 - \max[q_1^{4(I)} q_3^{2(I)}, q_2^{2(I)} q_3^{2(I)}, q_3^{2(I)} q_4^{4(I)}, q_3^{2(I)} q_5^{2(I)},$$
$$q_4^{2(I)}, q_2^{4(I)} q_3^{2(I)} q_5^{4(I)}, (q_3^{2(I)} \amalg q_4^{2(I)})(q_1^{4(I)} \amalg q_2^{2(I)} \amalg q_4^{4(I)} \amalg q_5^{4(I)})$$
$$(q_1^{4(I)} \amalg q_2^{4(I)} \amalg q_4^{4(I)} \amalg q_5^{2(I)})] = 1 - \bar{L}^{*2(I)}_{\phi_2}(\boldsymbol{Q}^{(I)}_{\phi_2}) \geq 1 - \max[q_1^{4(I)}$$
$$q_3^{2(I)}, q_2^{2(I)} q_3^{2(I)}, q_3^{2(I)} q_4^{4(I)}, q_3^{2(I)} q_5^{2(I)}, q_4^{2(I)}, q_2^{4(I)} q_3^{2(I)} q_5^{4(I)},$$
$$(q_1^{4(I)} \amalg q_2^{2(I)} \amalg q_4^{4(I)} \amalg q_5^{2(I)})(q_3^{2(I)} \amalg q_4^{2(I)}), (q_3^{2(I)} \amalg q_4^{2(I)})$$
$$(q_1^{4(I)} \amalg q_2^{2(I)} \amalg q_4^{4(I)} \amalg q_5^{4(I)})(q_1^{4(I)} \amalg q_2^{4(I)} \amalg q_4^{4(I)} \amalg q_5^{2(I)})]$$
$$= 1 - \bar{B}^{*2(I)}_{\phi_2}(\boldsymbol{Q}^{(I)}_{\phi_2})$$

$$1 - \bar{\ell}'^{3(I)}_{\phi_2}(\boldsymbol{Q}^{(I)}_{\phi_2})$$
$$= 1 - \max[q_4^{2(I)}, q_1^{4(I)} q_3^{4(I)}, q_2^{2(I)} q_3^{4(I)}, q_2^{4(I)} q_3^{2(I)}, q_3^{2(I)} q_5^{4(I)}, q_3^{4(I)} q_4^{4(I)},$$
$$q_3^{4(I)} q_5^{2(I)}] \geq 1 - \max[q_4^{2(I)}, q_1^{4(I)} q_3^{4(I)}, q_2^{2(I)} q_3^{4(I)}, q_2^{4(I)} q_3^{2(I)}, q_3^{2(I)} q_5^{4(I)},$$
$$q_3^{4(I)} q_4^{4(I)}, q_3^{4(I)} q_5^{2(I)}, (q_1^{4(I)} \amalg q_2^{2(I)} \amalg q_3^{2(I)} \amalg q_4^{4(I)} \amalg q_5^{2(I)})$$
$$(q_3^{4(I)} \amalg q_4^{2(I)})(q_1^{4(I)} \amalg q_2^{4(I)} \amalg q_4^{4(I)} \amalg q_5^{4(I)})] = 1 - \bar{L}^{*3(I)}_{\phi_2}(\boldsymbol{Q}^{(I)}_{\phi_2}) \geq 1$$
$$- \max[q_4^{2(I)}, q_1^{4(I)} q_3^{4(I)}, q_2^{2(I)} q_3^{4(I)}, q_2^{4(I)} q_3^{2(I)}, q_3^{2(I)} q_5^{4(I)}, q_3^{4(I)} q_4^{4(I)},$$
$$q_3^{4(I)} q_5^{2(I)}, (q_1^{4(I)} \amalg q_2^{2(I)} \amalg q_4^{4(I)} \amalg q_5^{2(I)})(q_3^{2(I)} \amalg q_4^{2(I)}), (q_3^{2(I)} \amalg q_4^{2(I)})$$
$$(q_1^{4(I)} \amalg q_2^{2(I)} \amalg q_4^{4(I)} \amalg q_5^{4(I)})(q_1^{4(I)} \amalg q_2^{4(I)} \amalg q_4^{4(I)} \amalg q_5^{2(I)}), (q_1^{4(I)}$$
$$\amalg q_2^{2(I)} \amalg q_3^{2(I)} \amalg q_4^{4(I)} \amalg q_5^{2(I)})(q_3^{4(I)} \amalg q_4^{2(I)})(q_1^{4(I)} \amalg q_2^{4(I)} \amalg q_4^{4(I)}$$
$$\amalg q_5^{4(I)})] = 1 - \bar{B}^{*3(I)}_{\phi_2}(\boldsymbol{Q}^{(I)}_{\phi_2})$$

$$
\begin{aligned}
&1 - \bar{\ell}'^{4(I)}_{\phi_2}(\boldsymbol{Q}^{(I)}_{\phi_2}) \\
&\quad = 1 - \max[q_4^{2(I)}, q_1^{4(I)} q_3^{4(I)}, q_2^{2(I)} q_3^{4(I)}, q_2^{4(I)} q_3^{2(I)}, q_3^{2(I)} q_5^{4(I)}, q_3^{4(I)} \\
&\quad q_4^{4(I)}, q_3^{4(I)} q_5^{2(I)}, q_2^{4(I)} q_3^{4(I)} q_5^{4(I)}] \geq 1 - \max[q_4^{2(I)}, q_1^{4(I)} q_3^{4(I)}, q_2^{2(I)} q_3^{4(I)}, \\
&\quad q_2^{4(I)} q_3^{2(I)}, q_3^{2(I)} q_5^{4(I)}, q_3^{4(I)} q_4^{4(I)}, q_3^{4(I)} q_5^{2(I)}, q_2^{4(I)} q_3^{4(I)} q_5^{4(I)}, \\
&\quad (q_3^{4(I)} \amalg q_4^{2(I)})(q_1^{4(I)} \amalg q_2^{4(I)} \amalg q_4^{4(I)} \amalg q_5^{4(I)})(q_1^{4(I)} \amalg q_2^{2(I)} \amalg q_3^{2(I)} \amalg \\
&\quad q_4^{4(I)} \amalg q_5^{4(I)})(q_1^{4(I)} \amalg q_2^{4(I)} \amalg q_3^{2(I)} \amalg q_4^{4(I)} \amalg q_5^{2(I)})] = 1 - \bar{L}^{*4(I)}_{\phi_2} \\
&\quad (\boldsymbol{Q}^{(I)}_{\phi_2}) \geq 1 - \max[q_4^{2(I)}, q_1^{4(I)} q_3^{4(I)}, q_2^{2(I)} q_3^{4(I)}, q_2^{4(I)} q_3^{2(I)}, q_3^{2(I)} q_5^{4(I)}, \\
&\quad q_3^{4(I)} q_4^{4(I)}, q_3^{4(I)} q_5^{2(I)}, q_2^{4(I)} q_3^{4(I)} q_5^{4(I)}, (q_1^{4(I)} \amalg q_2^{2(I)} \amalg q_4^{4(I)} \amalg q_5^{2(I)}) \\
&\quad (q_3^{2(I)} \amalg q_4^{2(I)}), (q_3^{2(I)} \amalg q_4^{2(I)})(q_1^{4(I)} \amalg q_2^{2(I)} \amalg q_4^{4(I)} \amalg q_5^{4(I)}) \\
&\quad (q_1^{4(I)} \amalg q_2^{4(I)} \amalg q_4^{4(I)} \amalg q_5^{2(I)}), (q_1^{4(I)} \amalg q_2^{2(I)} \amalg q_3^{2(I)} \amalg q_4^{4(I)} \amalg q_5^{2(I)}) \\
&\quad (q_3^{4(I)} \amalg q_4^{2(I)})(q_1^{4(I)} \amalg q_2^{4(I)} \amalg q_4^{4(I)} \amalg q_5^{4(I)}), (q_3^{4(I)} \amalg q_4^{2(I)}) \\
&\quad (q_1^{4(I)} \amalg q_2^{4(I)} \amalg q_4^{4(I)} \amalg q_5^{4(I)})(q_1^{4(I)} \amalg q_2^{2(I)} \amalg q_3^{2(I)} \amalg q_4^{4(I)} \amalg q_5^{4(I)}) \\
&\quad (q_1^{4(I)} \amalg q_2^{4(I)} \amalg q_3^{2(I)} \amalg q_4^{4(I)} \amalg q_5^{2(I)})] = 1 - \bar{B}^{*4(I)}_{\phi_2}(\boldsymbol{Q}^{(I)}_{\phi_2}). \qquad (3.90)
\end{aligned}
$$

By using the minimal path and cut vectors of $\phi_1$ and $\phi_2$ in Tables 2.7–2.10 and the availabilities and unavailabilities of Table 3.6, we arrive at the bounds based on Corollary 3.11 given in Table 3.7 for the availabilities and unavailabilities linked to $\phi_1$ and $\phi_2$.

First it should be noted that stationarity is achieved after half a year, as can be seen from the fact that the upper bounds do not depend on interval length. We see that the bounds are rather informative for $I = [0.5, 0.6]$, corresponding to an interval of 36 days. For $I = [0.5, 1]$, corresponding to an interval of half a year, bounds are less informative. Note, however, that from the theory we know that the lower bounds are far better than the upper bounds especially for long intervals. As can be seen from Equation (3.90), the reason for the upper bounds of the system availabilities of $\phi_2$ being close to 1 is that all component unavailabilities are 0 or close to 0.

Again, to illustrate the application of a modular decomposition consider the module of $\phi_2$ given by Equation (2.4) and organizing structure function given by Equation (2.5). The minimal path vectors and minimal cut vectors for the module are given respectively in Tables 2.11 and

Table 3.7 Bounds for $p_{\phi_1}^{j(I)}$, $q_{\phi_1}^{j(I)}$, $p_{\phi_2}^{j(I)}$ and $q_{\phi_2}^{j(I)}$.

| $k$ | $j$ | $I$ | bounds for $p_{\phi_k}^{j(I)}$ | | bounds for $q_{\phi_k}^{j(I)}$ | |
|---|---|---|---|---|---|---|
| | | | Lower | Upper | Lower | Upper |
| 1 | 4 | [0.5, 1] | 0.0002 | 0.9995 | 0.0000 | 0.0004 |
| | 4 | [0.5, 0.6] | 0.1832 | 0.9995 | 0.0000 | 0.0004 |
| | 2 | [0.5, 1] | 0.1767 | 0.9995 | 0.0000 | 0.0004 |
| | 2 | [0.5, 0.6] | 0.7195 | 0.9995 | 0.0000 | 0.0004 |
| 2 | 4 | [0.5, 1] | 0.0000 | 0.9994 | 0.0000 | 0.0011 |
| | 4 | [0.5, 0.6] | 0.0440 | 0.9994 | 0.0000 | 0.0011 |
| | 3 | [0.5, 1] | 0.0000 | 0.9994 | 0.0000 | 0.0011 |
| | 3 | [0.5, 0.6] | 0.0440 | 0.9994 | 0.0000 | 0.0011 |
| | 2 | [0.5, 1] | 0.1767 | 0.9995 | 0.0000 | 0.0004 |
| | 2 | [0.5, 0.6] | 0.7056 | 0.9995 | 0.0000 | 0.0004 |
| | 1 | [0.5, 1] | 0.1767 | 0.9995 | 0.0000 | 0.0004 |
| | 1 | [0.5, 0.6] | 0.7056 | 0.9995 | 0.0000 | 0.0004 |

2.12, whereas the minimal path vectors and minimal cut vectors for the organizing structure are given respectively in Tables 2.13 and 2.14.

Again Theorem 3.6, Equations (3.12) and (3.13) of Theorem 3.7, Corollaries 3.9 and 3.11, Theorem 3.14 and Theorems 3.22–3.24 can be applied, concentrating on bounds on the availabilities. Since, according to Theorem 3.6, the lower bound of Equation (3.6) reduces to the corresponding one of Equation (3.8) when the marginal performance processes of the components are independent in $I$, we treat only the latter one. It should be mentioned that, since from Table 2.11 there is only one minimal path set for the module both to levels 1 and 4, it follows from Theorem 3.22 that for these levels the corresponding lower bound of Equation (3.8) is exactly correct. Then, by using the minimal path and cut vectors in Tables 2.11–2.14, the corresponding lower bounds are given by

$$\begin{aligned}
\ell_\chi^{\prime 1(I)}(\boldsymbol{P}_{\phi_2}^{(I)}) &= \ell_\chi^{**1(I)}(\boldsymbol{P}_{\phi_2}^{(I)}) = L_\chi^{*1(I)}(\boldsymbol{P}_{\phi_2}^{(I)}) \\
&= B_\chi^{*1(I)}(\boldsymbol{P}_{\phi_2}^{(I)}) = p_1^{4(I)} p_2^{2(I)} p_5^{2(I)} = p_\chi^{1(I)}(\boldsymbol{P}_{\phi_2}^{(I)}) \\
\ell_\chi^{\prime 2(I)}(\boldsymbol{P}_{\phi_2}^{(I)}) &= p_1^{4(I)} \max[p_2^{2(I)} p_5^{4(I)}, p_2^{4(I)} p_5^{2(I)}] \\
&\leq p_1^{4(I)} \max[p_2^{2(I)} p_5^{4(I)}, p_2^{4(I)} p_5^{2(I)}, p_2^{2(I)} p_5^{2(I)} \\
&\quad (p_2^{4(I)} \amalg p_5^{4(I)})] = L_\chi^{*2(I)}(\boldsymbol{P}_{\phi_2}^{(I)}) = B_\chi^{*2(I)}(\boldsymbol{P}_{\phi_2}^{(I)})
\end{aligned}$$

$$\ell_{\chi}^{\prime 4(I)}(\boldsymbol{P}_{\phi_2}^{(I)}) = \ell_{\chi}^{**4(I)}(\boldsymbol{P}_{\phi_2}^{(I)}) = L_{\chi}^{*4(I)}(\boldsymbol{P}_{\phi_2}^{(I)}) = B_{\chi}^{*4(I)}(\boldsymbol{P}_{\phi_2}^{(I)})$$
$$= p_1^{4(I)} p_2^{4(I)} p_5^{4(I)} = p_{\chi}^{4(I)}(\boldsymbol{P}_{\phi_2}^{(I)})$$
$$\ell_{\psi}^{\prime 1(I)}(\boldsymbol{B}_{\psi}^{*(I)}(\boldsymbol{P}_{\phi_2}^{(I)})) = \max[p_4^{4(I)} p_1^{4(I)} p_2^{2(I)} p_5^{2(I)}, p_3^{2(I)} p_4^{2(I)}]$$
$$\ell_{\psi}^{\prime 2(I)}(\boldsymbol{B}_{\psi}^{*(I)}(\boldsymbol{P}_{\phi_2}^{(I)})) = \max[p_3^{2(I)} p_4^{2(I)}, p_4^{4(I)} p_1^{4(I)} \max[p_2^{2(I)} p_5^{4(I)}, p_2^{4(I)} p_5^{2(I)}, p_2^{2(I)} p_5^{2(I)}(p_2^{4(I)} \amalg p_5^{4(I)})]]$$
$$\ell_{\psi}^{\prime 3(I)}(\boldsymbol{B}_{\psi}^{*(I)}(\boldsymbol{P}_{\phi_2}^{(I)})) = \max[p_3^{2(I)} p_4^{4(I)} p_1^{4(I)} p_2^{2(I)} p_5^{2(I)}, p_4^{4(I)} p_1^{4(I)} p_2^{4(I)} p_5^{4(I)}, p_3^{4(I)} p_4^{2(I)}]$$
$$\ell_{\psi}^{\prime 4(I)}(\boldsymbol{B}_{\psi}^{*(I)}(\boldsymbol{P}_{\phi_2}^{(I)})) = \max[p_4^{4(I)} p_1^{4(I)} p_2^{4(I)} p_5^{4(I)}, p_3^{4(I)} p_4^{2(I)}, p_3^{2(I)} p_4^{4(I)} p_1^{4(I)} \max[p_2^{2(I)} p_5^{4(I)}, p_2^{4(I)} p_5^{2(I)}, p_2^{2(I)} p_5^{2(I)}(p_2^{4(I)} \amalg p_5^{4(I)})]]$$
$$\ell_{\psi}^{**1(I)}(\boldsymbol{B}_{\psi}^{*(I)}(\boldsymbol{P}_{\phi_2}^{(I)})) = (p_3^{2(I)} \amalg p_1^{4(I)} p_2^{2(I)} p_5^{2(I)})(p_3^{2(I)} \amalg p_4^{4(I)}) p_4^{2(I)}$$
$$\ell_{\psi}^{**2(I)}(\boldsymbol{B}_{\psi}^{*(I)}(\boldsymbol{P}_{\phi_2}^{(I)})) = (p_3^{2(I)} \amalg p_4^{4(I)}) p_4^{2(I)} (p_3^{2(I)} \amalg p_1^{4(I)} \max[p_2^{2(I)} p_5^{4(I)}, p_2^{4(I)} p_5^{2(I)}, p_2^{2(I)} p_5^{2(I)}(p_2^{4(I)} \amalg p_5^{4(I)})])$$
$$\ell_{\psi}^{**3(I)}(\boldsymbol{B}_{\psi}^{*(I)}(\boldsymbol{P}_{\phi_2}^{(I)})) = p_4^{2(I)}(p_3^{4(I)} \amalg p_1^{4(I)} p_2^{2(I)} p_5^{2(I)})(p_3^{4(I)} \amalg p_4^{4(I)})(p_3^{2(I)} \amalg p_1^{4(I)} p_2^{4(I)} p_5^{4(I)})$$
$$\ell_{\psi}^{**4(I)}(\boldsymbol{B}_{\psi}^{*(I)}(\boldsymbol{P}_{\phi_2}^{(I)})) = p_4^{2(I)}(p_3^{4(I)} \amalg p_4^{4(I)})(p_3^{2(I)} \amalg p_1^{4(I)} p_2^{4(I)} p_5^{4(I)})(p_3^{4(I)} \amalg p_1^{4(I)} \max[p_2^{2(I)} p_5^{4(I)}, p_2^{4(I)} p_5^{2(I)}, p_2^{2(I)} p_5^{2(I)} (p_2^{4(I)} \amalg p_5^{4(I)})])$$
$$B_{\psi}^{*j(I)}(\boldsymbol{B}_{\psi}^{*(I)}(\boldsymbol{P}_{\phi_2}^{(I)})) = L_{\psi}^{*j(I)}(\boldsymbol{B}_{\psi}^{*(I)}(\boldsymbol{P}_{\phi_2}^{(I)}))$$
$$= \max[\ell_{\psi}^{\prime j(I)}(\boldsymbol{B}_{\psi}^{*(I)}(\boldsymbol{P}_{\phi_2}^{(I)})), \ell_{\psi}^{**j(I)}(\boldsymbol{B}_{\psi}^{*(I)}(\boldsymbol{P}_{\phi_2}^{(I)}))],$$
$$j = 1, \ldots, 4. \tag{3.91}$$

The upper bounds are given by

$$u_{\chi}^{\prime\prime 1(I)} = \min[p_1^{4(I)}, p_2^{2(I)}, p_5^{2(I)}]$$
$$1 - \bar{\ell}_{\chi}^{\prime 1(I)}(\boldsymbol{Q}_{\phi_2}^{(I)}) = 1 - \max[q_1^{4(I)}, q_2^{2(I)}, q_5^{2(I)}] \geq 1 - q_1^{4(I)} \amalg q_2^{2(I)} \amalg q_5^{2(I)}$$
$$= 1 - \bar{L}_{\chi}^{*1(I)}(\boldsymbol{Q}_{\phi_2}^{(I)}) = 1 - \bar{B}_{\chi}^{*1(I)}(\boldsymbol{Q}_{\phi_2}^{(I)})$$

$$1 - \bar{\ell}'^{2(I)}_{\chi}(\boldsymbol{Q}^{(I)}_{\phi_2})$$
$$= 1 - \max[q_1^{4(I)}, q_2^{2(I)}, q_5^{2(I)}, q_2^{4(I)}q_5^{4(I)}] \geq 1 - \max[q_1^{4(I)}, q_2^{2(I)}, q_5^{2(I)},$$
$$(q_1^{4(I)} \amalg q_2^{2(I)} \amalg q_5^{4(I)})(q_1^{4(I)} \amalg q_2^{4(I)} \amalg q_5^{2(I)})] = 1 - \bar{L}^{*2(I)}_{\chi}(\boldsymbol{Q}^{(I)}_{\phi_2})$$
$$\geq 1 - \max[q_1^{4(I)} \amalg q_2^{2(I)} \amalg q_5^{2(I)}, (q_1^{4(I)} \amalg q_2^{2(I)} \amalg q_5^{4(I)})$$
$$(q_1^{4(I)} \amalg q_2^{4(I)} \amalg q_5^{2(I)})] = 1 - \bar{B}^{*2(I)}_{\chi}(\boldsymbol{Q}^{(I)}_{\phi_2})$$

$$u''^{4(I)}_{\chi} = \min[p_1^{4(I)}, p_2^{4(I)}, p_5^{4(I)}]$$

$$1 - \bar{\ell}'^{4(I)}_{\chi}(\boldsymbol{Q}^{(I)}_{\phi_2})$$
$$= 1 - \max[q_1^{4(I)}, q_2^{4(I)}, q_5^{4(I)}] \geq 1 - q_1^{4(I)} \amalg q_2^{4(I)} \amalg q_5^{4(I)}$$
$$= 1 - \bar{L}^{*4(I)}_{\chi}(\boldsymbol{Q}^{(I)}_{\phi_2}) = 1 - \bar{B}^{*4(I)}_{\chi}(\boldsymbol{Q}^{(I)}_{\phi_2})$$

$$1 - \bar{\ell}'^{1(I)}_{\psi}(\bar{\boldsymbol{B}}^{*(I)}_{\psi}(\boldsymbol{Q}^{(I)}_{\phi_2}))$$
$$= 1 - \max[q_3^{2(I)}(q_1^{4(I)} \amalg q_2^{2(I)} \amalg q_5^{2(I)}), q_3^{2(I)}q_4^{4(I)}, q_4^{2(I)}] \geq 1 - \max$$
$$[q_4^{2(I)}, (q_1^{4(I)} \amalg q_2^{2(I)} \amalg q_4^{4(I)} \amalg q_5^{2(I)})(q_3^{2(I)} \amalg q_4^{2(I)})] = 1 - \bar{L}^{*1(I)}_{\psi}$$
$$(\bar{\boldsymbol{B}}^{*(I)}_{\psi}(\boldsymbol{Q}^{(I)}_{\phi_2})) = 1 - \bar{B}^{*1(I)}_{\psi}(\bar{\boldsymbol{B}}^{*(I)}_{\psi}(\boldsymbol{Q}^{(I)}_{\phi_2}))$$

$$1 - \bar{\ell}'^{2(I)}_{\psi}(\bar{\boldsymbol{B}}^{*(I)}_{\psi}(\boldsymbol{Q}^{(I)}_{\phi_2}))$$
$$= 1 - \max[q_3^{2(I)}q_4^{4(I)}, q_4^{2(I)}, q_3^{2(I)}(q_1^{4(I)} \amalg q_2^{2(I)} \amalg q_5^{2(I)}), q_3^{2(I)}$$
$$(q_1^{4(I)} \amalg q_2^{2(I)} \amalg q_5^{4(I)})(q_1^{4(I)} \amalg q_2^{4(I)} \amalg q_5^{2(I)})] \geq 1 - \max[q_4^{2(I)}, (q_3^{2(I)} \amalg$$
$$q_4^{2(I)})(q_4^{4(I)} \amalg \max[q_1^{4(I)} \amalg q_2^{2(I)} \amalg q_5^{2(I)}, (q_1^{4(I)} \amalg q_2^{2(I)} \amalg q_5^{4(I)})$$
$$(q_1^{4(I)} \amalg q_2^{4(I)} \amalg q_5^{2(I)})])] = 1 - \bar{L}^{*2(I)}_{\psi}(\bar{\boldsymbol{B}}^{*(I)}_{\psi}(\boldsymbol{Q}^{(I)}_{\phi_2}))$$
$$= 1 - \bar{B}^{*2(I)}_{\psi}(\bar{\boldsymbol{B}}^{*(I)}_{\psi}(\boldsymbol{Q}^{(I)}_{\phi_2}))$$

$$1 - \bar{\ell}'^{3(I)}_{\psi}(\bar{\boldsymbol{B}}^{*(I)}_{\psi}(\boldsymbol{Q}^{(I)}_{\phi_2}))$$
$$= 1 - \max[q_4^{2(I)}, q_3^{4(I)}(q_1^{4(I)} \amalg q_2^{2(I)} \amalg q_5^{2(I)}), q_3^{4(I)}q_4^{4(I)}, q_3^{2(I)}$$
$$(q_1^{4(I)} \amalg q_2^{4(I)} \amalg q_5^{4(I)})] \geq 1 - \max[q_4^{2(I)}, q_3^{4(I)}(q_1^{4(I)} \amalg q_2^{2(I)} \amalg q_5^{2(I)}),$$
$$q_3^{4(I)}q_4^{4(I)}, q_3^{2(I)}(q_1^{4(I)} \amalg q_2^{4(I)} \amalg q_5^{4(I)}), (q_1^{4(I)} \amalg q_2^{2(I)} \amalg q_3^{2(I)} \amalg q_4^{4(I)} \amalg$$

$q_5^{2(I)})(q_1^{4(I)} \amalg q_2^{4(I)} \amalg q_4^{4(I)} \amalg q_5^{4(I)})(q_3^{4(I)} \amalg q_4^{2(I)})] = 1 - \bar{L}_\psi^{*3(I)}$
$(\bar{\boldsymbol{B}}_\psi^{*(I)}(\boldsymbol{Q}_{\phi_2}^{(I)})) \geq 1-\max[q_4^{2(I)}, q_3^{4(I)}(q_1^{4(I)} \amalg q_2^{2(I)} \amalg q_5^{2(I)}), q_3^{4(I)}q_4^{4(I)},$
$q_3^{2(I)}(q_1^{4(I)} \amalg q_2^{4(I)} \amalg q_5^{4(I)}), (q_1^{4(I)} \amalg q_2^{2(I)} \amalg q_3^{2(I)} \amalg q_4^{4(I)} \amalg q_5^{2(I)})$
$(q_1^{4(I)} \amalg q_2^{4(I)} \amalg q_4^{4(I)} \amalg q_5^{4(I)})(q_3^{4(I)} \amalg q_4^{2(I)}), (q_3^{2(I)} \amalg q_4^{2(I)})$
$(q_4^{4(I)} \amalg \max[q_1^{4(I)} \amalg q_2^{2(I)} \amalg q_5^{2(I)}, (q_1^{4(I)} \amalg q_2^{2(I)} \amalg q_5^{4(I)})$
$(q_1^{4(I)} \amalg q_2^{4(I)} \amalg q_5^{2(I)})])] = 1 - \bar{B}_\psi^{*3(I)}(\bar{\boldsymbol{B}}_\psi^{*(I)}(\boldsymbol{Q}_{\phi_2}^{(I)}))$

$1 - \bar{\ell}_\psi^{\prime 4(I)}(\bar{\boldsymbol{B}}_\psi^{*(I)}(\boldsymbol{Q}_{\phi_2}^{(I)}))$
$= 1 - \max[q_4^{2(I)}, q_3^{4(I)}q_4^{4(I)}, q_3^{2(I)}(q_1^{4(I)} \amalg q_2^{4(I)} \amalg q_5^{4(I)}), q_3^{4(I)}$
$(q_1^{4(I)} \amalg q_2^{2(I)} \amalg q_5^{2(I)}), q_3^{4(I)}(q_1^{4(I)} \amalg q_2^{2(I)} \amalg q_5^{4(I)})(q_1^{4(I)} \amalg q_2^{4(I)}$
$\amalg q_5^{2(I)})] \geq 1 - \max[q_4^{2(I)}, q_3^{4(I)}q_4^{4(I)}, q_3^{2(I)}(q_1^{4(I)} \amalg q_2^{4(I)} \amalg q_5^{4(I)}),$
$q_3^{4(I)}(q_1^{4(I)} \amalg q_2^{2(I)} \amalg q_5^{2(I)}), q_3^{4(I)}(q_1^{4(I)} \amalg q_2^{2(I)} \amalg q_5^{4(I)})$
$(q_1^{4(I)} \amalg q_2^{4(I)} \amalg q_5^{2(I)}), (q_1^{4(I)} \amalg q_2^{4(I)} \amalg q_4^{4(I)} \amalg q_5^{4(I)})(q_3^{4(I)} \amalg q_4^{2(I)})$
$(q_3^{2(I)} \amalg q_4^{4(I)} \amalg \max[q_1^{4(I)} \amalg q_2^{2(I)} \amalg q_5^{2(I)}, (q_1^{4(I)} \amalg q_2^{2(I)} \amalg q_5^{4(I)})$
$(q_1^{4(I)} \amalg q_2^{4(I)} \amalg q_5^{2(I)})])] = 1 - \bar{L}_\psi^{*4(I)}(\bar{\boldsymbol{B}}_\psi^{*(I)}(\boldsymbol{Q}_{\phi_2}^{(I)})) \geq 1 - \max[q_4^{2(I)},$
$q_3^{4(I)}q_4^{4(I)}, q_3^{2(I)}(q_1^{4(I)} \amalg q_2^{4(I)} \amalg q_5^{4(I)}), q_3^{4(I)}(q_1^{4(I)} \amalg q_2^{2(I)} \amalg q_5^{2(I)}),$
$q_3^{4(I)}(q_1^{4(I)} \amalg q_2^{2(I)} \amalg q_5^{4(I)})(q_1^{4(I)} \amalg q_2^{4(I)} \amalg q_5^{2(I)}), (q_1^{4(I)} \amalg q_2^{4(I)} \amalg$
$q_4^{4(I)} \amalg q_5^{4(I)})(q_3^{4(I)} \amalg q_4^{2(I)})(q_3^{2(I)} \amalg q_4^{4(I)} \amalg \max[q_1^{4(I)} \amalg q_2^{2(I)} \amalg q_5^{2(I)},$
$(q_1^{4(I)} \amalg q_2^{2(I)} \amalg q_5^{4(I)})(q_1^{4(I)} \amalg q_2^{4(I)} \amalg q_5^{2(I)})]), (q_3^{2(I)} \amalg q_4^{2(I)})$
$(q_4^{4(I)} \amalg \max[q_1^{4(I)} \amalg q_2^{2(I)} \amalg q_5^{2(I)}, (q_1^{4(I)} \amalg q_2^{2(I)} \amalg q_5^{4(I)})$
$(q_1^{4(I)} \amalg q_2^{4(I)} \amalg q_5^{2(I)})])] = 1 - \bar{B}_\psi^{*4(I)}(\bar{\boldsymbol{B}}_\psi^{*(I)}(\boldsymbol{Q}_{\phi_2}^{(I)})).$

# 4

# An offshore gas pipeline network

## 4.1 Description of the system

The offshore gas pipeline network treated in this chapter, covering the paper by Natvig and Mørch (2003), is the most complex of the ones considered in Mørch (1991). It constitutes the main parts of the network in the North Sea, as of the end of the 1980s, transporting gas to Emden in Germany. The network, along with its modules, is shown in Figure 4.1. As can be seen from this figure, the network consists altogether of 32 components and seven nontrivial modules. $a_1$ and $a_2$ are pipelines from the production field at Statfjord, $c_1$ from the Heimdal and Troll fields, $c_2$ from the Sleipner field and finally $e$ from the Ekofisk field. All these oil and gas fields are in the Norwegian sector of the North Sea west of southern Norway. $k$ is the pipeline from what is called H7 to Emden. The compressor components of the network are $f_1, h_1$ and $j_1$.

There are ten passages in the network, all supposed to function perfectly. For instance, $P_f(45)$ is the passage of module $f$ having a capacity of $45\,\text{MSm}^3/d$ (million standard cubic meters per day). Similarly, $P_{b_4}(4.2)$ is the passage of component $b_4$ having a capacity of $4.2\,\text{MSm}^3/d$. The module passages $P_f$, $P_h$ and $P_j$ are used whenever the corresponding modules cannot transport and compress the incoming amounts of gas, by transporting, within their capacities, the surplus amount of gas. The seven component passages are only used when

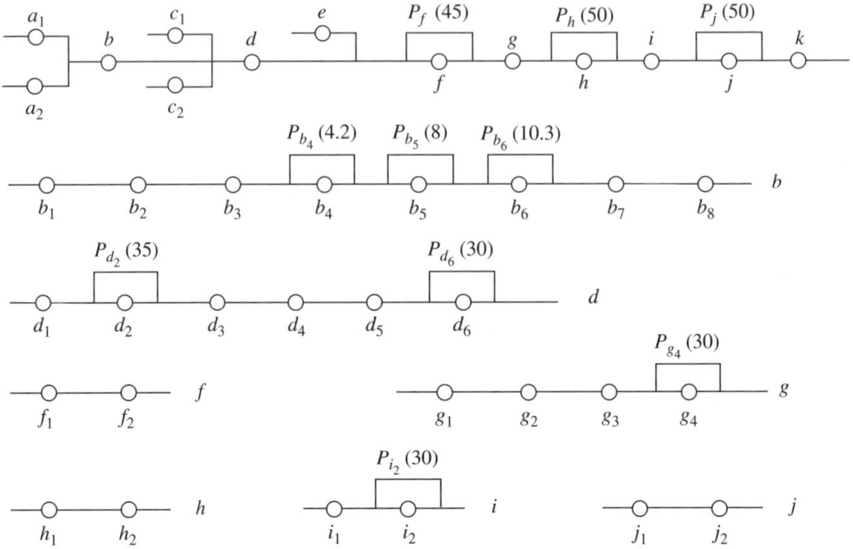


*Figure 4.1* *Gas pipeline network with corresponding modules* $b, d, f, g, h, i, j$.

the corresponding components are not in the perfect functioning state, chosen somewhat arbitrarily to be $M = 16$.

The gas pipeline from Ekofisk to Emden, being part of this network, is called Norpipe. Today there are two more pipelines, Europipe 1 and Europipe 2, providing Norwegian offshore gas to Emden. In addition, Zeepipe provides gas to Zeebrugge in Belgium and Franpipe gas to Dunkerque in France. Hence, the total Norwegian offshore gas pipeline network of today is more complex than the one considered in the present chapter.

Let us look closer at the compressor components and start with $f_1$. $f_1$ consists of four compressors each with a capacity of transporting and compressing 11 $MSm^3/d$. The states of $f_1$ are defined in Table 4.1. Since the expected maintenance time for a compressor is assumed longer than the expected repair time, it is seen from the table that a failed compressor leads to a higher component state than a maintained one.

Furthermore, we assume that at most three compressors can be used at the same time, implying that the maximum capacity of $f_1$ is 33 $MSm^3/d$, which is achieved when the component state is 14, 15 or 16. The compressor components $h_1$ and $j_1$ both consist of three compressors each with a capacity of transporting and compressing 4.5 $MSm^3/d$. The states of $h_1$ and $j_1$ are defined in Table 4.2.

Table 4.1 The states of compressor component $f_1$.

| State | The number of operative compressors | The number of failed compressors | The number of maintained compressors |
|---|---|---|---|
| 0 | 0 | 3 | 1 |
| 1 | 0 | 4 | 0 |
| 2 | 1 | 2 | 1 |
| 3 | 1 | 3 | 0 |
| 4 | 2 | 1 | 1 |
| 5 | 2 | 2 | 0 |
| 14 | 3 | 0 | 1 |
| 15 | 3 | 1 | 0 |
| 16 | 4 | 0 | 0 |

Table 4.2 The states of compressor components $h_1$ and $j_1$.

| State | The number of operative compressors | The number of failed compressors | The number of maintained compressors |
|---|---|---|---|
| 0 | 0 | 2 | 1 |
| 1 | 0 | 3 | 0 |
| 2 | 1 | 1 | 1 |
| 3 | 1 | 2 | 0 |
| 14 | 2 | 0 | 1 |
| 15 | 2 | 1 | 0 |
| 16 | 3 | 0 | 0 |

Now at most two compressors can be used at the same time, implying that the maximum capacity of both $h_1$ and $j_1$ is $9\,\mathrm{MSm}^3/d$, which is achieved when the component state is 14, 15 or 16.

For the remaining 29 components gas can only be transported when the component is in the perfect functioning state 16. The corresponding transport capacities are given in Table 4.3.

For all these components the set of states is of the form $\{0, 1, 2, \ldots, r, 16\}$, where $0 \le r < 16$. The states $0, 1, 2, \ldots, r$ are

Table 4.3 Transport capacities in $\mathrm{MSm}^3/d$ for the remaining 29 components.

| $a_1$ | $a_2$ | $b_1$ | $b_2$ | $b_3$ | $b_4$ | $b_5$ | $b_6$ | $b_7$ | $b_8$ | $c_1$ | $c_2$ | $d_1$ | $d_2$ | $d_3$ |
|---|---|---|---|---|---|---|---|---|---|---|---|---|---|---|
| 5 | 20 | 20 | 20 | 16 | 16 | 16 | 16 | 16 | 16 | 48 | 42 | 50 | 50 | 50 |

| $d_4$ | $d_5$ | $d_6$ | $e$ | $f_2$ | $g_1$ | $g_2$ | $g_3$ | $g_4$ | $h_2$ | $i_1$ | $i_2$ | $j_2$ | $k$ |
|---|---|---|---|---|---|---|---|---|---|---|---|---|---|
| 50 | 50 | 50 | 11 | 33 | 59 | 59 | 59 | 59 | 9 | 59 | 59 | 9 | 59 |

failure states ranked according to expected repair times. To arrive at the structure function, only perfect functioning or not is of relevance for these components.

Now let $a$ be the module consisting of the components $a_1, a_2$. We let the set of states be $S_a = \{0, 1, 16\}$, where state 1 is achieved when the state of $a_2$ is less than 16, whereas the state of $a_1$ is 16. The state 16 is achieved when at least the state of $a_2$ is 16. From Figure 4.1 and Table 4.3 one easily arrives at Table 4.4, summing up the states of $a$.

The corresponding structure function is given by

$$a = \phi_a(a_1, a_2) = I(a_1 = 16)I(a_2 < 16) + 16I(a_2 = 16), \tag{4.1}$$

where $I(A)$ equals 1 if $A$ is true and 0 otherwise. For notational convenience, in Equation (4.1) and in the following corresponding equations we denote the state of a module or component by its name.

From Figure 4.1 and Table 4.3 one gets Table 4.5, summing up the states of module $b$.

The corresponding structure function is given by

$$\begin{aligned} b &= \phi_b(b_1, b_2, b_3, b_4, b_5, b_6, b_7, b_8) \\ &= I(\min(b_1, b_2, b_3, b_7, b_8) = 16)\{I(b_4 < 16) \\ &\quad + I(b_4 = 16)[2 + I(b_5 = 16)[1 + 13I(b_6 = 16)]]\}. \end{aligned} \tag{4.2}$$

Combining the modules $a$ and $b$ into module $ab$, from Tables 4.4 and 4.5 one arrives at Table 4.6, summing up the states of this module.

Table 4.4 The states of module $a$.

| State | Capacity |
|---|---|
| 0 | 0 |
| 1 | 5 |
| 16 | 20–25 |

Table 4.5 The states of module $b$.

| State | Capacity |
|---|---|
| 0 | 0 |
| 1 | 4.2 |
| 2 | 8.0 |
| 3 | 10.3 |
| 16 | 16 |

Table 4.6 The states of module $ab$.

| State | Capacity |
|---|---|
| 0 | 0 |
| 1 | 4.2 |
| 2 | 5 |
| 3 | 8.0 |
| 4 | 10.3 |
| 16 | 16 |

The corresponding structure function is given by

$$\begin{aligned} ab = \phi_{ab}(a, b) &= I(\min(a, b) > 0) \\ &\times \{I(a = 1)[1 + I(b \geq 2)] + I(a = 16)[I(b = 1) \\ &+ 3I(b = 2) + 4I(b = 3) + 16I(b = 16)]\}. \end{aligned} \quad (4.3)$$

Let $c$ be the module consisting of the components $c_1$ and $c_2$. We let the set of states be $S_c = \{0, 6, 7, 16\}$. Here state 6 is achieved when the state of $c_1$ is less than 16, whereas the state of $c_2$ is 16. The state 7 is achieved when it is the other way round, whereas the state 16 is achieved when the state of both $c_1$ and $c_2$ is 16. From Figure 4.1 and Table 4.3 one gets Table 4.7, summing up the states of $c$.

The corresponding structure function is given by

$$\begin{aligned} c = \phi_c(c_1, c_2) &= 6I(c_1 < 16)I(c_2 = 16) + 7I(c_1 = 16)I(c_2 < 16) \\ &+ 16I(c_1 = 16)I(c_2 = 16). \end{aligned} \quad (4.4)$$

From Figure 4.1 and Table 4.3 one arrives at Table 4.8, summing up the states of module $d$.

The corresponding structure function is given by

$$\begin{aligned} d = \phi_d(d_1, d_2, d_3, d_4, d_5, d_6) &= I(\min(d_1, d_3, d_4, d_5) = 16) \\ &\times \{3 + I(d_2 < 16)I(d_6 = 16) + 13I(d_2 = 16)I(d_6 = 16)\}. \end{aligned} \quad (4.5)$$

Table 4.7 The states of module $c$.

| State | Capacity |
|---|---|
| 0 | 0 |
| 6 | 42 |
| 7 | 48 |
| 16 | 90 |

Table 4.8 The states of module $d$.

| State | Capacity |
|---|---|
| 0 | 0 |
| 3 | 30 |
| 4 | 35 |
| 16 | 50 |

Table 4.9 The states of module $abcde$.

| State | Capacity |
|---|---|
| 0 | 0 |
| 1 | 4.2–19 |
| 2 | 21.3–27 |
| 3 | 30 |
| 4 | 35 |
| 5 | 41–42 |
| 7 | 46–48 |
| 8 | 50–53 |
| 16 | 57.2–61 |

Combining the modules $ab$, $c$, $d$ and component $e$ into module $abcde$, from Figure 4.1 and Tables 4.3, 4.6, 4.7 and 4.8 one gets Table 4.9, summing up the states of module $abcde$.

The corresponding structure function is given by

$$
\begin{aligned}
abcde &= \phi_{abcde}(ab, c, d, e) \\
&= I(d=0)I(e=16) + I(d=3)\{I(c=0) \\
&\quad \times [I(e<16)I(ab>0) + I(e=16)(I(ab<4) + 2I(ab \geq 4))] \\
&\quad + I(c>0)[3 + 2I(e=16)]\} + I(d=4)\{I(c=0) \\
&\quad \times [I(e<16)I(ab>0) + I(e=16)(I(ab<4) + 2I(ab \geq 4))] \\
&\quad + I(c>0)[4 + 3I(e=16)]\} + I(d=16)\{I(c=0) \\
&\quad \times [I(e<16)I(ab>0) + I(e=16)(I(ab<4) + 2I(ab \geq 4))] \\
&\quad + I(c=6)[5 + I(e<16)(2I(1 \leq ab \leq 2) + 3I(ab \geq 3)) \\
&\quad + I(e=16)(3 + 8I(ab \geq 1))] + I(c=7)[7 \\
&\quad + I(e<16)I(ab>0) + 9I(e=16)] \\
&\quad + I(c=16)[8 + 8I(e=16)]\}. 
\end{aligned} \tag{4.6}
$$

Table 4.10 The states of module $f$.

| State | Capacity/Compression |
|---|---|
| 0 | 0 |
| 2 | 11 |
| 4 | 22 |
| 16 | 33 |

From Figure 4.1 and Tables 4.1 and 4.3 one arrives at Table 4.10, summing up the states of module $f$.

The corresponding structure function is given by

$$f = \phi_f(f_1, f_2) = I(f_2 = 16)[2I(2 \le f_1 \le 3) + 4I(4 \le f_1 \le 5) + 16I(14 \le f_1 \le 16)]. \tag{4.7}$$

Similarly, from Figure 4.1 and Tables 4.2 and 4.3 we obtain Table 4.11, summing up the states of modules $h$ and $j$.

The corresponding structure functions are given by

$$h = \phi_h(h_1, h_2) = I(h_2 = 16)[I(2 \le h_1 \le 3) + 16I(14 \le h_1 \le 16)] \tag{4.8}$$

$$j = \phi_j(j_1, j_2) = I(j_2 = 16)[I(2 \le j_1 \le 3) + 16I(14 \le j_1 \le 16)]. \tag{4.9}$$

From Figure 4.1 and Table 4.3 we arrive at Table 4.12, summing up the states of modules $g$ and $i$.

Table 4.11 The states of modules $h$ and $j$.

| State | Capacity/Compression |
|---|---|
| 0 | 0 |
| 1 | 4.5 |
| 16 | 9 |

Table 4.12 The states of modules $g$ and $i$.

| State | Capacity |
|---|---|
| 0 | 0 |
| 3 | 30 |
| 16 | 59 |

Table 4.13 System states.

| State | Capacity | Compression |
|---|---|---|
| 0 | 0 | – |
| 1 | 4.2–19 | – |
| 2 | 21.3–27 | – |
| 3 | 30 | – |
| 4 | 35 | – |
| 5 | 41–42 | – |
| 6 | 45 | – |
| 7 | 46–48 | – |
| 8 | 50 | 0 |
| 9 | 50–53 | 4.5 |
| 10 | 50–53 | 9 |
| 11 | 50–53 | $9^+$ |
| 12 | 54.5 | 4.5 |
| 13 | 54.5 | 9 |
| 14 | 56 | 9 |
| 15 | 57.2–59 | 9 |
| 16 | 57.2–59 | $9^+$ |

The corresponding structure functions are given by

$$g = \phi_g(g_1, g_2, g_3, g_4) = I(\min(g_1, g_2, g_3) = 16)[3 + 13I(g_4 = 16)] \tag{4.10}$$

$$i = \phi_i(i_1, i_2) = I(i_1 = 16)[3 + 13I(i_2 = 16)]. \tag{4.11}$$

At last we are in the position of considering the system as a whole. From Figure 4.1 and Tables 4.9–4.12 we obtain Table 4.13, summing up the system states. In this table mainly the amount of gas compressed by module $j$, closest to Emden, is taken into account, but only when the amount of transported gas is at least $50\,\mathrm{MSm}^3/d$. In addition a '+' in the compression column indicates the advantageous situation where all compressor modules $f, h$ and $j$ are in state 16. Of course this simplifying approach is not necessary, but as is seen, even this approach leads to a rather complex system structure function.

Some careful thinking leads to the following system structure function

$$\begin{aligned}
&\phi(abcde, f, g, h, i, j, k)\\
&\quad = I(k = 16)I(\min(abcde, g, i) > 0)\\
&\quad \{I(abcde \leq 3)abcde + I(abcde = 4)[3 + I(g = 16)I(i = 16)]\\
&\quad + I(abcde = 5)[3 + 2I(g = 16)I(i = 16)]\\
&\quad + I(abcde = 7)[3 + I(g = 16)I(i = 16)(3I(f = 0) + 4I(f > 0))]
\end{aligned}$$

$$
\begin{aligned}
&+ I(abcde = 8)[3 + I(g = 16)I(i = 16)\{3I(f = 0)\\
&+ I(2 \le f \le 4)(5 + I(j = 1) + 2I(j = 16))\\
&+ I(f = 16)[I(h \le 1)(5 + I(j = 1) + 2I(j = 16))\\
&+ I(h = 16)(5 + I(j = 1) + 3I(j = 16))]\}]\\
&+ I(abcde = 16)[3 + I(g = 16)I(i = 16)\{3I(f = 0)\\
&+ I(f = 2)[I(h = 0)(5 + I(j = 1) + 2I(j = 16))\\
&+ I(h = 1)(5 + 4I(j = 1) + 5I(j = 16))\\
&+ I(h = 16)(5 + 4I(j = 1) + 6I(j = 16))]\\
&+ I(f = 4)[I(h = 0)(5 + I(j = 1) + 2I(j = 16))\\
&+ I(h = 1)(5 + 4I(j = 1) + 5I(j = 16))\\
&+ I(h = 16)(5 + 4I(j = 1) + 7I(j = 16)]\\
&+ I(f = 16)[I(h = 0)(5 + I(j = 1) + 2I(j = 16))\\
&+ I(h = 1)(5 + 4I(j = 1) + 5I(j = 16))\\
&+ I(h = 16)(5 + 4I(j = 1) + 8I(j = 16))]\}]\}. \qquad (4.12)
\end{aligned}
$$

## 4.2 Bounds for system availabilities and unavailabilities

The availabilities and unavailabilities of the 32 components of the offshore gas pipeline system are determined by the computer package MUSTAFA (MUltiSTAte Fault tree Analysis) developed by Høgåsen (1990) applying the formula given by Equation (3.75) for the component transition probabilities. The only input needed is the intensity matrices of the components, implicitly giving the corresponding sets of states, along with the time interval $I$.

The intensity matrices of $a_1$ and the compressor component $f_1$ are given below. For the interested reader, aiming for instance at checking the results by a simulation study, the remaining intensity matrices are given in Appendix B.

$$
\boldsymbol{A}_{a_1} = \begin{bmatrix} -4 & 0 & 0 & 0 & 4 \\ 0 & -8.3721 & 0 & 0 & 8.3721 \\ 0 & 0 & -9 & 0 & 9 \\ 0 & 0 & 0 & -22.25 & 22.25 \\ 0.00003 & 0.0019 & 0.028 & 0.0061 & -0.03603 \end{bmatrix}
$$

Note that $S_{a_1} = \{0, 1, 2, 3, 16\}$, and that the failure states are ranked according to inverse repair rates or, equivalently, the expected repair times.

$$A_{f_1} = \begin{bmatrix} -1783.35 & 0 & 1711.35 & 72 & 0 & 0 & 0 & 0 & 0 \\ 0 & -2281.8 & 0 & 2281.8 & 0 & 0 & 0 & 0 & 0 \\ 9.6 & 0 & -1222.5 & 0 & 1140.9 & 72 & 0 & 0 & 0 \\ 0 & 9.6 & 0 & -1720.95 & 0 & 1711.35 & 0 & 0 & 0 \\ 0 & 0 & 19.2 & 0 & -661.65 & 0 & 570.45 & 72 & 0 \\ 0 & 0 & 0 & 19.2 & 0 & -1160.1 & 0 & 1140.9 & 0 \\ 0 & 0 & 0 & 0 & 28.8 & 0 & -100.8 & 0 & 72 \\ 0 & 0 & 0 & 0 & 0 & 28.8 & 0 & -599.25 & 570.45 \\ 0 & 0 & 0 & 0 & 0 & 0 & 20 & 38.4 & -58.4 \end{bmatrix}$$

Note that the maintenance completion rate of 72 completions per year is smaller than the repair rate of a single compressor of 570.45 repairs per year as assumed when ranking the states of $f_1$ in Table 4.1, and that there are four repairmen. Note also that a compressor is only sent for maintenance when all four compressors are operative, and that in this case the fourth compressor is in hot standby.

By using the computer package MUSTAFA, which finds the minimal path and cut vectors to all levels from Equations (4.1)–(4.12) and then applies the best bounds of Corollary 3.11, we arrive at the bounds shown in Table 4.14, which are not based on a modular decomposition. Note that for the case $I = [0.5, 0.5]$ we know from Section 3.1 that $q_{\phi}^{j[0.5,0.5]} = 1 - p_{\phi}^{j[0.5,0.5]}$. Hence, the bounds for $q_{\phi}^{j[0.5,0.5]}$ follow immediately from the ones of $p_{\phi}^{j[0.5,0.5]}$. Furthermore, note that the upper bounds both for $p_{\phi}^{j[0.5,0.6]}$ and $p_{\phi}^{j[0.5,0.5]}$, and also for $q_{\phi}^{j[0.5,0.6]}$ and $q_{\phi}^{j[0.5,0.5]}$, are almost completely identical. This is due to the fact that the upper bounds for a fixed time point $t$ in Corollary 3.11 do not depend on $t$ for $t \geq 0.5$. Hence, stationarity is reached after half a year. We see that all bounds are very informative for $I = [0.5, 0.5]$ and for the unavailabilities also for $I = [0.5, 0.6]$, corresponding to an interval length of 36 days. They are less informative for the availabilities for $I = [0.5, 0.6]$. Note, however, that from the theory we know that all lower bounds are far better than the corresponding upper bounds especially for long intervals. Also, to be on the conservative side, the lower bounds for the availabilities are the most interesting.

It should be mentioned that none of the intervals in Table 4.14 reduce to a single exact value. This follows from Theorem 3.24 since there is more than one single minimal path and minimal cut vector to all levels. Exactly correct values are obtained in Mørch (1991) for some simpler networks.

The computer package MUSTAFA can also find a modular decomposition of a system. Then, bounds for the availabilities and unavailabilities for the modules can be established from the availabilities and unavailabilities of the components. Furthermore, bounds for the availabilities and unavailabilities of the system can be established from these bounds for

Table 4.14 Bounds for $p_\phi^{j(I)}$ and $q_\phi^{j(I)}$ not based on a modular decomposition.

| $j$ | $I$ | Bounds for $p_\phi^{j(I)}$ | | Bounds for $q_\phi^{j(I)}$ | |
|---|---|---|---|---|---|
| | | Lower | Upper | Lower | Upper |
| 16 | [0.5, 0.6] | 0.0897 | 0.9600 | 0.0120 | 0.0836 |
| | [0.5, 0.5] | 0.9164 | 0.9601 | 0.0399 | 0.0836 |
| 15 | [0.5, 0.6] | 0.1672 | 0.9600 | 0.0120 | 0.0738 |
| | [0.5, 0.5] | 0.9262 | 0.9601 | 0.0399 | 0.0738 |
| 14 | [0.5, 0.6] | 0.1706 | 0.9600 | 0.0120 | 0.0737 |
| | [0.5, 0.5] | 0.9263 | 0.9601 | 0.0399 | 0.0737 |
| 13 | [0.5, 0.6] | 0.2422 | 0.9600 | 0.0120 | 0.0686 |
| | [0.5, 0.5] | 0.9314 | 0.9601 | 0.0399 | 0.0686 |
| 12 | [0.5, 0.6] | 0.3439 | 0.9600 | 0.0120 | 0.0635 |
| | [0.5, 0.5] | 0.9365 | 0.9601 | 0.0399 | 0.0635 |
| 11 | [0.5, 0.6] | 0.3439 | 0.9968 | 0.0011 | 0.0254 |
| | [0.5, 0.5] | 0.9746 | 0.9968 | 0.0032 | 0.0254 |
| 10 | [0.5, 0.6] | 0.3904 | 0.9968 | 0.0011 | 0.0230 |
| | [0.5, 0.5] | 0.9770 | 0.9968 | 0.0032 | 0.0230 |
| 9 | [0.5, 0.6] | 0.4113 | 0.9968 | 0.0011 | 0.0228 |
| | [0.5, 0.5] | 0.9772 | 0.9968 | 0.0032 | 0.0228 |
| 8 | [0.5, 0.6] | 0.4492 | 0.9968 | 0.0011 | 0.0211 |
| | [0.5, 0.5] | 0.9789 | 0.9968 | 0.0032 | 0.0211 |
| 7 | [0.5, 0.6] | 0.4566 | 0.9968 | 0.0011 | 0.0207 |
| | [0.5, 0.5] | 0.9793 | 0.9968 | 0.0032 | 0.0207 |
| 6 | [0.5, 0.6] | 0.4720 | 0.9968 | 0.0011 | 0.0199 |
| | [0.5, 0.5] | 0.9801 | 0.9968 | 0.0032 | 0.0199 |
| 5 | [0.5, 0.6] | 0.4802 | 0.9968 | 0.0011 | 0.0195 |
| | [0.5, 0.5] | 0.9805 | 0.9968 | 0.0032 | 0.0195 |

(*continued overleaf*)

Table 4.14 (*continued*)

| $j$ | $I$ | Bounds for $p_\phi^{j(I)}$ | | Bounds for $q_\phi^{j(I)}$ | |
|---|---|---|---|---|---|
| | | Lower | Upper | Lower | Upper |
| 4 | [0.5, 0.6] | 0.4810 | 0.9968 | 0.0011 | 0.0195 |
| | [0.5, 0.5] | 0.9805 | 0.9968 | 0.0032 | 0.0195 |
| 3 | [0.5, 0.6] | 0.4952 | 0.9968 | 0.0011 | 0.0189 |
| | [0.5, 0.5] | 0.9811 | 0.9968 | 0.0032 | 0.0189 |
| 2 | [0.5, 0.6] | 0.4952 | 0.9968 | 0.0011 | 0.0189 |
| | [0.5, 0.5] | 0.9811 | 0.9968 | 0.0032 | 0.0189 |
| 1 | [0.5, 0.6] | 0.7354 | 0.9975 | 0.0011 | 0.0112 |
| | [0.5, 0.5] | 0.9888 | 0.9975 | 0.0025 | 0.0112 |

Table 4.15 Bounds for $p_\phi^{j(I)}$ and $q_\phi^{j(I)}$ based on a modular decomposition.

| $j$ | $I$ | Bounds for $p_\phi^{j(I)}$ | | Bounds for $q_\phi^{j(I)}$ | |
|---|---|---|---|---|---|
| | | Lower | Upper | Lower | Upper |
| 16 | [0.5, 0.6] | 0.0897 | 0.9246 | 0.0148 | 0.0836 |
| | [0.5, 0.5] | 0.9164 | 0.9247 | 0.0753 | 0.0836 |
| 15 | [0.5, 0.6] | 0.1672 | 0.9344 | 0.0148 | 0.0738 |
| | [0.5, 0.5] | 0.9262 | 0.9345 | 0.0655 | 0.0738 |
| 14 | [0.5, 0.6] | 0.1706 | 0.9346 | 0.0148 | 0.0737 |
| | [0.5, 0.5] | 0.9263 | 0.9347 | 0.0653 | 0.0737 |
| 13 | [0.5, 0.6] | 0.2422 | 0.9397 | 0.0148 | 0.0686 |
| | [0.5, 0.5] | 0.9314 | 0.9398 | 0.0602 | 0.0686 |
| 12 | [0.5, 0.6] | 0.3439 | 0.9449 | 0.0148 | 0.0635 |
| | [0.5, 0.5] | 0.9365 | 0.9450 | 0.0550 | 0.0635 |

Table 4.15 (*continued*)

| $j$ | $I$ | Bounds for $p_{\phi}^{j(I)}$ | | Bounds for $q_{\phi}^{j(I)}$ | |
|---|---|---|---|---|---|
| | | Lower | Upper | Lower | Upper |
| 11 | [0.5, 0.6] | 0.3439 | 0.9828 | 0.0034 | 0.0257 |
| | [0.5, 0.5] | 0.9743 | 0.9829 | 0.0171 | 0.0257 |
| 10 | [0.5, 0.6] | 0.3439 | 0.9828 | 0.0034 | 0.0232 |
| | [0.5, 0.5] | 0.9768 | 0.9829 | 0.0171 | 0.0232 |
| 9 | [0.5, 0.6] | 0.4113 | 0.9828 | 0.0034 | 0.0228 |
| | [0.5, 0.5] | 0.9772 | 0.9829 | 0.0171 | 0.0228 |
| 8 | [0.5, 0.6] | 0.4492 | 0.9845 | 0.0033 | 0.0211 |
| | [0.5, 0.5] | 0.9789 | 0.9846 | 0.0154 | 0.0211 |
| 7 | [0.5, 0.6] | 0.4566 | 0.9845 | 0.0033 | 0.0207 |
| | [0.5, 0.5] | 0.9793 | 0.9846 | 0.0154 | 0.0207 |
| 6 | [0.5, 0.6] | 0.4720 | 0.9853 | 0.0033 | 0.0199 |
| | [0.5, 0.5] | 0.9801 | 0.9854 | 0.0146 | 0.0199 |
| 5 | [0.5, 0.6] | 0.4802 | 0.9853 | 0.0033 | 0.0195 |
| | [0.5, 0.5] | 0.9805 | 0.9854 | 0.0146 | 0.0195 |
| 4 | [0.5, 0.6] | 0.4810 | 0.9853 | 0.0033 | 0.0195 |
| | [0.5, 0.5] | 0.9805 | 0.9854 | 0.0146 | 0.0195 |
| 3 | [0.5, 0.6] | 0.4952 | 0.9859 | 0.0033 | 0.0189 |
| | [0.5, 0.5] | 0.9811 | 0.9860 | 0.0140 | 0.0189 |
| 2 | [0.5, 0.6] | 0.4952 | 0.9859 | 0.0033 | 0.0189 |
| | [0.5, 0.5] | 0.9811 | 0.9860 | 0.0140 | 0.0189 |
| 1 | [0.5, 0.6] | 0.7354 | 0.9889 | 0.0027 | 0.0112 |
| | [0.5, 0.5] | 0.9888 | 0.9890 | 0.0110 | 0.0112 |

the availabilities and unavailabilities for the modules. Using this approach we arrive at the best bounds shown in Table 4.15.

First note that the comments made on Table 4.14 are still valid for Table 4.15. Secondly, note that the bounds for a fixed time point $t = 0.5$ are more informative in Table 4.15 than in Table 4.14. This is mainly in accordance with Corollary 3.21. The same is also true to some extent for the interval $I = [0.5, 0.6]$, but this is not true in general, which is also mainly in accordance with Corollary 3.21. For instance, Table 4.14 gives $0.3904 \leq p_\phi^{10[0.5,0.6]} \leq 0.9968$, whereas Table 4.15 gives $0.3439 \leq p_\phi^{10[0.5,0.6]} \leq 0.9828$. Finally, it should be concluded that with the computer package MUSTAFA, more complex systems than the one treated here can be attacked.

# 5

# Bayesian assessment of system availabilities

## 5.1 Basic ideas

In the present chapter, covering the paper Natvig *et al.* (2010a) and following a Bayesian approach, the ambition is to arrive at the posterior distributions of the system availabilities and unavailabilities, to the various levels in a fixed time interval, based on both prior information and data on both the components and the system. We argue that a realistic approach is to start out by describing our uncertainty on the component availabilities and unavailabilities to the various levels in a fixed time interval, based on both prior information and data on the components, by the moments up to order $m$ of their marginal distributions. From these moments analytic bounds on the corresponding moments of the system availabilities and unavailabilities to the various levels in a fixed time interval are arrived at. Applying these bounds and prior system information we may then fit prior distributions of the system availabilities and unavailabilities to the various levels in a fixed time interval. These can, in turn, be updated by relevant data on the system. This generalizes results given in Natvig and Eide (1987) considering a binary monotone system of binary components at a fixed point of time. Furthermore, considering the simple network system given in Figure 1.1, we show that the analytic bounds can be slightly improved by straightforward simulation techniques.

Suppose that we run $K_i$ independent experiments for component $i$ registering $x_i^{(k)}(s)$ $\forall s \in \tau(I)$ in the $k$th experiment, $k = 1, \ldots, K_i, i = 1, \ldots, n$. Let, for $j = 1, \ldots, M, i = 1, \ldots, n$

$$D_i^{1j(I)} = \sum_{k=1}^{K_i} I[x_i^{(k)}(s) \geq j \ \ \forall s \in \tau(I)]$$

$$D_i^{2j(I)} = \sum_{k=1}^{K_i} I[x_i^{(k)}(s) < j \ \ \forall s \in \tau(I)],$$

and for $j = 0, \ldots, M, i = 1, \ldots, n$

$$D_i^{j(I)} = \sum_{k=1}^{K_i} I[\min_{s \in \tau(I)} x_i^{(k)}(s) = j].$$

Let, for $r = 1, 2$, $\boldsymbol{D}_i^{r(I)} = (D_i^{r1(I)}, \ldots, D_i^{rM(I)})$, $\boldsymbol{D}^{r(I)} = (\boldsymbol{D}_1^{r(I)}, \ldots, \boldsymbol{D}_n^{r(I)})$. Furthermore, let $\boldsymbol{D}_i^{(I)} = (D_i^{0(I)}, \ldots, D_i^{M(I)})$ and $\boldsymbol{D}^{(I)} = (\boldsymbol{D}_1^{(I)}, \ldots, \boldsymbol{D}_n^{(I)})$.

Suppose also that we run $K$ independent experiments on the system level registering $\phi(\boldsymbol{x}^{(k)}(s))$ $\forall s \in \tau(I)$ in the $k$th experiment, $k = 1, \ldots, K$. Let, for $j = 1, \ldots, M$

$$D_\phi^{1j(I)} = \sum_{k=1}^{K} I[\phi(\boldsymbol{x}^{(k)}(s)) \geq j \ \ \forall s \in \tau(I)]$$

$$D_\phi^{2j(I)} = \sum_{k=1}^{K} I[\phi(\boldsymbol{x}^{(k)}(s)) < j \ \ \forall s \in \tau(I)],$$

and for $j = 0, \ldots, M$

$$D_\phi^{j(I)} = \sum_{k=1}^{K} I[\min_{s \in \tau(I)} \phi(\boldsymbol{x}^{(k)}(s)) = j].$$

Let, for $r = 1, 2$, $\boldsymbol{D}_\phi^{r(I)} = (D_\phi^{r1(I)}, \ldots, D_\phi^{rM(I)})$. Furthermore, let $\boldsymbol{D}_\phi^{(I)} = (D_\phi^{0(I)}, \ldots, D_\phi^{M(I)})$. When $I = [t, t]$, we drop $I$ from the notation in all these data variables and data vectors.

Assume that the prior distribution respectively of the component availability and unavailability matrices, before running any experiment on the component level, $\pi(\boldsymbol{P}_\phi^{(I)})$ and $\pi(\boldsymbol{Q}_\phi^{(I)})$, can be written as

$$\pi(\boldsymbol{P}_\phi^{(I)}) = \prod_{i=1}^{n} \pi_i(\boldsymbol{p}_i^{(I)}) \qquad \pi(\boldsymbol{Q}_\phi^{(I)}) = \prod_{i=1}^{n} \pi_i(\boldsymbol{q}_i^{(I)}),$$

where, for $i = 1, \ldots, n$

$$\boldsymbol{p}_i^{(I)} = \{p_i^{j(I)}\}_{j=1,\ldots,M} \qquad \boldsymbol{q}_i^{(I)} = \{q_i^{j(I)}\}_{j=1,\ldots,M}$$

are the component availability and unavailability vectors and $\pi_i(\boldsymbol{p}_i^{(I)})$ the prior marginal distribution of $\boldsymbol{p}_i^{(I)}$, and $\pi_i(\boldsymbol{q}_i^{(I)})$ the prior marginal distribution of $\boldsymbol{q}_i^{(I)}$. Hence, we assume that the components have independent prior component availability vectors and independent prior component unavailability vectors.

Note that before the experiments are carried through, $D_i^{1j(I)}$ is binomially distributed with parameters $K_i$ and $p_i^{j(I)}$, and $D_i^{2j(I)}$ binomially distributed with parameters $K_i$ and $q_i^{j(I)}$. We assume that given $\boldsymbol{P}_\phi^{(I)}$, $\boldsymbol{D}_1^{1(I)}, \ldots, \boldsymbol{D}_n^{1(I)}$ are independent and that given $\boldsymbol{Q}_\phi^{(I)}$, $\boldsymbol{D}_1^{2(I)}, \ldots, \boldsymbol{D}_n^{2(I)}$ are independent. Hence, since we have assumed that the components have independent prior availability vectors, using Bayes's theorem the posterior distribution of the component availability matrix, $\pi(\boldsymbol{P}_\phi^{(I)}|\boldsymbol{D}^{1(I)})$, can be written as

$$\begin{aligned}
\pi(\boldsymbol{P}_\phi^{(I)}|\boldsymbol{D}^{1(I)}) &= \frac{\pi(\boldsymbol{D}^{1(I)}|\boldsymbol{P}_\phi^{(I)})\pi(\boldsymbol{P}_\phi^{(I)})}{\int \pi(\boldsymbol{D}^{1(I)}|\boldsymbol{P}_\phi^{(I)})\pi(\boldsymbol{P}_\phi^{(I)})\,\mathrm{d}\boldsymbol{P}_\phi^{(I)}} \\
&= \frac{\prod_{i=1}^n \pi_i(\boldsymbol{D_i}^{1(I)}|\boldsymbol{p}_i^{(I)})\pi_i(\boldsymbol{p}_i^{(I)})}{\prod_{i=1}^n \int \pi_i(\boldsymbol{D_i}^{1(I)}|\boldsymbol{p}_i^{(I)})\pi_i(\boldsymbol{p}_i^{(I)})\,\mathrm{d}\boldsymbol{p}_i^{(I)}} \\
&= \prod_{i=1}^{n} \frac{\pi_i(\boldsymbol{D_i}^{1(I)}|\boldsymbol{p}_i^{(I)})\pi_i(\boldsymbol{p}_i^{(I)})}{\pi_i(\boldsymbol{D}_i^{1(I)})} = \prod_{i=1}^{n} \pi_i(\boldsymbol{p}_i^{(I)}|\boldsymbol{D}_i^{1(I)}),
\end{aligned}$$

where $\pi_i(\boldsymbol{p}_i^{(I)}|\boldsymbol{D}_i^{1(I)})$ is the posterior marginal distribution of $\boldsymbol{p}_i^{(I)}$. Similarly, the posterior distribution of the component unavailability matrix, $\pi(\boldsymbol{Q}_\phi^{(I)}|\boldsymbol{D}^{2(I)})$, can be written as

$$\pi(\boldsymbol{Q}_\phi^{(I)}|\boldsymbol{D}^{2(I)}) = \prod_{i=1}^{n} \pi_i(\boldsymbol{q}_i^{(I)}|\boldsymbol{D}_i^{2(I)}).$$

Hence, the posterior component availability vectors are independent given $\boldsymbol{D}^{1(I)}$ and the posterior component unavailability vectors are independent given $\boldsymbol{D}^{2(I)}$.

Now specialize $I = [t, t]$ and assume that the component states $X_1, \ldots, X_n$ are independent given $\boldsymbol{P}_\phi$. Let

$$\boldsymbol{p}_\phi = \{p_\phi^j\}_{j=1,\ldots,M}$$

be the system reliability vector. Since, in this case, $\boldsymbol{p}_\phi$ is a function of $\boldsymbol{P}_\phi$, the distribution $\pi(\boldsymbol{p}_\phi(\boldsymbol{P}_\phi)|\boldsymbol{D}^1)$ can then be arrived at. Based on prior knowledge on the system level this may be adjusted to the prior distribution of the system reliability vector $\pi_0(\boldsymbol{p}_\phi(\boldsymbol{P}_\phi)|\boldsymbol{D}^1)$. Note that before the experiments are carried through, $D_\phi^{1j}$ is binomially distributed with parameters $K$ and $p_\phi^j$. Including the data $\boldsymbol{D}_\phi^1$, we end up with the posterior distribution of the system reliability vector $\pi(\boldsymbol{p}_\phi(\boldsymbol{P}_\phi)|\boldsymbol{D}^1, \boldsymbol{D}_\phi^1)$ for $j = 1, \ldots, M$.

When considering the case $I = [t, t]$, we can, instead of $\boldsymbol{P}_\phi$, consider the parameter matrix

$$\boldsymbol{R}_\phi = \{r_i^j\}_{\substack{i=1,\ldots,n \\ j=0,\ldots,M}},$$

and assume that the components have independent prior vectors

$$\boldsymbol{r}_i = \{r_i^j\}_{j=0,\ldots,M}, \quad i = 1, \ldots, n,$$

each having a Dirichlet distribution being the natural conjugate prior. Furthermore, we assume that given $\boldsymbol{R}_\phi$, $\boldsymbol{D}_1, \ldots, \boldsymbol{D}_n$ are independent. Note that before the experiments are carried through, $\boldsymbol{D}_i$ is multinomially distributed with parameters $K_i$ and $\boldsymbol{r}_i$. Hence, the posterior marginal distribution of $\boldsymbol{r}_i$ given the data $\boldsymbol{D}_i$, $\pi_i(\boldsymbol{r}_i|\boldsymbol{D}_i)$, is Dirichlet. Furthermore, we have

$$\pi(\boldsymbol{R}_\phi|\boldsymbol{D}) = \prod_{i=1}^{n} \pi_i(\boldsymbol{r}_i|\boldsymbol{D}_i).$$

Hence, life can be made easy at the component level. Assume that the component states $X_1, \ldots, X_n$ are independent given $\boldsymbol{R}_\phi$. Let

$$\boldsymbol{r}_\phi = \{r_\phi^j\}_{j=0,\ldots,M}.$$

We attempt to derive the corresponding distribution, $\pi(\boldsymbol{r}_\phi(\boldsymbol{R}_\phi)|\boldsymbol{D})$. If this is successful, based on prior knowledge on the system level, it is adjusted to $\pi_0(\boldsymbol{r}_\phi(\boldsymbol{R}_\phi)|\boldsymbol{D})$. This may be possible for simple systems.

Note that before the experiments are carried through, $\boldsymbol{D}_\phi$ is multinomially distributed with parameters $K$ and $\boldsymbol{r}_\phi$. Hence, if $\pi_0(\boldsymbol{r}_\phi(\boldsymbol{R}_\phi)|\boldsymbol{D})$, as in a dream, ended up as a Dirichlet distribution, the posterior distribution $\pi(\boldsymbol{r}_\phi(\boldsymbol{R}_\phi)|\boldsymbol{D}, \boldsymbol{D}_\phi)$ would also be a Dirichlet distribution. Do not forget this was a dream and also based on independent components given $\boldsymbol{R}_\phi$! Life will not be easy at the system level even when $I = [t, t]$.

For an arbitrary $I$ let

$$\boldsymbol{p}_\phi^{(I)} = \{p_\phi^{j(I)}\}_{j=1,\ldots,M} \qquad \boldsymbol{q}_\phi^{(I)} = \{q_\phi^{j(I)}\}_{j=1,\ldots,M}.$$

Then $\boldsymbol{p}_\phi^{(I)}$ is not a function of just $\boldsymbol{P}_\phi^{(I)}$, and $\boldsymbol{q}_\phi^{(I)}$ not a function of just $\boldsymbol{Q}_\phi^{(I)}$. Hence, the approach above for the case $I = [t, t]$ cannot be extended.

## 5.2 Moments for posterior component availabilities and unavailabilities

Based on the experiences of the previous section we reduce our ambitions. We start by specifying marginal moments $E\{(p_i^{j(I)})^s\}$ and $E\{(q_i^{j(I)})^s\}$ for $s = 1, \ldots, m + K_i$, $j = 1, \ldots, M$ of $\pi_i(\boldsymbol{p}_i^{(I)})$ and $\pi_i(\boldsymbol{q}_i^{(I)})$, $i = 1, \ldots, n$. We will first illustrate how these can be updated to give posterior moments $E\{(p_i^{j(I)})^s|D_i^{1j(I)}\}$ and $E\{(q_i^{j(I)})^s|D_i^{2j(I)}\}$ for $s = 1, \ldots, m$, $j = 1, \ldots, M$ by using Lemma 1 in Mastran and Singpurwalla (1978). Note that we lose information by conditioning on $D_i^{1j(I)}$ instead of $\boldsymbol{D}_i^{1(I)}$ and $D_i^{2j(I)}$ instead of $\boldsymbol{D}_i^{2(I)}$. However, such improved conditioning does not work with this approach. We have

$$\begin{aligned}
&E\{(p_i^{j(I)})^s|D_i^{1j(I)}\} \\
&\quad\propto \int_0^1 (p_i^{j(I)})^s (p_i^{j(I)})^{D_i^{1j(I)}} (1 - p_i^{j(I)})^{K_i - D_i^{1j(I)}} \pi_i(p_i^{j(I)})\, \mathrm{d}p_i^{j(I)} \\
&\quad= \int_0^1 (p_i^{j(I)})^{s+D_i^{1j(I)}} \sum_{r=0}^{K_i - D_i^{1j(I)}} \binom{K_i - D_i^{1j(I)}}{r} \\
&\qquad\times (-1)^r (p_i^{j(I)})^r \pi_i(p_i^{j(I)})\, \mathrm{d}p_i^{j(I)}.
\end{aligned}$$

Hence,

$$E\{(p_i^{j(I)})^s|D_i^{1j(I)}\} = \frac{\sum_{r=0}^{K_i-D_i^{1j(I)}} \binom{K_i-D_i^{1j(I)}}{r}(-1)^r E\{(p_i^{j(I)})^{s+D_i^{1j(I)}+r}\}}{\sum_{r=0}^{K_i-D_i^{1j(I)}} \binom{K_i-D_i^{1j(I)}}{r}(-1)^r E\{(p_i^{j(I)})^{D_i^{1j(I)}+r}\}}.$$

A similar expression is valid for $E\{(q_i^{j(I)})^s | D_i^{2j(I)}\}$. The advantage of using this lemma is that it is applicable for general prior distributions $\pi_i(p_i^{j(I)})$ and $\pi_i(q_i^{j(I)})$. A serious drawback is, however, that to arrive at $E\{(p_i^{j(I)})^s | D_i^{1j(I)}\}$ and $E\{(q_i^{j(I)})^s | D_i^{2j(I)}\}$ for $s = 1, \ldots, m$, $j = 1, \ldots, M$ one must specify marginal moments up to order $m + K_i$ of the corresponding prior distributions $\pi_i(p_i^{j(I)})$ and $\pi_i(q_i^{j(I)})$. This may be completely unrealistic unless $K_i$ is small.

A more realistic alternative is given in the following. Let, for $j = 0, \ldots, M, i = 1, \ldots, n$

$$r_i^{j(I)} = p_i^{j(I)} - p_i^{j+1(I)} = P[\min_{t \in \tau(I)} X_i(t) = j].$$

Assume that the components have independent prior vectors

$$\boldsymbol{r}_i^I = \{r_i^{j(I)}\}_{j=0,\ldots,M}, \quad i = 1, \ldots, n,$$

each having a Dirichlet distribution being the natural conjugate prior. Note that before the experiments are carried through, $\boldsymbol{D}_i^{(I)}$ is multinomially distributed with parameters $K_i$ and $\boldsymbol{r}_i^{(I)}$. Hence, the posterior marginal distribution of $\boldsymbol{r}_i^{(I)}$ given the data $\boldsymbol{D}_i^{(I)}$, $\pi_i(\boldsymbol{r}_i^{(I)} | \boldsymbol{D}_i^{(I)})$, is Dirichlet. We now have

$$p_i^{j(I)} = \sum_{\ell=j}^{M} r_i^{\ell(I)} \quad D_i^{1j(I)} = \sum_{\ell=j}^{M} D_i^{\ell(I)}. \tag{5.1}$$

Hence, the posterior marginal distribution of $p_i^{j(I)}$ given the data $D_i^{1j(I)}$ is beta. Accordingly, we lose no information by conditioning on $D_i^{1j(I)}$ instead of $\boldsymbol{D}_i^{1(I)}$. We now assume that the prior distribution $\pi_i(p_i^{j(I)})$ is beta with parameters $a_i^{j(I)}$ and $b_i^{j(I)}$. It then follows that $\pi_i(p_i^{j(I)} | D_i^{1j(I)})$ is beta with parameters $a_i^{j(I)} + D_i^{1j(I)}$ and $b_i^{j(I)} + K_i - D_i^{1j(I)}$. We have

$$\begin{aligned}
&E\{(p_i^{j(I)})^s | D_i^{1j(I)}\} \\
&= \int_0^1 (p_i^{j(I)})^s \frac{\Gamma(a_i^{j(I)} + b_i^{j(I)} + K_i)}{\Gamma(a_i^{j(I)} + D_i^{1j(I)})\Gamma(b_i^{j(I)} + K_i - D_i^{1j(I)})} \\
&\quad (p_i^{j(I)})^{a_i^{j(I)} + D_i^{1j(I)} - 1}(1 - p_i^{j(I)})^{b_i^{j(I)} + K_i - D_i^{1j(I)} - 1}\, dp_i^{j(I)} \\
&= \frac{\Gamma(a_i^{j(I)} + b_i^{j(I)} + K_i)\Gamma(a_i^{j(I)} + D_i^{1j(I)} + s)}{\Gamma(a_i^{j(I)} + D_i^{1j(I)})\Gamma(a_i^{j(I)} + b_i^{j(I)} + K_i + s)}
\end{aligned}$$

$$\int_0^1 \frac{\Gamma(a_i^{j(I)} + b_i^{j(I)} + K_i + s)}{\Gamma(a_i^{j(I)} + D_i^{1j(I)} + s)\Gamma(b_i^{j(I)} + K_i - D_i^{1j(I)})}$$
$$(p_i^{j(I)})^{a_i^{j(I)} + D_i^{1j(I)} + s - 1}(1 - p_i^{j(I)})^{b_i^{j(I)} + K_i - D_i^{1j(I)} - 1} \, \mathrm{d}p_i^{j(I)}$$
$$= \frac{\Gamma(a_i^{j(I)} + b_i^{j(I)} + K_i)\Gamma(a_i^{j(I)} + D_i^{1j(I)} + s)}{\Gamma(a_i^{j(I)} + D_i^{1j(I)})\Gamma(a_i^{j(I)} + b_i^{j(I)} + K_i + s)},$$

the integral being equal to 1 since we are integrating up a beta density with parameters $a_i^{j(I)} + D_i^{1j(I)} + s$ and $b_i^{j(I)} + K_i - D_i^{1j(I)}$. A similar expression is valid for $E\{(q_i^{j(I)})^s | D_i^{2j(I)}\}$.

## 5.3 Bounds for moments for system availabilities and unavailabilities

From the marginal moments $E\{(p_i^{\ell(I)})^s | D_i^{1\ell(I)}\}$ and $E\{(q_i^{\ell(I)})^s | D_i^{2\ell(I)}\}$, we derive lower bounds on the marginal moments $E\{(p_\phi^{j(I)})^s | \boldsymbol{D}^{1(I)}\}$ and upper bounds on the marginal moments $E\{(p_\phi^{j(I)})^s | \boldsymbol{D}^{2(I)}\}$ for $s = 1, \ldots, m$, $\ell = 1, \ldots, M$, $j = 1, \ldots, M$. Similarly, we derive lower bounds on the marginal moments $E\{(q_\phi^{j(I)})^s | \boldsymbol{D}^{2(I)}\}$ and upper bounds on the marginal moments $E\{(q_\phi^{j(I)})^s | \boldsymbol{D}^{1(I)}\}$. Note that we now do not necessarily need the marginal performance processes of the components to be independent in $I$. From these bounds and prior knowledge on the system level we may fit $\pi_0(\boldsymbol{p}_\phi^{(I)})$ and $\pi_0(\boldsymbol{q}_\phi^{(I)})$. This may finally be updated to give $\pi(\boldsymbol{p}_\phi^{(I)} | \boldsymbol{D}_\phi^{1(I)})$ and $\pi(\boldsymbol{q}_\phi^{(I)} | \boldsymbol{D}_\phi^{2(I)})$.

What we will concentrate on is how to establish the bounds on the marginal moments of system availabilities and unavailabilities from the marginal moments of component availabilities and unavailabilities. To simplify notation we drop the reference to the data $(\boldsymbol{D}^{1(I)}, \boldsymbol{D}^{2(I)})$ from experiments on the component level.

Let us just for a while return to the case $I = [t, t]$ and assume that the component states $X_1, \ldots, X_n$ are independent given $\boldsymbol{R}_\phi$. Then, parallel to Equation (2.17), we get

$$p_\phi^j(\boldsymbol{R}_\phi) = \sum_{\boldsymbol{x}} I[\phi(\boldsymbol{x}) \geq j] \prod_{i=1}^n r_i^{x_i}.$$

Hence, since we assume that the components have independent prior vectors $\boldsymbol{r}_i$ for $i = 1, \ldots, n$, generalizing a result in Natvig and

Eide (1987), we get

$$E\{p_\phi^j(\boldsymbol{R}_\phi)\} = \sum_{\boldsymbol{x}} I[\phi(\boldsymbol{x}) \geq j] \prod_{i=1}^{n} E\{r_i^{x_i}\} = p_\phi^j(E\{\boldsymbol{R}_\phi\}),$$

where

$$E\{\boldsymbol{R}_\phi\} = \{E\{r_i^j\}\}_{\substack{i=1,\ldots,n \\ j=0,\ldots,M}}.$$

Accordingly, one can arrive at an exact expression for $E\{p_\phi^j(\boldsymbol{R}_\phi)\}$ for not too large systems. The point is, however, that there seems to be no easy way to extend the approach above to give exact expressions for higher order moments of $p_\phi^j(\boldsymbol{R}_\phi)$. Hence, even when $I = [t, t]$ and component states are independent given $\boldsymbol{R}_\phi$, one needs bounds on higher order moments of $p_\phi^j(\boldsymbol{R}_\phi)$.

We need the following theorem proved in Natvig and Eide (1987).

**Theorem 5.1:** If $Y_1, \ldots, Y_n$ are associated random variables such that $0 \leq Y_i \leq 1$, $i = 1, \ldots, n$, then, for $\alpha > 0$

$$E\left\{\left(\prod_{i=1}^{n} Y_i\right)^{\alpha}\right\} \geq \prod_{i=1}^{n} E\{(Y_i)^{\alpha}\} \tag{5.2}$$

$$E\left\{\coprod_{i=1}^{n} Y_i\right\} \leq \coprod_{i=1}^{n} E\{Y_i\}. \tag{5.3}$$

In the special case of independent random variables $Y_1$ and $Y_2$ with $0 \leq Y_i \leq 1$, $i = 1, 2$, we have

$$E\left\{\left(\coprod_{i=1}^{2} Y_i\right)^{2}\right\} \geq \coprod_{i=1}^{2} E\{(Y_i)^2\}. \tag{5.4}$$

**Proof:** For the case $Y_1, \ldots, Y_n$ binary and $\alpha = 1$, Equations (5.2) and (5.3) are proved in Theorem 3.1, page 32 of Barlow and Proschan (1975a). The proof, however, also works when $0 \leq Y_i \leq 1$, $i = 1, \ldots, n$. Using this fact and Property $P_3$ of associated random variables we get

$$E\left\{\left(\prod_{i=1}^{n} Y_i\right)^{\alpha}\right\} = E\{\prod_{i=1}^{n}(Y_i)^{\alpha}\} \geq \prod_{i=1}^{n} E\{(Y_i)^{\alpha}\},$$

and Equation (5.2) is proved. Equation (5.3) is proved in the same way. Finally, Equation (5.4) follows since

$$\left(\coprod_{i=1}^{2} Y_i\right)^2 = \coprod_{i=1}^{2} Y_i^2 + 2(Y_1^2 - Y_1)(Y_2^2 - Y_2) \geq \coprod_{i=1}^{2} Y_i^2,$$

and that $Y_1$ and $Y_2$ are assumed to be independent.

Equation (5.4) reveals the unpleasant fact that a symmetry in Equations (5.2) and (5.3) seems only possible for $\alpha = 1$, when $Y_1, \ldots, Y_n$ are not binary.

Introduce the $m \times n \times M$ arrays of component availability and unavailability moments

$$E\{(\boldsymbol{P}_\phi^{(I)})^m\} = \{E\{(p_i^{j(I)})^s\}\}_{\substack{s=1,\ldots,m \\ i=1,\ldots,n \\ j=1,\ldots,M}} \tag{5.5}$$

$$E\{(\boldsymbol{Q}_\phi^{(I)})^m\} = \{E\{(q_i^{j(I)})^s\}\}_{\substack{s=1,\ldots,m \\ i=1,\ldots,n \\ j=1,\ldots,M}} \,. \tag{5.6}$$

**Theorem 5.2:** Let $(C, \phi)$ be an MMS. Furthermore, for $j \in \{1, \ldots, M\}$ let $\boldsymbol{y}_k^j = (y_{1k}^j, \ldots, y_{nk}^j)$, $k = 1, \ldots, n_\phi^j$ be its minimal path vectors to level $j$ and $\boldsymbol{z}_k^j = (z_{1k}^j, \ldots, z_{nk}^j)$, $k = 1, \ldots, m_\phi^j$ be its minimal cut vectors to level $j$ and

$$C_\phi^j(\boldsymbol{y}_k^j), k = 1, \ldots, n_\phi^j \text{ and } D_\phi^j(\boldsymbol{z}_k^j), k = 1, \ldots, m_\phi^j$$

the corresponding minimal path and cut sets to level $j$. Assume that the component availability vectors $\boldsymbol{p}_i^{(I)}, i = 1, \ldots, n$ and the component unavailability vectors $\boldsymbol{q}_i^{(I)}$, $i = 1, \ldots, n$ are independent. Let

$$\ell_\phi^{\prime j(I)m}(E\{(\boldsymbol{P}_\phi^{(I)})^m\}) = \max_{1 \leq k \leq n_\phi^j} \prod_{i \in C_\phi^j(\boldsymbol{y}_k^j)} E\{(p_i^{y_{ik}^j(I)})^m\}$$

$$u_\phi^{\prime j(I)m}(E\{(\boldsymbol{Q}_\phi^{(I)})^m\}) = \min_{1 \leq k \leq m_\phi^j} \sum_{r=o}^{m} \binom{m}{r} (-1)^r \prod_{i \in D_\phi^j(\boldsymbol{z}_k^j)} E\{(q_i^{z_{ik}^j+1(I)})^r\}$$

$$\bar{\ell}_{\phi}^{\prime j(I)m}(E\{(\boldsymbol{Q}_{\phi}^{(I)})^m\}) = \max_{1\leq k\leq m_{\phi}^{j}} \prod_{i\in D_{\phi}^{j}(\boldsymbol{z}_k^j)} E\{(q_i^{z_{ik}^j+1(I)})^m\}$$

$$\bar{u}_{\phi}^{\prime j(I)m}(E\{(\boldsymbol{P}_{\phi}^{(I)})^m\}) = \min_{1\leq k\leq n_{\phi}^{j}} \sum_{r=o}^{m} \binom{m}{r}(-1)^r \prod_{i\in C_{\phi}^{j}(\boldsymbol{y}_k^j)} E\{(p_i^{y_{ik}^j(I)})^r\}.$$

If the joint performance process for the system's components is associated in $I$, or the marginal performance processes of the components are independent in $I$, then, for $m = 1, 2, \ldots$

$$\ell_{\phi}^{\prime j(I)m}(E\{(\boldsymbol{P}_{\phi}^{(I)})^m\}) \leq E\{(p_{\phi}^{j(I)})^m\}$$
$$E\{(p_{\phi}^{j(I)})^m\} \leq u_{\phi}^{\prime j(I)m}(E\{(\boldsymbol{Q}_{\phi}^{(I)})^m\}) \tag{5.7}$$
$$\bar{\ell}_{\phi}^{\prime j(I)m}(E\{(\boldsymbol{Q}_{\phi}^{(I)})^m\}) \leq E\{(q_{\phi}^{j(I)})^m\}$$
$$E\{(q_{\phi}^{j(I)})^m\} \leq \bar{u}_{\phi}^{\prime j(I)m}(E\{(\boldsymbol{P}_{\phi}^{(I)})^m\}). \tag{5.8}$$

**Proof:** From Equation (3.8) we have

$$E\{(p_{\phi}^{j(I)})^m\} \geq E\left\{\max_{1\leq k\leq n_{\phi}^{j}} \prod_{i\in C_{\phi}^{j}(\boldsymbol{y}_k^j)} (p_i^{y_{ik}^j(I)})^m\right\} \geq E\left\{\prod_{i\in C_{\phi}^{j}(\boldsymbol{y}_k^j)} (p_i^{y_{ik}^j(I)})^m\right\}$$
$$= \prod_{i\in C_{\phi}^{j}(\boldsymbol{y}_k^j)} E\{(p_i^{y_{ik}^j(I)})^m\}, \quad 1 \leq k \leq n_{\phi}^{j},$$

having used the independence of the component availability vectors. Since the inequality holds for all $1 \leq k \leq n_{\phi}^{j}$, the lower bound of Equation (5.7) follows. Similarly, from Equation (3.8)

$$E\{(p_{\phi}^{j(I)})^m\} \leq E\left\{\left(\min_{1\leq k\leq m_{\phi}^{j}}\left[1 - \prod_{i\in D_{\phi}^{j}(\boldsymbol{z}_k^j)} q_i^{z_{ik}^j+1(I)}\right]\right)^m\right\}$$
$$\leq \min_{1\leq k\leq m_{\phi}^{j}} E\left\{\left(1 - \prod_{i\in D_{\phi}^{j}(\boldsymbol{z}_k^j)} q_i^{z_{ik}^j+1(I)}\right)^m\right\}$$

$$= \min_{1 \le k \le m_\phi^j} E\left\{ \sum_{r=o}^{m} \binom{m}{r} (-1)^r \prod_{i \in D_\phi^j(\boldsymbol{z}_k^j)} (q_i^{z_{ik}^j+1(I)})^r \right\}$$

$$= \min_{1 \le k \le m_\phi^j} \sum_{r=o}^{m} \binom{m}{r} (-1)^r \prod_{i \in D_\phi^j(\boldsymbol{z}_k^j)} E\{(q_i^{z_{ik}^j+1(I)})^r\},$$

having used the independence of the component unavailability vectors. Hence, the upper bound of Equation (5.7) is proved. The bounds of Equation (5.8) follow completely similarly from Equation (3.9).

Note that respectively $p_i^{j([t_1,t_1])}$ and $p_i^{j([t_2,t_2])}$, and $q_i^{j([t_1,t_1])}$ and $q_i^{j([t_2,t_2])}$ are dependent for $t_1 \in \tau(I)$, $t_2 \in \tau(I)$, $t_1 \neq t_2$. Hence,

$$E(p_i^{j([t,t])} | \boldsymbol{D}^{1(I)}) \neq E(p_i^{j([t,t])} | \boldsymbol{D}^1)$$

$$E(q_i^{j([t,t])} | \boldsymbol{D}^{2(I)}) \neq E(q_i^{j([t,t])} | \boldsymbol{D}^2).$$

This means that we cannot apply the best upper bounds in Equations (3.8) and (3.9).

**Theorem 5.3:** Let $(C, \phi)$ be an MMS. Furthermore, for $j \in \{1, \ldots, M\}$ let $\boldsymbol{y}_k^j = (y_{1k}^j, \ldots, y_{nk}^j)$, $k = 1, \ldots, n_\phi^j$ be its minimal path vectors to level $j$ and $\boldsymbol{z}_k^j = (z_{1k}^j, \ldots, z_{nk}^j)$, $k = 1, \ldots, m_\phi^j$ be its minimal cut vectors to level $j$ and

$$C_\phi^j(\boldsymbol{y}_k^j), k = 1, \ldots, n_\phi^j \text{ and } D_\phi^j(\boldsymbol{z}_k^j), k = 1, \ldots, m_\phi^j$$

the corresponding minimal path and cut sets to level $j$. Assume that the component availability vectors $\boldsymbol{p}_i^{(I)}, i = 1, \ldots, n$ and the component unavailability vectors $\boldsymbol{q}_i^{(I)}, i = 1, \ldots, n$ are independent. Let

$$\ell_\phi^{**j(I)m}(E\{(\boldsymbol{P}_\phi^{(I)})^m\}) = \prod_{k=1}^{m_\phi^j} \sum_{r=o}^{m} \binom{m}{r} (-1)^r \prod_{i \in D_\phi^j(\boldsymbol{z}_k^j)} \sum_{s=o}^{r} \binom{r}{s} (-1)^s E\{(p_i^{z_{ik}^j+1(I)})^s\}$$

$$u_\phi^{**j(I)1}(E\{(\boldsymbol{Q}_\phi^{(I)})^1\}) = \coprod_{k=1}^{n_\phi^j} \prod_{i \in C_\phi^j(\boldsymbol{y}_k^j)} (1 - E\{q_i^{y_{ik}^j(I)}\})$$

$$\bar{\ell}_{\phi}^{**j(I)m}(E\{(\boldsymbol{Q}_{\phi}^{(I)})^m\}) = \prod_{k=1}^{n_{\phi}^j} \sum_{r=o}^{m} \binom{m}{r} (-1)^r$$

$$\prod_{i \in C_{\phi}^j(\boldsymbol{y}_k^j)} \sum_{s=o}^{r} \binom{r}{s} (-1)^s E\{(q_i^{y_{ik}^j(I)})^s\}$$

$$\bar{u}_{\phi}^{**j(I)1}(E\{(\boldsymbol{P}_{\phi}^{(I)})^1\}) = \coprod_{k=1}^{m_{\phi}^j} \prod_{i \in D_{\phi}^j(\boldsymbol{z}_k^j)} (1 - E\{p_i^{z_{ik}^j+1(I)}\}).$$

If the marginal performance processes of the components are independent in $I$, then, for $m = 1, 2, \ldots$

$$\ell_{\phi}^{**j(I)m}(E\{(\boldsymbol{P}_{\phi}^{(I)})^m\}) \leq E\{(p_{\phi}^{j(I)})^m\} \tag{5.9}$$

$$E\{p_{\phi}^{j(I)}\} \leq u_{\phi}^{**j(I)1}(E\{(\boldsymbol{Q}_{\phi}^{(I)})^1\}) \tag{5.10}$$

$$\bar{\ell}_{\phi}^{**j(I)m}(E\{(\boldsymbol{Q}_{\phi}^{(I)})^m\}) \leq E\{(q_{\phi}^{j(I)})^m\} \tag{5.11}$$

$$E\{q_{\phi}^{j(I)}\} \leq \bar{u}_{\phi}^{**j(I)1}(E\{(\boldsymbol{P}_{\phi}^{(I)})^1\}). \tag{5.12}$$

**Proof:** From Equation (3.12) we have

$$E\{(p_{\phi}^{j(I)})^m\} \geq E\left\{\left(\prod_{k=1}^{m_{\phi}^j} \coprod_{i \in D_{\phi}^j(\boldsymbol{z}_k^j)} p_i^{z_{ik}^j+1(I)}\right)^m\right\}$$

$$\geq \prod_{k=1}^{m_{\phi}^j} E\left\{\left(\coprod_{i \in D_{\phi}^j(\boldsymbol{z}_k^j)} p_i^{z_{ik}^j+1(I)}\right)^m\right\},$$

having applied Equation (5.2). The random variables

$$\coprod_{i \in D_{\phi}^j(\boldsymbol{z}_k^j)} p_i^{z_{ik}^j+1(I)}, \quad k = 1, \ldots, m_{\phi}^j,$$

are associated by Properties $P_5$ and $P_3$ of associated random variables, having used the independence of the component availability vectors.

Continuing the derivation we get

$$= \prod_{k=1}^{m_\phi^j} E\left\{\sum_{r=o}^{m}\binom{m}{r}(-1)^r \prod_{i\in D_\phi^j(\mathbf{z}_k^j)} \sum_{s=o}^{r}\binom{r}{s}(-1)^s (p_i^{z_{ik}^j+1(I)})^s\right\}$$

$$= \prod_{k=1}^{m_\phi^j}\sum_{r=o}^{m}\binom{m}{r}(-1)^r \prod_{i\in D_\phi^j(\mathbf{z}_k^j)} \sum_{s=o}^{r}\binom{r}{s}(-1)^s E\{(p_i^{z_{ik}^j+1(I)})^s\},$$

again having applied the independence of the component availability vectors. Hence, Equation (5.9) follows. Similarly, from Equation (3.12)

$$E\{p_\phi^{j(I)}\} \le E\left\{\coprod_{k=1}^{n_\phi^j} \prod_{i\in C_\phi^j(\mathbf{y}_k^j)} (1 - q_i^{y_{ik}^j(I)})\right\}$$

$$\le \coprod_{k=1}^{n_\phi^j} \prod_{i\in C_\phi^j(\mathbf{y}_k^j)} (1 - E\{q_i^{y_{ik}^j(I)}\}),$$

having used Equation (5.3), noting that the random variables

$$\prod_{i\in C_\phi^j(\mathbf{y}_k^j)} (1 - q_i^{y_{ik}^j(I)}),\ k = 1, \ldots, n_\phi^j,$$

are associated by the same argument as above, and the independence of the component unavailability vectors. Hence, Equation (5.10) is proved. The bounds of Equations (5.11) and (5.12) follow completely similarly from Equation (3.13).

Due to the lack of symmetry in Theorem 5.1, we have not been able to obtain corresponding upper bounds for $E\{(p_\phi^{j(I)})^m\}$ and $E\{(q_\phi^{j(I)})^m\}$, $m = 2, 3, \ldots$ in this theorem. As for Theorem 5.2 we cannot apply the best upper bounds in Equations (3.12) and (3.13).

**Corollary 5.4:** Make the same assumptions as in Theorem 5.3 and let

$$L_\phi^{*j(I)m}(E\{(\mathbf{P}_\phi^{(I)})^m\})$$
$$= \max[\ell_\phi^{\prime j(I)m}(E\{(\mathbf{P}_\phi^{(I)})^m\}), \ell_\phi^{**j(I)m}(E\{(\mathbf{P}_\phi^{(I)})^m\})]$$

$$U_\phi^{*j(I)1}(E\{(\boldsymbol{Q}_\phi^{(I)})^1\})$$
$$= \min[u_\phi^{\prime j(I)1}(E\{(\boldsymbol{Q}_\phi^{(I)})^1\}), u_\phi^{**j(I)1}(E\{(\boldsymbol{Q}_\phi^{(I)})^1\})]$$
$$\bar{L}_\phi^{*j(I)m}(E\{(\boldsymbol{Q}_\phi^{(I)})^m\})$$
$$= \max[\bar{\ell}_\phi^{\prime j(I)m}(E\{(\boldsymbol{Q}_\phi^{(I)})^m\}), \bar{\ell}_\phi^{**j(I)m}(E\{(\boldsymbol{Q}_\phi^{(I)})^m\})]$$
$$\bar{U}_\phi^{*j(I)1}(E\{(\boldsymbol{P}_\phi^{(I)})^1\})$$
$$= \min[\bar{u}_\phi^{\prime j(I)1}(E\{(\boldsymbol{P}_\phi^{(I)})^1\}), \bar{u}_\phi^{**j(I)1}(E\{(\boldsymbol{P}_\phi^{(I)})^1\})].$$

Then, for $m = 1, 2, \ldots$

$$L_\phi^{*j(I)m}(E\{(\boldsymbol{P}_\phi^{(I)})^m\}) \leq E\{(p_\phi^{j(I)})^m\} \tag{5.13}$$
$$E\{p_\phi^{j(I)}\} \leq U_\phi^{*j(I)1}(E\{(\boldsymbol{Q}_\phi^{(I)})^1\}) \tag{5.14}$$
$$\bar{L}_\phi^{*j(I)m}(E\{(\boldsymbol{Q}_\phi^{(I)})^m\}) \leq E\{(q_\phi^{j(I)})^m\} \tag{5.15}$$
$$E\{q_\phi^{j(I)}\} \leq \bar{U}_\phi^{*j(I)1}(E\{(\boldsymbol{P}_\phi^{(I)})^1\}). \tag{5.16}$$

**Corollary 5.5:** Make the same assumptions as in Theorem 5.3 and let

$$B_\phi^{*j(I)m}(E\{(\boldsymbol{P}_\phi^{(I)})^m\}) = \max_{j\leq k\leq M}[L_\phi^{*k(I)m}(E\{(\boldsymbol{P}_\phi^{(I)})^m\})]$$
$$C_\phi^{*j(I)1}(E\{(\boldsymbol{Q}_\phi^{(I)})^1\}) = \min_{1\leq k\leq j}[U_\phi^{*k(I)1}(E\{(\boldsymbol{Q}_\phi^{(I)})^1\})]$$
$$D_\phi^{*j(I)m}(E\{(\boldsymbol{Q}_\phi^{(I)})^m\}) = \min_{1\leq k\leq j}[u_\phi^{\prime k(I)m}(E\{(\boldsymbol{Q}_\phi^{(I)})^m\})]$$
$$\bar{B}_\phi^{*j(I)m}(E\{(\boldsymbol{Q}_\phi^{(I)})^m\}) = \max_{1\leq k\leq j}[\bar{L}_\phi^{*k(I)m}(E\{(\boldsymbol{Q}_\phi^{(I)})^m\})]$$
$$\bar{C}_\phi^{*j(I)1}(E\{(\boldsymbol{P}_\phi^{(I)})^1\}) = \min_{j\leq k\leq M}[\bar{U}_\phi^{*k(I)1}(E\{(\boldsymbol{P}_\phi^{(I)})^1\})]$$
$$\bar{D}_\phi^{*j(I)m}(E\{(\boldsymbol{P}_\phi^{(I)})^m\}) = \min_{j\leq k\leq M}[\bar{u}_\phi^{\prime k(I)m}(E\{(\boldsymbol{P}_\phi^{(I)})^m\})].$$

Then, for $m = 1, 2, \ldots$

$$L_\phi^{*j(I)m}(E\{(\boldsymbol{P}_\phi^{(I)})^m\}) \leq B_\phi^{*j(I)m}(E\{(\boldsymbol{P}_\phi^{(I)})^m\}) \leq E\{(p_\phi^{j(I)})^m\} \tag{5.17}$$
$$E\{p_\phi^{j(I)}\} \leq C_\phi^{*j(I)1}(E\{(\boldsymbol{Q}_\phi^{(I)})^1\}) \leq U_\phi^{*j(I)1}(E\{(\boldsymbol{Q}_\phi^{(I)})^1\}) \tag{5.18}$$
$$E\{(p_\phi^{j(I)})^m\} \leq D_\phi^{*j(I)m}(E\{(\boldsymbol{Q}_\phi^{(I)})^m\}) \leq u_\phi^{\prime j(I)m}(E\{(\boldsymbol{Q}_\phi^{(I)})^m\}) \tag{5.19}$$

$$\bar{L}_{\phi}^{*j(I)m}(E\{(\boldsymbol{Q}_{\phi}^{(I)})^m\}) \leq \bar{B}_{\phi}^{*j(I)m}(E\{(\boldsymbol{Q}_{\phi}^{(I)})^m\}) \leq E\{(q_{\phi}^{j(I)})^m\} \tag{5.20}$$

$$E\{q_{\phi}^{j(I)}\} \leq \bar{C}_{\phi}^{*j(I)1}(E\{(\boldsymbol{P}_{\phi}^{(I)})^1\}) \leq \bar{U}_{\phi}^{*j(I)1}(E\{(\boldsymbol{P}_{\phi}^{(I)})^1\}) \tag{5.21}$$

$$E\{(q_{\phi}^{j(I)})^m\} \leq \bar{D}_{\phi}^{*j(I)m}(E\{(\boldsymbol{P}_{\phi}^{(I)})^m\}) \leq \bar{u}_{\phi}^{\prime j(I)m}(E\{(\boldsymbol{P}_{\phi}^{(I)})^m\}). \tag{5.22}$$

It is important to note that the bounds for $E\{(p_{\phi}^{j(I)})\}$ and $E\{(q_{\phi}^{j(I)})\}$ given in this section equal the bounds in Section 3.2 by replacing $\boldsymbol{P}_{\phi}^{(I)}$ by $E\{(\boldsymbol{P}_{\phi}^{(I)})^1\}$ and $\boldsymbol{Q}_{\phi}^{(I)}$ by $E\{(\boldsymbol{Q}_{\phi}^{(I)})^1\}$. However, this is not true for higher order moments.

## 5.4 A simulation approach and an application to the simple network system

An objection against the bounds in Theorems 5.2 and 5.3 and Corollaries 5.4 and 5.5 is that they are based on knowing all minimal path and cut vectors of the system. Referring to the discussion at the end of Section 3.1, it is natural to try to both improve the bounds and reduce the computational complexity by introducing modular decompositions.

Looking at the bounds for $E\{(p_{\phi}^{j(I)})^m\}$ and $E\{(q_{\phi}^{j(I)})^m\}$ for $m = 2, 3, \ldots$, it seems that only the lower bounds of Theorem 5.2 are of the form that fits into the machinery of Section 3.3. We now get the following theorem

**Theorem 5.6:** Let $(C, \phi)$ be an MMS with modular decomposition given by Definition 1.6. Make the same assumptions as in Theorem 5.3. Then, for $j = 1, \ldots, M$, $m = 1, 2, \ldots$

$$\ell_{\phi}^{\prime j(I)}(E\{(\boldsymbol{P}_{\phi}^{(I)})^m\}) = \ell_{\psi}^{\prime j(I)}(\boldsymbol{\ell}_{\psi}^{\prime(I)}(E\{(\boldsymbol{P}_{\phi}^{(I)})^m\}) \leq E\{(p_{\phi}^{j(I)})^m\} \tag{5.23}$$

$$\bar{\ell}_{\phi}^{\prime j(I)}(E\{(\boldsymbol{Q}_{\phi}^{(I)})^m\}) = \bar{\ell}_{\psi}^{\prime j(I)}(\bar{\boldsymbol{\ell}}_{\psi}^{\prime(I)}(E\{(\boldsymbol{Q}_{\phi}^{(I)})^m\}) \leq E\{(q_{\phi}^{j(I)})^m\}. \tag{5.24}$$

**Proof:** Equation (5.23) follows from the lower bound of Equation (5.7) and from Equation (3.45), by replacing $\boldsymbol{P}_{\phi}^{(I)}$ by $E\{(\boldsymbol{P}_{\phi}^{(I)})^m\}$ in the last expression. Note that in this case the latter array given by Equation (5.5) can be replaced by an $n \times M$ matrix by fixing $s$ at $m$. Hence, an $n \times M$ matrix is replaced by an $n \times M$ matrix. Equation (5.24) follows by applying Equation (5.23) on the dual structure and dual level, remembering Equations (3.2), (3.5) and (3.28).

Hence, our analytical bounds are not improved by using a modular decomposition. On the other hand, the computational complexity is reduced since we have to find minimal path and cut vectors only for each module and for the organizing structure.

All analytical bounds on the marginal moments $E\{(p_\phi^{j(I)})^m|\boldsymbol{D}^{1(I)}\}$ and $E\{(q_\phi^{j(I)})^m|\boldsymbol{D}^{2(I)}\}$ for $m = 1, \ldots, j = 1, \ldots, M$ given in Section 5.3 can be improved by straightforward simulation techniques. Let us illustrate this on the lower bounds in Equation (5.7). As in the proof of Theorem 5.2, with full notation, we have, from Equation (3.8)

$$E\{(p_\phi^{j(I)})^m|\boldsymbol{D}^{1(I)}\} \geq E\left\{ \max_{1\leq k\leq n_\phi^j} \prod_{i\in C_\phi^j(\boldsymbol{y}_k^j)} (p_i^{y_{ik}^j(I)})^m|\boldsymbol{D}^{1(I)}\right\}. \tag{5.25}$$

For $i = 1, \ldots, n$ we simulate from the posterior marginal distribution of $\boldsymbol{r}_i^{(I)}$ given the data $\boldsymbol{D}_i^{(I)}, \pi_i(\boldsymbol{r}_i^{(I)}|\boldsymbol{D}_i^{(I)})$, which is assumed to be Dirichlet. We then calculate $p_i^{j(I)}$ from Equation (5.1) for $i = 1, \ldots, n$, $j = 1, \ldots, M$. For each round of $n$ simulations the quantity $\max_{1\leq k\leq n_\phi^j} \prod_{i\in C_\phi^j(\boldsymbol{y}_k^j)} (p_i^{y_{ik}^j(I)})^m$ is calculated, and the right-hand side of Equation (5.25) is estimated by the average of the simulated quantities. Theoretically, as seen from the proof, this improves the lower bound of Equation (5.7) of Theorem 5.2. Similarly, we obtain a simulated lower bound which improves the lower bound of Equation (5.9) of Theorem 5.3. In practice, the analytic bounds may be marginally better due to simulation uncertainty.

This simulation technique can also be applied to arrive at improved bounds using modular decompositions. From Theorem 3.14, for instance, we get the following inequalities as starting points for the simulations.

**Corollary 5.7:** Let $(C, \phi)$ be an MMS with modular decomposition given by Definition 1.6. Assume the marginal performance processes of the components to be independent in the time interval $I$. Then, for $j = 1, \ldots, M$

$$E\{(B_\psi^{*j(I)}(\boldsymbol{B}_\psi^{*(I)}(\boldsymbol{P}_\phi^{(I)})))^m|\boldsymbol{D}^{1(I)}\} \leq E\{(p_\phi^j)^m|\boldsymbol{D}^{1(I)}\}$$
$$E\{(p_\phi^j)^m|\boldsymbol{D}^{2(I)}\} \leq E\{(1-\bar{B}_\psi^{*j(I)}(\bar{\boldsymbol{B}}_\psi^{*(I)}(\boldsymbol{Q}_\phi^{(I)})))^m|\boldsymbol{D}^{2(I)}\} \tag{5.26}$$

$$E\{(\bar{B}_{\psi}^{*j(I)}(\bar{\boldsymbol{B}}_{\psi}^{*(I)}(\boldsymbol{Q}_{\phi}^{(I)})))^{m}|\boldsymbol{D}^{2(I)}\} \leq E\{(q_{\phi}^{j})^{m}|\boldsymbol{D}^{2(I)}\}$$
$$E\{(q_{\phi}^{j})^{m}|\boldsymbol{D}^{1(I)}\} \leq E\{(1-B_{\psi}^{*j(I)}(\boldsymbol{B}_{\psi}^{*(I)}(\boldsymbol{P}_{\phi}^{(I)})))^{m}|\boldsymbol{D}^{1(I)}\}. \quad (5.27)$$

To illustrate the theory consider the simple network system depicted in Figure 1.1. Following the argument leading to Equation (5.1) we assume that the posterior marginal distribution of $p_i^{j(I)}$ given the data $D_i^{1j(I)}$ is beta with parameters $\alpha E\{p_i^{j(I)}\}$ and $\alpha(1 - E\{p_i^{j(I)}\})$. Hence,

$$\mathrm{Var}\{p_i^{j(I)}\} = E\{p_i^{j(I)}\}(1 - E\{p_i^{j(I)}\})/(\alpha + 1),$$

and the second order moment is given by

$$E\{(p_i^{j(I)})^2\} = E\{p_i^{j(I)}\}(1 + E\{p_i^{j(I)}\}\alpha)/(\alpha + 1).$$

$E\{p_i^{j(I)}\}$ is chosen to be equal to the value of $p_i^{j(I)}$ calculated by a standard deterministic analysis as indicated at the beginning of Section 3.7. In these calculations the marginal performance processes of the two modules are assumed independent in the time interval $I$ and we also assume that the two components of each module fail and are repaired/replaced independently of each other. All components have the same instantaneous failure rate $\lambda = 0.001$ and repair/replacement rate $\mu = 0.01$.

In Table 5.1 the analytical lower bounds from Corollary 5.5, $B_{\phi}^{*j(I)m}(E\{(\boldsymbol{P}_{\phi}^{(I)})^m\})$, for $E\{(p_{\phi}^{j(I)})^m\}$ for $m = 1, 2$ and the corresponding simulated lower bounds, both not using and using modular decompositions, are calculated for the time interval $I$ equal to [100, 110], [100, 200], [1000, 1100], the parameter $\alpha$ equal to 1, 10, 1000 and for system level $j$ equal to 1, 2, 3. Looking at the lower bounds for $E\{(p_{\phi}^{j(I)})\}$ there are just minor differences between the analytical and the simulated bounds that are not based on modular decompositions, except for $\alpha = 1$, $j = 2$ and the two longest intervals [100, 200] and [1000, 1100], where the improvements are quite small. Correspondingly, there are just minor improvements of the simulated bounds that are based on modular decompositions compared to the ones that are not, except for $\alpha = 10$, 1000, $j = 2$ and the two longest intervals [100, 200] and [1000, 1100], where the improvements are quite small. Furthermore, the analytical bounds do not depend on the $\alpha$ parameter, which is natural

Table 5.1 Lower bounds of the simple network system of Figure 1.1.

| | | | Analytical | | Sim. minus m.d. | | Sim. plus m.d. | |
|---|---|---|---|---|---|---|---|---|
| $I$ | $\alpha$ | $j$ | 1. m. | 2. m. | 1. m. | 2. m. | 1. m. | 2. m. |
| [100, 110] | 1 | 1 | 0.9902 | 0.9833 | 0.9902 | 0.9833 | 0.9902 | 0.9833 |
| [100, 110] | 1 | 2 | 0.9710 | 0.9507 | 0.9726 | 0.9546 | 0.9730 | 0.9551 |
| [100, 110] | 1 | 3 | 0.7481 | 0.6487 | 0.7481 | 0.6487 | 0.7481 | 0.6487 |
| [100, 110] | 10 | 1 | 0.9902 | 0.9807 | 0.9902 | 0.9807 | 0.9902 | 0.9807 |
| [100, 110] | 10 | 2 | 0.9710 | 0.9433 | 0.9713 | 0.9446 | 0.9724 | 0.9465 |
| [100, 110] | 10 | 3 | 0.7481 | 0.5751 | 0.7481 | 0.5752 | 0.7481 | 0.5752 |
| [100, 110] | 1000 | 1 | 0.9902 | 0.9805 | 0.9902 | 0.9805 | 0.9902 | 0.9805 |
| [100, 110] | 1000 | 2 | 0.9710 | 0.9428 | 0.9710 | 0.9428 | 0.9722 | 0.9451 |
| [100, 110] | 1000 | 3 | 0.7481 | 0.5598 | 0.7481 | 0.5598 | 0.7481 | 0.5598 |
| [100, 200] | 1 | 1 | 0.9555 | 0.9263 | 0.9555 | 0.9262 | 0.9555 | 0.9262 |
| [100, 200] | 1 | 2 | 0.8723 | 0.7947 | 0.8857 | 0.8226 | 0.8884 | 0.8256 |
| [100, 200] | 1 | 3 | 0.5219 | 0.3821 | 0.5217 | 0.3819 | 0.5217 | 0.3819 |
| [100, 200] | 10 | 1 | 0.9555 | 0.9142 | 0.9555 | 0.9142 | 0.9555 | 0.9142 |
| [100, 200] | 10 | 2 | 0.8723 | 0.7641 | 0.8751 | 0.7731 | 0.8834 | 0.7865 |
| [100, 200] | 10 | 3 | 0.5219 | 0.2903 | 0.5219 | 0.2903 | 0.5219 | 0.2903 |
| [100, 200] | 1000 | 1 | 0.9555 | 0.9130 | 0.9555 | 0.9130 | 0.9555 | 0.9130 |
| [100, 200] | 1000 | 2 | 0.8723 | 0.7610 | 0.8723 | 0.7611 | 0.8819 | 0.7778 |
| [100, 200] | 1000 | 3 | 0.5219 | 0.2726 | 0.5219 | 0.2726 | 0.5219 | 0.2726 |
| [1000, 1100] | 1 | 1 | 0.9380 | 0.8986 | 0.9381 | 0.8987 | 0.9381 | 0.8987 |
| [1000, 1100] | 1 | 2 | 0.8254 | 0.7256 | 0.8460 | 0.7663 | 0.8500 | 0.7705 |
| [1000, 1100] | 1 | 3 | 0.4578 | 0.3157 | 0.4580 | 0.3159 | 0.4580 | 0.3159 |
| [1000, 1100] | 10 | 1 | 0.9380 | 0.8818 | 0.9380 | 0.8818 | 0.9380 | 0.8818 |
| [1000, 1100] | 10 | 2 | 0.8254 | 0.6857 | 0.8296 | 0.6987 | 0.8422 | 0.7178 |
| [1000, 1100] | 10 | 3 | 0.4578 | 0.2265 | 0.4578 | 0.2266 | 0.4578 | 0.2266 |
| [1000, 1100] | 1000 | 1 | 0.9380 | 0.8800 | 0.9380 | 0.8799 | 0.9380 | 0.8799 |
| [1000, 1100] | 1000 | 2 | 0.8254 | 0.6813 | 0.8254 | 0.6815 | 0.8400 | 0.7057 |
| [1000, 1100] | 1000 | 3 | 0.4578 | 0.2098 | 0.4578 | 0.2098 | 0.4578 | 0.2098 |

since $E\{p_i^{j(I)}\}$ is independent of this parameter. That these lower bounds are decreasing in the length of the interval $I$ and the system state $j$ is just a reflection of the fact that these properties hold for $E\{(p_\phi^{j(I)})\}$.

Turning to the lower bounds for $E\{(p_\phi^{j(I)})^2\}$ there are again just minor differences between the analytical and the simulated bounds that are not based on modular decompositions, again except for $\alpha = 1, j = 2$ and the two longest intervals $[100, 200]$ and $[1000, 1100]$, where the improvements are quite small. Correspondingly, there are just minor improvements of the simulated bounds that are based on modular decompositions compared to the ones that are not, except for $\alpha = 10, 1000$, $j = 2$ and the two longest intervals $[100, 200]$ and $[1000, 1100]$, where the improvements are quite small. Furthermore, these bounds are decreasing in the $\alpha$ parameter, which is natural since $E\{(p_i^{j(I)})^2\}$ is decreasing in this parameter. That these lower bounds are decreasing in the length of the interval $I$ and the system state $j$ is just a reflection of the fact that these properties also hold for $E\{(p_\phi^{j(I)})^2\}$. Finally, it should be noted that combining the lower bounds for $E\{(p_\phi^{j(I)})\}$ and $E\{(p_\phi^{j(I)})^2\}$ does not lead to a lower bound for $\text{Var}\{(p_\phi^{j(I)})\}$. However, for the analytical lower bounds it is revealing that this leads to positive variances. For the corresponding simulated lower bounds this is obviously the case. Finally, it should be mentioned that in contrast to the analytical bounds, the simulated bounds are improved by using modular decompositions.

# 6

# Measures of importance of system components

## 6.1 Introduction

There seem to be two main reasons for coming up with a measure of the importance of system components. Reason 1: it permits the analyst to determine which components merit the most additional research and development to improve overall system reliability at minimal cost or effort. Reason 2: it may suggest the most efficient way to diagnose system failure by generating a repair checklist for an operator to follow. It should be noted that no measure of importance can be expected to be universally best irrespective of usage purpose. In this chapter we will concentrate on what could be considered all-round measures of component importance.

In Natvig and Gåsemyr (2009) dynamic and stationary measures of importance of a component in a binary system were considered. To arrive at explicit results the performance processes of the components were assumed to be independent and the system to be coherent. In particular, the Barlow–Proschan and the Natvig measures were treated in detail and a series of new results and approaches were given. For the case of components not undergoing repair, it was shown that both measures are sensible. Reasonable measures of component importance for repairable systems represent a challenge. A basic idea here is also to take a so-called dual term into account. For a binary coherent system, according to the extended Barlow–Proschan measure, a component is important if there are high probabilities both that its failure is the cause of system failure and that its repair is the cause of system repair. Even with this extension, results for the stationary Barlow–Proschan measure are not satisfactory. For a binary coherent system, according to the extended Natvig measure, a component is important if both by failing it strongly reduces the expected system uptime and by being repaired it strongly reduces the expected system downtime. With this extension the results for the stationary Natvig measure seem very sensible. In Natvig *et al.* (2009) a thorough numerical analysis of the Natvig measures for repairable systems is reported along with an application to an offshore oil and gas production system. The analysis is based on advanced simulation methods presented in Huseby *et al.* (2010). In the present chapter, covering the paper Natvig (2011), most results from Natvig and Gåsemyr (2009) are generalized to multistate strongly coherent systems. For such systems little has been published until now on measures of component importance even in the nonrepairable case.

We now consider the relation between the stochastic performance of the system $(C, \phi)$ and the stochastic performances of the components. As before, $X_i(t)$ is the random state of the $i$th component at time $t$, $i = 1, \dots, n$ with the corresponding random vector $\boldsymbol{X}(t) = (X_1(t), \dots, X_n(t))$. Now if $\phi$ is a multistate structure function, $\phi(\boldsymbol{X}(t))$ is the corresponding random system state at time $t$. Assume also that the stochastic processes $\{X_i(t), t \in [0, \infty)\}, i = 1, \dots, n$, are mutually independent. For the dynamic approach of the present chapter this is a necessary assumption in order to arrive at explicit results.

The chapter is organized as follows. In Section 6.2 the Birnbaum, Barlow–Proschan and Natvig measures of component importance in nonrepairable systems are considered. The Birnbaum and Barlow–Proschan measures of component importance in repairable systems, the latter with its dual extension, are treated in Section 6.3. The corresponding Natvig measure with its dual extension is treated in Section 6.4. Finally, some concluding remarks are given in Section 6.5.

## 6.2 Measures of component importance in nonrepairable systems

In this section we restrict our attention to the case where the components, and hence the system, cannot be repaired. In order to avoid a rather complex notation we will, in the following, assume that $S_i = S, i = 1, \dots, n$. Furthermore, assume that $X_i(t), i = 1, \dots, n$ for $t \in [0, \infty)$, are Markov processes in continuous time and that at time $t = 0$ all components are in the perfect functioning state $M$; i.e. $\boldsymbol{X}(0) = \boldsymbol{M}$. Introduce the notation

$$\begin{aligned}
p_i^j(t) &= P(X_i(t) \geq j), \quad j = 1, \ldots, M \\
r_i^j(t) &= P(X_i(t) = j), \quad j = 0, \ldots, M \\
\boldsymbol{r}(t) &= (r_1^1(t), \ldots, r_1^M(t), r_2^1(t), \ldots, r_n^M(t)) \\
p_i^{(k,\ell)}(t, t+u) &= P(X_i(t+u) = \ell \mid X_i(t) = k), \quad 0 \leq \ell < k \leq M \\
\lambda_i^{(k,\ell)}(t) &= \lim_{h \to 0} p_i^{(k,\ell)}(t, t+h)/h, \quad 0 \leq \ell < k \leq M
\end{aligned}$$

$$P[\phi(\boldsymbol{X}(t)) \geq j] = P[I(\phi(\boldsymbol{X}(t)) \geq j) = 1] = p_\phi^j(\boldsymbol{r}(t)),$$

where $I(\cdot)$ is the indicator function. $p_i^j(t)$ and $p_\phi^j(\boldsymbol{r}(t))$ are respectively the reliability to level $j$ of the $i$th component and the system at time $t$.

In order to make things not too complex we assume that

$$\lambda_i^{(k,\ell)}(t) = 0, \quad 0 \leq \ell < k - 1 \leq M - 1.$$

Hence, each component deteriorates by going through all states from the perfect functioning state until the complete failure state. Let the $i$th component have an absolutely continuous distribution $F_i^k(t)$ of time spent in state $k$, before jumping downwards to state $k - 1$, with density $f_i^k(t)$ and $\bar{F}_i^k(t) = 1 - F_i^k(t)$. It is assumed that all these times spent in the various states are independent. Finally, introduce the $M$-dimensional row vectors

$$\boldsymbol{e}^k = (1_k, \boldsymbol{0}), \; k = 1, \ldots, M \quad \boldsymbol{e}^0 = \boldsymbol{0}.$$

### 6.2.1 The Birnbaum measure

We now have the following generalization of $I_B^{(i)}(t)$, the Birnbaum (1969) measure of the importance of the $i$th component at time $t$

$$\begin{aligned}
I_B^{(i,k,j)}(t) = P[&\text{The system is in a state at time } t \text{ in which the functioning} \\
&\text{in state } k \text{ instead of state } k-1 \text{ of the } i\text{th component} \\
&\text{is critical for the system being in states } \{j, \ldots, M\}] \\
= P[&I(\phi(k_i, \boldsymbol{X}(t)) \geq j) - I(\phi((k-1)_i, \boldsymbol{X}(t)) \geq j) = 1] \\
= p_\phi^j&((\boldsymbol{e}^k)_i, \boldsymbol{r}(t)) - p_\phi^j((\boldsymbol{e}^{k-1})_i, \boldsymbol{r}(t)), i = 1, \ldots, n, k \\
= 1, &\ldots, M, j = 1, \ldots, M. \qquad (6.1)
\end{aligned}$$

This is the probability that the system is in the states $\{j, \ldots, M\}$ if the $i$th component is in state $k$ and is not if the $i$th component is in state $k-1$.

Now, from the proof of Theorem 2.20, having used an argument from Theorem 4.1 in El-Neweihi *et al.* (1978) based on the fact that $\sum_{k=0}^{M} r_i^k(t) = 1, i = 1, \ldots, n$, we have

$$p_\phi^j(\boldsymbol{r}(t)) = \sum_{k=1}^{M} r_i^k(t)[p_\phi^j((\boldsymbol{e}^k)_i, \boldsymbol{r}(t)) - p_\phi^j((\boldsymbol{e}^0)_i, \boldsymbol{r}(t))] + p_\phi^j((\boldsymbol{e}^0)_i, \boldsymbol{r}(t))$$

$$= \sum_{k=1}^{M} p_i^k(t)[p_\phi^j((\boldsymbol{e}^k)_i, \boldsymbol{r}(t)) - p_\phi^j((\boldsymbol{e}^{k-1})_i, \boldsymbol{r}(t))] + p_\phi^j((\boldsymbol{e}^0)_i, \boldsymbol{r}(t)).$$

Thus, for $i = 1, \ldots, n, k = 1, \ldots, M, j = 1, \ldots, M$

$$\frac{\partial p_\phi^j(\boldsymbol{r}(t))}{\partial r_i^k(t)} = p_\phi^j((\boldsymbol{e}^k)_i, \boldsymbol{r}(t)) - p_\phi^j((\boldsymbol{e}^0)_i, \boldsymbol{r}(t)) \tag{6.2}$$

$$\frac{\partial p_\phi^j(\boldsymbol{r}(t))}{\partial p_i^k(t)} = p_\phi^j((\boldsymbol{e}^k)_i, \boldsymbol{r}(t)) - p_\phi^j((\boldsymbol{e}^{k-1})_i, \boldsymbol{r}(t)) = I_B^{(i,k,j)}(t). \tag{6.3}$$

Note that for the binary case, when $M = 1$, we have the well-known result

$$I_B^{(i,1,1)}(t) = I_B^{(i)}(t) = \frac{\partial p_\phi^1(\boldsymbol{r}(t))}{\partial r_i^1(t)} = \frac{\partial p_\phi^1(\boldsymbol{r}(t))}{\partial p_i^1(t)}. \tag{6.4}$$

Inspired by Griffith (1980) let, for $j \in \{1, \ldots, M\}$

$a_j$ = utility attached to the system being in the states $\{j, \ldots, M\}$,

where $a_M \geq a_{M-1} \geq \cdots \geq a_1$. Furthermore, let

$a_j^c$ = utility attached to the system being in the states $\{0, \ldots, j-1\}$,

where $a_M^c \geq a_{M-1}^c \geq \cdots \geq a_1^c = 0$ and $a_j \geq a_j^c$. Finally, let

$0 \leq a_j - a_j^c = c_j$ = the loss of utility attached to the system leaving the states $\{j, \ldots, M\}$.

Assume $\sum_{j=1}^{M} c_j = 1$. We now suggest the following generalized Birnbaum measure, $I_B^{(i,j)}(t)$, and generalized weighted Birnbaum measure, $I_B^{*(i)}(t)$, of the importance of the $i$th component at time $t$

$$I_B^{(i,j)}(t) = \sum_{k=1}^{M} I_B^{(i,k,j)}(t) / \sum_{r=1}^{n}\sum_{k=1}^{M} I_B^{(r,k,j)}(t) \tag{6.5}$$

$$I_B^{*(i)}(t) = \sum_{j=1}^{M} c_j I_B^{(i,j)}(t). \tag{6.6}$$

We obviously have

$$\sum_{i=1}^{n} I_B^{(i,j)}(t) = 1, \qquad 0 \le I_B^{(i,j)}(t) \le 1$$

$$\sum_{i=1}^{n} I_B^{*(i)}(t) = 1, \qquad 0 \le I_B^{*(i)}(t) \le 1. \tag{6.7}$$

These Birnbaum measures reflect Reason 1. However, there are two main objections to these measures. Firstly, they give the importance at fixed points of time, leaving the analyst at the system development phase to determine which points are important. Secondly, the measures do not depend on the performance of the $i$th component, whether good or bad, although the ranking of the importances of the components depends on the performances of all components.

### 6.2.2 The Barlow–Proschan measure

These objections cannot be raised to the following generalization of $I_{B-P}^{(i)}$, the time-independent Barlow and Proschan (1975b) measure of the importance of the $i$th component

$$\begin{aligned} I_{B-P}^{(i,j)} &= P[\text{The jump downwards of the } i\text{th component coincides with} \\ &\qquad \text{the system leaving the states } \{j, \ldots, M\}] \\ &= \int_0^\infty \sum_{k=1}^{M} I_B^{(i,k,j)}(t) r_i^k(t) \lambda_i^{(k,k-1)}(t)\,\mathrm{d}t \\ &= \int_0^\infty \sum_{k=1}^{M} [p_\phi^j((\boldsymbol{e}^k)_i, \boldsymbol{r}(t)) - p_\phi^j((\boldsymbol{e}^{k-1})_i, \boldsymbol{r}(t))] r_i^k(t) \lambda_i^{(k,k-1)}(t)\,\mathrm{d}t, \end{aligned}$$

$$i = 1, \ldots, n,\ j \in \{1, \ldots, M\}. \tag{6.8}$$

Note that for the binary case we have

$$I_{B-P}^{(i,1)} = I_{B-P}^{(i)}. \tag{6.9}$$

Since the system leaving the states $\{j, \ldots, M\}$ coincides with the jump downwards of exactly one component, we have

$$\sum_{i=1}^{n} I_{B-P}^{(i,j)} = 1. \tag{6.10}$$

We now suggest the following generalized weighted Barlow–Proschan measure, $I_{B-P}^{*(i)}$, of the importance of the $i$th component

$$I_{B-P}^{*(i)} = \sum_{j=1}^{M} c_j I_{B-P}^{(i,j)}. \tag{6.11}$$

We then have

$$\sum_{i=1}^{n} I_{B-P}^{*(i)} = 1, \qquad 0 \leq I_{B-P}^{*(i)} \leq 1. \tag{6.12}$$

Both the generalized and the generalized weighted Barlow–Proschan measures of the importance of the $i$th component are weighted averages of the generalized Birnbaum measure, $I_B^{(i,k,j)}(t)$. According to these measures a component is more important the more likely it is to be the direct cause of system deterioration, indicating that it takes care of both Reasons 1 and 2.

### 6.2.3 The Natvig measure

Intuitively it seems that components which, by deteriorating, strongly reduce the expected remaining system time in the better states are very important. This seems at least true during the system development phase. This is the motivation for the following generalization of $I_N^{(i)}$, the Natvig (1979) measure of the importance of the $i$th component. In order to formalize this, we introduce, for $i = 1, \ldots, n$, $k \in \{0, \ldots, M-1\}$

$T_{i,k}$ = the time of the jump of the $i$th component into state $k$.

$T'_{i,k}$ = the fictive time of the jump of the $i$th component into state $k$ after a fictive minimal repair of the component at $T_{i,k}$; i.e. it is repaired to have the same distribution of remaining time in state $k+1$ as it had just before jumping downwards to state $k$.

As in Natvig (1982b) a stochastic representation of this generalized measure is obtained by considering the random variables for $i = 1, \ldots, n, k \in \{1, \ldots, M\}, j \in \{1, \ldots, M\}$

$$Z_{i,k,j} = Y^1_{i,k,j} - Y^0_{i,k,j}, \tag{6.13}$$

where

$Y^1_{i,k,j}$ = system time in the states $\{j, \ldots, M\}$ in the interval $[T_{i,k-1}, T'_{i,k-1}]$ just *after* the jump downwards from state $k$ to state $k-1$ of the $i$th component, which, however, immediately undergoes a fictive minimal repair.

$Y^0_{i,k,j}$ = system time in the states $\{j, \ldots, M\}$ in the interval $[T_{i,k-1}, T'_{i,k-1}]$ just *after* the jump downwards from state $k$ to state $k-1$ of the $i$th component, assuming that the component stays in the latter state throughout this interval.

Thus, $Z_{i,k,j}$ can be interpreted as the fictive increase in system time in the states $\{j, \ldots, M\}$ in the interval $[T_{i,k-1}, T'_{i,k-1}]$ due to a fictive minimal repair of the $i$th component when jumping downwards from state $k$ to state $k-1$. Note that since the minimal repair is fictive, we have chosen to calculate the effect of this repair over the entire interval $[T_{i,k-1}, T'_{i,k-1}]$ even though this interval may extend beyond the time of the next jump of the $i$th component. Note that the fictive minimal repair periods; i.e. the intervals of the form $[T_{i,k-1}, T'_{i,k-1}]$, may sometimes overlap. Thus, at a given point of time we may have contributions from more than one fictive minimal repair. This was efficiently dealt with by the simulation methods presented in Huseby *et al.* (2010) in the binary case. Taking the expectation, we get, for $i = 1, \ldots, n, j \in \{1, \ldots, M\}$, the following generalized Natvig measure, $I_N^{(i,j)}$, and generalized weighted Natvig measure, $I_N^{*(i)}$, of the importance of the $i$th component

$$I_N^{(i,j)} = \sum_{k=1}^{M} EZ_{i,k,j} / \sum_{r=1}^{n} \sum_{k=1}^{M} EZ_{r,k,j} \tag{6.14}$$

$$I_N^{*(i)} = \sum_{j=1}^{M} c_j I_N^{(i,j)}, \tag{6.15}$$

tacitly assuming $EZ_{i,k,j} < \infty$, $i = 1, \ldots, n$, $k \in \{1, \ldots, M\}$, $j \in \{1, \ldots, M\}$. We obviously have

$$\sum_{i=1}^{n} I_N^{(i,j)} = 1, \qquad 0 \le I_N^{(i,j)} \le 1 \tag{6.16}$$

$$\sum_{i=1}^{n} I_N^{*(i)} = 1, \qquad 0 \le I_N^{*(i)} \le 1. \tag{6.17}$$

We will now prove the following theorem.

**Theorem 6.1:**

$$EZ_{i,k,j} = \int_0^\infty \int_0^\infty I_B^{(i,k,j)}(u+w)\bar{F}_i^k(w)(-\ln \bar{F}_i^k(w))\,\mathrm{d}w$$
$$r_i^{k+1}(u)\lambda_i^{(k+1,k)}(u)\,\mathrm{d}u,\ k \in \{1, \ldots, M-1\}$$
$$EZ_{i,M,j} = \int_0^\infty I_B^{(i,M,j)}(w)\bar{F}_i^M(w)(-\ln \bar{F}_i^M(w))\,\mathrm{d}w.$$

**Proof:** For $k \in \{1, \ldots, M-1\}$ from Equation (6.13) we have

$$EZ_{i,k,j} = EY_{i,k,j}^1 - EY_{i,k,j}^0 = \int_0^\infty \int_0^\infty \int_0^\infty [p_\phi^j((\mathbf{0},$$
$$(1 - \bar{F}_i^k(z+v)/\bar{F}_i^k(z))_{k-1}, (\bar{F}_i^k(z+v)/\bar{F}_i^k(z))_k)_i, \boldsymbol{r}(u+z+v))$$
$$- p_\phi^j((\mathbf{0}, 1_{k-1}, 0_k)_i, \boldsymbol{r}(u+z+v))]\,\mathrm{d}v\, f_i^k(z)\,\mathrm{d}z\, r_i^{k+1}(u)\lambda_i^{(k+1,k)}(u)\,\mathrm{d}u.$$

From Equation (6.3) by pivot decomposition this reduces to

$$= \int_0^\infty \int_0^\infty \int_0^\infty [(\bar{F}_i^k(z+v)/\bar{F}_i^k(z))p_\phi^j((\mathbf{0}, 0_{k-1}, 1_k)_i, \boldsymbol{r}(u+z+v))$$
$$+ (1 - \bar{F}_i^k(z+v)/\bar{F}_i^k(z))p_\phi^j((\mathbf{0}, 1_{k-1}, 0_k)_i, \boldsymbol{r}(u+z+v))$$
$$- p_\phi^j((\mathbf{0}, 1_{k-1}, 0_k)_i, \boldsymbol{r}(u+z+v))]\,\mathrm{d}v\, f_i^k(z)\,\mathrm{d}z\, r_i^{k+1}(u)\lambda_i^{(k+1,k)}(u)\,\mathrm{d}u$$
$$= \int_0^\infty \int_0^\infty \int_0^\infty \frac{\bar{F}_i^k(z+v)}{\bar{F}_i^k(z)} I_B^{(i,k,j)}(u+z+v)\,\mathrm{d}v\, f_i^k(z)\,\mathrm{d}z$$
$$r_i^{k+1}(u)\lambda_i^{(k+1,k)}(u)\,\mathrm{d}u$$
$$= \int_0^\infty \int_0^\infty \int_0^\infty \frac{\bar{F}_i^k(z+v)}{\bar{F}_i^k(z)} I_B^{(i,k,j)}(u+z+v) f_i^k(z)\,\mathrm{d}z\,\mathrm{d}v$$
$$r_i^{k+1}(u)\lambda_i^{(k+1,k)}(u)\,\mathrm{d}u$$

$$= \int_0^\infty \int_0^\infty I_B^{(i,k,j)}(u+w)\bar{F}_i^k(w) \int_0^w \frac{f_i^k(z)}{\bar{F}_i^k(z)}\, \mathrm{d}z\, \mathrm{d}w\, r_i^{k+1}(u)\lambda_i^{(k+1,k)}(u)\, \mathrm{d}u$$

$$= \int_0^\infty \int_0^\infty I_B^{(i,k,j)}(u+w)\bar{F}_i^k(w)(-\ln \bar{F}_i^k(w))\, \mathrm{d}w\, r_i^{k+1}(u)\lambda_i^{(k+1,k)}(u)\, \mathrm{d}u.$$

We similarly get

$$\begin{aligned} EZ_{i,M,j} &= EY_{i,M,j}^1 - EY_{i,M,j}^0 \\ &= \int_0^\infty \int_0^\infty [p_\phi^j((\mathbf{0}, (1 - \bar{F}_i^M(z+v)/\bar{F}_i^M(z))_{M-1}, \\ &\quad (\bar{F}_i^M(z+v)/\bar{F}_i^M(z))_M)_i, \boldsymbol{r}(z+v)) \\ &\quad - p_\phi^j((\mathbf{0}, 1_{M-1}, 0_M)_i, \boldsymbol{r}(z+v))]\, \mathrm{d}v\, f_i^M(z)\, \mathrm{d}z. \end{aligned}$$

By pivot decomposition this reduces to

$$\begin{aligned} &= \int_0^\infty \int_0^\infty [(\bar{F}_i^M(z+v)/\bar{F}_i^M(z))p_\phi^j((\mathbf{0}, 0_{M-1}, 1_M)_i, \boldsymbol{r}(z+v)) \\ &+ (1 - \bar{F}_i^M(z+v)/\bar{F}_i^M(z))p_\phi^j((\mathbf{0}, 1_{M-1}, 0_M)_i, \boldsymbol{r}(z+v)) \\ &- p_\phi^j((\mathbf{0}, 1_{M-1}, 0_M)_i, \boldsymbol{r}(z+v))]\, \mathrm{d}v\, f_i^M(z)\, \mathrm{d}z \\ &= \int_0^\infty \int_0^\infty \frac{\bar{F}_i^M(z+v)}{\bar{F}_i^M(z)} I_B^{(i,M,j)}(z+v)\, \mathrm{d}v\, f_i^M(z)\, \mathrm{d}z \\ &= \int_0^\infty \int_0^\infty \frac{\bar{F}_i^M(z+v)}{\bar{F}_i^M(z)} I_B^{(i,M,j)}(z+v) f_i^M(z)\, \mathrm{d}z\, \mathrm{d}v \\ &= \int_0^\infty I_B^{(i,M,j)}(w)\bar{F}_i^M(w) \int_0^w \frac{f_i^M(z)}{\bar{F}_i^M(z)}\, \mathrm{d}z\, \mathrm{d}w \\ &= \int_0^\infty I_B^{(i,M,j)}(w)\bar{F}_i^M(w)(-\ln \bar{F}_i^M(w))\, \mathrm{d}w. \end{aligned}$$

Hence, as for the generalized and generalized weighted Barlow–Proschan measures $EZ_{i,k,j}$ for $k \in \{1, \ldots, M\}$ is a weighted average of the generalized Birnbaum measure, $I_B^{(i,k,j)}(t)$. In a way the generalized weighted Natvig measure can be considered a more complex cousin of the generalized weighted Barlow–Proschan measure.

## 6.3 The Birnbaum and Barlow–Proschan measures of component importance in repairable systems and the latter's dual extension

In this and the subsequent section we consider the case where the components, and hence the system, can be repaired. Again, in order to make things not too complex, we assume that each component deteriorates by going through all states from the perfect functioning state until the complete failure state before being repaired to the perfect functioning state. Also at time $t = 0$ all components are in the perfect functioning state $M$. Let the $i$th component still have an absolutely continuous distribution $F_i^k(t)$ of time spent in the state $k$, before jumping downwards to state $k-1$, with density $f_i^k(t)$, $\bar{F}_i^k(t) = 1 - F_i^k(t)$ and mean $\mu_i^k$. Furthermore, let the $i$th component have an absolutely continuous repair time distribution $G_i(t)$ with density $g_i(t)$, $\bar{G}_i(t) = 1 - G_i(t)$ and mean $\mu_i^0$. It is still assumed that all these times spent in the various states are independent.

Introduce the notation

$$P(X_i(t) = j) = a_i^j(t), \quad j = 0, \ldots, M$$

$$\boldsymbol{a}(t) = (a_1^1(t), \ldots, a_1^M(t), a_2^1(t), \ldots, a_n^M(t))$$

$$P[\phi(\boldsymbol{X}(t)) \geq j] = P[I(\phi(\boldsymbol{X}(t)) \geq j) = 1] = p_\phi^j(\boldsymbol{a}(t)).$$

We denote $a_i^j(t)$ the availability of the $i$th component at level $j$ at time $t$ and $p_\phi^j(\boldsymbol{a}(t))$ the availability of the system to level $j$ at time $t$. The corresponding stationary availabilities for $i = 1, \ldots, n$ and $j \in \{0, \ldots, M\}$ are given by

$$a_i^j = \lim_{t\to\infty} a_i^j(t) = \frac{\mu_i^j}{\sum_{\ell=0}^M \mu_i^\ell}. \tag{6.18}$$

Introduce

$$\boldsymbol{a} = (a_1^1, \ldots, a_1^M, a_2^1, \ldots, a_n^M).$$

### 6.3.1 The Birnbaum measure

Now the generalized Birnbaum measure in repairable systems is expressed as

$$I_B^{(i,k,j)}(t) = p_\phi^j((\boldsymbol{e}^k)_i, \boldsymbol{a}(t)) - p_\phi^j((\boldsymbol{e}^{k-1})_i, \boldsymbol{a}(t)),$$

$$i = 1, \ldots, n, k \in \{1, \ldots, M\}, j \in \{1, \ldots, M\}. \tag{6.19}$$

Using Equation (6.19) the generalized Birnbaum measure and the generalized weighted Birnbaum measure are still given by Equations (6.5) and (6.6) and the properties given in Equation (6.7) still hold. The corresponding stationary measures are given by

$$I_B^{(i,k,j)} = \lim_{t\to\infty} I_B^{(i,k,j)}(t) = p_\phi^j((\boldsymbol{e}^k)_i, \boldsymbol{a}) - p_\phi^j((\boldsymbol{e}^{k-1})_i, \boldsymbol{a}) \tag{6.20}$$

$$I_B^{(i,j)} = \sum_{k=1}^{M} I_B^{(i,k,j)} / \sum_{r=1}^{n}\sum_{k=1}^{M} I_B^{(r,k,j)} \tag{6.21}$$

$$I_B^{*(i)} = \sum_{j=1}^{M} c_j I_B^{(i,j)},$$

$$i = 1, \ldots, n, k \in \{1, \ldots, M\}, j \in \{1, \ldots, M\}. \tag{6.22}$$

We obviously have

$$\sum_{i=1}^{n} I_B^{(i,j)} = 1, \qquad 0 \le I_B^{(i,j)} \le 1$$

$$\sum_{i=1}^{n} I_B^{*(i)} = 1, \qquad 0 \le I_B^{*(i)} \le 1. \tag{6.23}$$

### 6.3.2 The Barlow–Proschan measure

For $i = 1, \ldots, n, k \in \{1, \ldots, M\}, j \in \{1, \ldots, M\}$, let

$N_i^{(k)}(t)$ = the number of jumps of the $i$th component from state $k$ to state $k-1$ in $[0, t]$.

$\tilde{N}_i^{(k,j)}(t)$ = the number of times the system leaves the states $\{j, \ldots, M\}$ in $[0, t]$ due to the jump of the $i$th component from state $k$ to $k-1$.

Finally, denote $EN_i^{(k)}(t)$ by $M_i^{(k)}(t)$. As in Barlow and Proschan (1975b) we have, for $i = 1, \ldots, n, k \in \{1, \ldots, M\}, j \in \{1, \ldots, M\}$

$$E\tilde{N}_i^{(k,j)}(t) = \int_0^t I_B^{(i,k,j)}(u)\, \mathrm{d}M_i^{(k)}(u), \tag{6.24}$$

where $I_B^{(i,k,j)}(u)$ is given by Equation (6.19). A generalized time-dependent Barlow–Proschan measure of the importance of the $i$th component in the time interval $[0, t]$ in repairable systems is given by

$$I_{B-P}^{(i,j)}(t) = \frac{\sum_{k=1}^{M} E\tilde{N}_i^{(k,j)}(t)}{\sum_{r=1}^{n}\sum_{k=1}^{M} E\tilde{N}_r^{(k,j)}(t)}. \tag{6.25}$$

The generalized weighted Barlow–Proschan measure is given by

$$I_{B-P}^{*(i)}(t) = \sum_{j=1}^{M} c_j I_{B-P}^{(i,j)}(t), \tag{6.26}$$

and we have the properties

$$\sum_{i=1}^{n} I_{B-P}^{(i,j)}(t) = 1, \qquad 0 \le I_{B-P}^{(i,j)}(t) \le 1$$

$$\sum_{i=1}^{n} I_{B-P}^{*(i)}(t) = 1, \qquad 0 \le I_{B-P}^{*(i)}(t) \le 1. \tag{6.27}$$

As in Barlow and Proschan (1975b), by a renewal theory argument we arrive at the corresponding stationary measures

$$I_{B-P}^{(i,j)} = \lim_{t\to\infty} I_{B-P}^{(i,j)}(t) = \frac{\sum_{k=1}^{M} I_B^{(i,k,j)}/(\sum_{\ell=0}^{M}\mu_i^{\ell})}{\sum_{r=1}^{n}\sum_{k=1}^{M} I_B^{(r,k,j)}/(\sum_{\ell=0}^{M}\mu_r^{\ell})}$$

$$I_{B-P}^{*(i)} = \sum_{j=1}^{M} c_j I_{B-P}^{(i,j)}. \tag{6.28}$$

$I_{B-P}^{(i,j)}$ is the stationary probability that the jump downwards of the $i$th component is the cause of the system leaving the states $\{j, \ldots, M\}$, given that the system has left these states, $j \in \{1, \ldots, M\}$. $I_{B-P}^{*(i)}$ is the weighted average of these probabilities.

**Theorem 6.2:** Let $i = 1, \ldots, n$, $j \in \{1, \ldots, M\}$. For a multistate series system; i.e. $\phi(\boldsymbol{x}) = \min_{1\le i\le n} x_i$, we have

$$I_{B-P}^{(i,j)} = \frac{1/\sum_{k=j}^{M}\mu_i^k}{\sum_{r=1}^{n} 1/\sum_{k=j}^{M}\mu_r^k},$$

whereas for a multistate parallel system; i.e. $\phi(\boldsymbol{x}) = \max_{1\le i\le n} x_i$, we have the dual expression

$$I_{B-P}^{(i,j)} = \frac{1/(\sum_{k=0}^{j-1}\mu_i^k)}{\sum_{r=1}^{n} 1/(\sum_{k=0}^{j-1}\mu_r^k)}.$$

**Proof:** From Equations (6.28), (6.20) and (6.18) we get, for the multistate series system

$$\begin{aligned} I_{B-P}^{(i,j)} &= \frac{I_B^{(i,j,j)}/(\sum_{\ell=0}^{M}\mu_i^\ell)}{\sum_{r=1}^{n} I_B^{(r,j,j)}/(\sum_{\ell=0}^{M}\mu_r^\ell)} \\ &= \frac{[\prod_{m\neq i}\sum_{k=j}^{M}\mu_m^k/(\sum_{\ell=0}^{M}\mu_m^\ell)]/(\sum_{\ell=0}^{M}\mu_i^\ell)}{\sum_{r=1}^{n}[\prod_{m\neq r}\sum_{k=j}^{M}\mu_m^k/(\sum_{\ell=0}^{M}\mu_m^\ell)]/(\sum_{\ell=0}^{M}\mu_r^\ell)} \\ &= \frac{\prod_{m\neq i}\sum_{k=j}^{M}\mu_m^k}{\sum_{r=1}^{n}\prod_{m\neq r}\sum_{k=j}^{M}\mu_m^k} = \frac{1/\sum_{k=j}^{M}\mu_i^k}{\sum_{r=1}^{n} 1/\sum_{k=j}^{M}\mu_r^k}. \end{aligned}$$

The proof for the multistate parallel system is completely analogous by noting that now

$$I_B^{(i,j,j)} = \prod_{m\neq i}\sum_{k=0}^{j-1}\mu_m^k \Big/ \left(\sum_{\ell=0}^{M}\mu_m^\ell\right).$$

According to this theorem, for a multistate series system the stationary Barlow–Proschan measure of importance of the $i$th component is decreasing in $\mu_i^k$, $k = j, \ldots, M$; i.e. the poorer the more important. This is sensible. However, this measure does not depend on component mean times to repair. This generalizes a result in the binary case shown in Natvig and Gåsemyr (2009) and is disappointing and an objection to the Barlow–Proschan measure for repairable systems. For a multistate parallel system, the stationary Barlow–Proschan measure depends both on component mean times to jumps downwards and to mean times to repair. This is not true in the binary case, as shown in Natvig and Gåsemyr (2009), where the one and only measure just depends on the mean times to repair. Note that these differences from results for the binary case are due to the asymmetric assumption that each component deteriorates by going through all states from the perfect functioning state until the complete failure state before being repaired to the perfect functioning state. For a multistate parallel system this measure of importance of the $i$th component is decreasing in $\mu_i^k$, $k = 1, \ldots, j-1$ and in the mean time to repair $\mu_i^0$; i.e. the better the more important. This is again sensible.

We have also arrived at the following theorem generalizing Theorem 4 in Natvig and Gåsemyr (2009)

**Theorem 6.3:** Let the $i$th component be in series (parallel) with the rest of the system; i.e. $\phi(\boldsymbol{x}) = \min(x_i, \phi(M_i, \boldsymbol{x}))$ ($\phi(\boldsymbol{x}) = \max(x_i, \phi(0_i, \boldsymbol{x}))$). Let, for $j \in \{1, \ldots, M\}$ and for an arbitrary component $k \neq i$ $\sum_{\ell=j}^M \mu_i^\ell \leq \mu_k^M$ ($\sum_{\ell=0}^{j-1} \mu_i^\ell \leq \mu_k^0$). Then $I_{B-P}^{(i,j)} \geq I_{B-P}^{(k,j)}$. Furthermore, for the numerator of the measure we have, respectively, when the $i$th component is in series and parallel with the rest of the system

$$\frac{\sum_{r=1}^M I_B^{(i,r,j)}}{\sum_{\ell=0}^M \mu_i^\ell} \geq \frac{\sum_{r=1}^M I_B^{(k,r,j)}}{\sum_{\ell=0}^M \mu_k^\ell} + \frac{p_\phi^j((\boldsymbol{e}^0)_k, \boldsymbol{a})}{\sum_{\ell=j}^M \mu_i^\ell}$$

$$\frac{\sum_{r=1}^M I_B^{(i,r,j)}}{\sum_{\ell=0}^M \mu_i^\ell} \geq \frac{\sum_{r=1}^M I_B^{(k,r,j)}}{\sum_{\ell=0}^M \mu_k^\ell} + \frac{1 - p_\phi^j((\boldsymbol{e}^M)_k, \boldsymbol{a})}{\sum_{\ell=0}^{j-1} \mu_i^\ell}.$$

**Proof:** When the $i$th component is in series with the rest of the system we have, from Equation (6.28) by applying Equations (6.20) and (6.18)

$$\begin{aligned}
\frac{\sum_{r=1}^M I_B^{(i,r,j)}}{\sum_{\ell=0}^M \mu_i^\ell} &= \frac{I_B^{(i,j,j)}}{\sum_{\ell=0}^M \mu_i^\ell} = \frac{p_\phi^j((\boldsymbol{e}^j)_i, \boldsymbol{a})}{\sum_{\ell=0}^M \mu_i^\ell} = \frac{p_\phi^j(\boldsymbol{a})}{\sum_{\ell=j}^M \mu_i^\ell} \\
&= \frac{\sum_{r=0}^M p_\phi^j((\boldsymbol{e}^r)_k, \boldsymbol{a}) \mu_k^r}{(\sum_{\ell=j}^M \mu_i^\ell)(\sum_{\ell=0}^M \mu_k^\ell)} \\
&= \frac{\sum_{r=0}^M [p_\phi^j((\boldsymbol{e}^r)_k, \boldsymbol{a}) - p_\phi^j((\boldsymbol{e}^{r-1})_k, \boldsymbol{a})] \sum_{\ell=r}^M \mu_k^\ell}{(\sum_{\ell=j}^M \mu_i^\ell)(\sum_{\ell=0}^M \mu_k^\ell)} \\
&= \frac{\sum_{r=1}^M I_B^{(k,r,j)} \sum_{\ell=r}^M \mu_k^\ell}{(\sum_{\ell=j}^M \mu_i^\ell)(\sum_{\ell=0}^M \mu_k^\ell)} + \frac{p_\phi^j((\boldsymbol{e}^0)_k, \boldsymbol{a})}{\sum_{\ell=j}^M \mu_i^\ell}.
\end{aligned}$$

Applying the assumption $\sum_{\ell=j}^M \mu_i^\ell \leq \mu_k^M$, the results follow. When the $i$th component is in parallel with the rest of the system we have

$$\frac{\sum_{r=1}^M I_B^{(i,r,j)}}{\sum_{\ell=0}^M \mu_i^\ell} = \frac{I_B^{(i,j,j)}}{\sum_{\ell=0}^M \mu_i^\ell} = \frac{1 - p_\phi^j((\boldsymbol{e}^{j-1})_i, \boldsymbol{a})}{\sum_{\ell=0}^M \mu_i^\ell} = \frac{1 - p_\phi^j(\boldsymbol{a})}{\sum_{\ell=0}^{j-1} \mu_i^\ell}$$

$$= \frac{1-\sum_{r=1}^{M} I_B^{(k,r,j)}(1-\sum_{\ell=0}^{r-1}\mu_k^\ell/(\sum_{\ell=0}^{M}\mu_k^\ell)) - p_\phi^j((\boldsymbol{e}^0)_k,\boldsymbol{a})}{\sum_{\ell=0}^{j-1}\mu_i^\ell}$$

$$= \frac{\sum_{r=1}^{M} I_B^{(k,r,j)}\sum_{\ell=0}^{r-1}\mu_k^\ell}{(\sum_{\ell=0}^{j-1}\mu_i^\ell)(\sum_{\ell=0}^{M}\mu_k^\ell)} + \frac{1-p_\phi^j((\boldsymbol{e}^M)_k,\boldsymbol{a})}{\sum_{\ell=0}^{j-1}\mu_i^\ell}.$$

Applying the assumption $\sum_{\ell=0}^{j-1}\mu_i^\ell \le \mu_k^0$, the results follow.

It is still discomforting that the assumption in the first inequality does not depend on component mean times to repair. The assumption in the second inequality does depend both on component mean times to jumps downwards and to mean times to repair. The assumptions are sensible since, according to them, a component in series (parallel) with the rest of the system is more important than any other component if it in some sense is poorer (better) than the other one.

### 6.3.3 The dual extension of the Barlow–Proschan measure

As an attempt to improve the Barlow–Proschan measures given in Equations (6.25), (6.26) and (6.28) for repairable systems, it is suggested that we take a dual term into account based on the probability that the repair of the $i$th component is the cause of system state improvement, given that a system state improvement has occurred. According to the dual term, a component is more important the more likely its repair is the direct cause of system improvement. For $i = 1, \ldots, n$, $j \in \{1, \ldots, M\}$, let

$V_i(t) =$ the number of jumps of the $i$th component from state 0 to state $M$ in $[0, t]$.

$\tilde{V}_i^{(j)}(t) =$ the number of times the system leaves the states $\{0, \ldots, j-1\}$ in $[0, t]$ due to the jump of the $i$th component from state 0 to $M$.

Denote $EV_i(t)$ by $R_i(t)$.

Note that

$$a_i^M(t) = P[V_i(t) - N_i^{(M)}(t) = 0] = E[V_i(t) - N_i^{(M)}(t) + 1]$$
$$= R_i(t) - M_i^{(M)}(t) + 1$$

$$a_i^k(t) = P[N_i^{(k+1)}(t) - N_i^{(k)}(t) = 1] = E[N_i^{(k+1)}(t) - N_i^{(k)}(t)]$$
$$= M_i^{(k+1)}(t) - M_i^{(k)}(t), k \in \{1, \ldots, M-1\}$$
$$a_i^0(t) = P[N_i^{(1)}(t) - V_i(t) = 1] = E[N_i^{(1)}(t) - V_i(t)] = M_i^{(1)}(t) - R_i(t).$$

Parallel to Equation (6.24) we get, for $i = 1, \ldots, n$, $j \in \{1, \ldots, M\}$

$$E\tilde{V}_i^{(j)}(t) = \int_0^t [p_\phi^j((\boldsymbol{e}^M)_i, \boldsymbol{a}(u)) - p_\phi^j((\boldsymbol{e}^0)_i, \boldsymbol{a}(u))]\, \mathrm{d}R_i(u)$$
$$= \int_0^t \sum_{k=1}^M I_B^{(i,k,j)}(u)\, \mathrm{d}R_i(u). \tag{6.29}$$

An extended version of Equation (6.25) is arrived at by applying Equations (6.24) and (6.29)

$$\bar{I}_{B-P}^{(i,j)}(t) = \frac{\sum_{k=1}^M E\tilde{N}_i^{(k,j)}(t) + E\tilde{V}_i^{j(t)}}{\sum_{r=1}^n [\sum_{k=1}^M E\tilde{N}_r^{(k,j)}(t) + E\tilde{V}_r^j(t)]}$$
$$= \frac{\int_0^t \sum_{k=1}^M I_B^{(i,k,j)}(u)\, \mathrm{d}(M_i^{(k)}(u) + R_i(u))}{\sum_{r=1}^n \int_0^t \sum_{k=1}^M I_B^{(r,k,j)}(u)\, \mathrm{d}(M_r^{(k)}(u) + R_r(u))}. \tag{6.30}$$

However, since from renewal theory

$$\lim_{t\to\infty} \frac{M_i^{(k)}(t)}{t} = \lim_{t\to\infty} \frac{R_i(t)}{t} = \frac{1}{\sum_{\ell=0}^M \mu_i^\ell},$$

it turns out that for the corresponding stationary measures we have

$$\bar{I}_{B-P}^{(i,j)} = \lim_{t\to\infty} \bar{I}_{B-P}^{(i,j)}(t) = I_{B-P}^{(i,j)}$$
$$\bar{I}_{B-P}^{*(i)} = \sum_{j=1}^M c_j \bar{I}_{B-P}^{(i,j)} = I_{B-P}^{*(i)}. \tag{6.31}$$

Hence, Theorems 6.2 and 6.3 are also valid for $\bar{I}_{B-P}^{(i,j)}$, which is disappointing since, under stationary, nothing is gained by introducing the extended measure of also taking the dual approach into account.

## 6.4 The Natvig measure of component importance in repairable systems and its dual extension

We start by introducing some basic random variables for $i = 1, \dots, n$, $k \in \{0, \dots, M\}$, $m = 1, 2, \dots$

$T_{i,k,m} =$ the time of the $m$th jump of the $i$th component into state $k$.

$D_{i,m} =$ the length of the $m$th repair time of the $i$th component.

We define $T_{i,M,0} = 0$ and have, for $m = 1, 2, \dots$

$$T_{i,M,m} = T_{i,0,m} + D_{i,m}.$$

### 6.4.1 The Natvig measure

Parallel to the nonrepairable case, we argue that components which, by deteriorating, strongly reduce the expected system time in the better states are very important. In order to formalize this, we introduce, for $i = 1, \dots, n$, $k \in \{0, \dots, M-1\}$, $m = 1, 2, \dots$

$T'_{i,k,m} =$ the fictive time of the $m$th jump of the $i$th component into state $k$ after a fictive minimal repair of the component at $T_{i,k,m}$; i.e. it is repaired to have the same distribution of remaining time in state $k+1$ as it had just before jumping downwards to state $k$.

As for the Barlow–Proschan measure, we consider a time interval $[0, t]$ and define, for $i = 1, \dots, n$, $k \in \{1, \dots, M\}$, $j \in \{1, \dots, M\}$, $m = 1, 2, \dots$

$Y^1_{i,k,j,m} =$ system time in the states $\{j, \dots, M\}$ in the interval $[\min(T_{i,k-1,m}, t), \min(T'_{i,k-1,m}, t)]$ just *after* the jump downwards from state $k$ to state $k-1$ of the $i$th component, which, however, immediately undergoes a fictive minimal repair.

$Y^0_{i,k,j,m} =$ system time in the states $\{j, \dots, M\}$ in the interval $[\min(T_{i,k-1,m}, t), \min(T'_{i,k-1,m}, t)]$ just *after* the jump downwards from state $k$ to state $k-1$ of the $i$th component, assuming that the component stays in the latter state throughout this interval.

In order to arrive at a stochastic representation similar to the nonrepairable case, see Equation (6.13), we introduce the following random variables

$$Z_{i,k,j,m} = Y^1_{i,k,j,m} - Y^0_{i,k,j,m}. \tag{6.32}$$

Thus, $Z_{i,k,j,m}$ can be interpreted as the fictive increase in system time in the states $\{j, \dots, M\}$ in the interval $[\min(T_{i,k-1,m}, t), \min(T'_{i,k-1,m}, t)]$ due to a fictive minimal repair of the $i$th component when jumping downwards from state $k$ to state $k-1$. Note that since the minimal repair is fictive, we have chosen to calculate the effect of this repair over the entire interval $[\min(T_{i,k-1,m}, t), \min(T'_{i,k-1,m}, t)]$ even though this interval may extend beyond the time of the next jump of the $i$th component.

In order to summarize the effects of all the fictive minimal repairs, we have chosen to simply add up these contributions. Taking the expectation, we get, for $i = 1, \dots, n$, $j \in \{1, \dots, M\}$

$$E\left[\sum_{m=1}^{\infty} I(T_{i,k,m} \le t) Z_{i,k,j,m}\right] \stackrel{d}{=} EY_{i,k,j}(t),\ k \in \{1, \dots, M-1\}$$

$$E\left[\sum_{m=1}^{\infty} I(T_{i,M,m-1} \le t) Z_{i,M,j,m}\right] \stackrel{d}{=} EY_{i,M,j}(t). \tag{6.33}$$

We then suggest the following generalized Natvig measure, $I_N^{(i,j)}(t)$, and generalized weighted Natvig measure, $I_N^{*(i)}(t)$, of the importance of the $i$th component in the time interval $[0, t]$ in repairable systems

$$I_N^{(i,j)}(t) = \sum_{k=1}^{M} EY_{i,k,j}(t) / \sum_{r=1}^{n} \sum_{k=1}^{M} EY_{r,k,j}(t) \tag{6.34}$$

$$I_N^{*(i)}(t) = \sum_{j=1}^{M} c_j I_N^{(i,j)}(t), \tag{6.35}$$

tacitly assuming $EY_{i,k,j}(t) < \infty$, $i = 1, \dots, n$, $k \in \{1, \dots, M\}$, $j \in \{1, \dots, M\}$. We obviously have

$$\sum_{i=1}^{n} I_N^{(i,j)}(t) = 1, \qquad 0 \le I_N^{(i,j)}(t) \le 1 \tag{6.36}$$

$$\sum_{i=1}^{n} I_N^{*(i)}(t) = 1, \qquad 0 \leq I_N^{*(i)}(t) \leq 1. \tag{6.37}$$

We will now prove the following theorem.

**Theorem 6.4:**

$$EY_{i,k,j}(t) = \int_0^t \int_u^t I_B^{(i,k,j)}(w)\bar{F}_i^k(w-u)(-\ln \bar{F}_i^k(w-u))\,\mathrm{d}w$$

$$\mathrm{d}M_i^{(k+1)}(u),\ k \in \{1, \ldots, M-1\}$$

$$EY_{i,M,j}(t) = \int_0^t I_B^{(i,M,j)}(w)\bar{F}_i^M(w)(-\ln \bar{F}_i^M(w))\,\mathrm{d}w$$

$$+ \int_0^t \int_u^t I_B^{(i,M,j)}(w)\bar{F}_i^M(w-u)(-\ln \bar{F}_i^M(w-u))\,\mathrm{d}w\,\mathrm{d}R_i(u).$$

To prove the theorem in a formal way we need the following lemma proved in Natvig and Gåsemyr (2009).

**Lemma 6.5:** Let $W_1, W_2, \ldots$ be an increasing sequence of positive random variables. Assume that $W_m - W_{m-1}$ are independent with an absolutely continuous distribution $H_m(u)$ and density $h_m(u)$, $m = 1, 2, \ldots$, where $W_0 \stackrel{d}{=} 0$. Let $\rho(u)$ be the jump intensity for the process $N(u) = \sum_{m=1}^{\infty} I(W_m \leq u)$, and let $N = N(t)$. For each $m = 1, 2, \ldots$ let $Y_m$ be a random variable which is independent of $W_1, \ldots, W_{m-1}$ given $W_m$, and suppose that $E(Y_m | W_m = u)$ does not depend on $m$. Finally, let $Y = \sum_{m=1}^{N} Y_m$. Then

$$EY = \int_0^t E(Y_m | W_m = u)\rho(u)\,\mathrm{d}u.$$

**Proof of Theorem 6.4:** We first apply Lemma 6.5 for $m = 1, 2, \ldots$ and $k \in \{1, \ldots, M-1\}$ with

$$W_m = T_{i,k,m}, \quad Y_m = Z_{i,k,j,m},$$

$$N = N(t) = \sum_{m=1}^{\infty} I(T_{i,k,m} \leq t) \stackrel{d}{=} N_k.$$

Since $E(Z_{i,k,j,m}|T_{i,k,m} = u)$ is shown not to depend on $m$, we get, from Equation (6.33)

$$EY_{i,k,j}(t) = E\left[\sum_{m=1}^{N_k} Z_{i,k,j,m}\right] = \sum_{m=1}^{N_k} EY_m = EY$$

$$= \int_0^t E(Z_{i,k,j,m}|T_{i,k,m} = u)\,\mathrm{d}M_i^{(k+1)}(u).$$

We then apply Lemma 6.5 for $m = 1, 2, \ldots$ with

$$W_m = T_{i,M,m},\quad Y_m = Z_{i,M,j,m+1},\quad N = N(t) = \sum_{m=1}^{\infty} I(T_{i,M,m} \le t) \stackrel{d}{=} N_M.$$

It will be shown that $E(Z_{i,M,j,m+1}|T_{i,M,m} = u)$ also does not depend on $m$. Hence, from Equation (6.33), remembering that $T_{i,M,0} = 0$

$$EY_{i,M,j}(t) = EZ_{i,M,j,1} + E\left[\sum_{m=1}^{\infty} I(T_{i,M,m} \le t) Z_{i,M,j,m+1}\right] = EZ_{i,M,j,1}$$

$$+ \sum_{m=1}^{N_M} EY_m = EZ_{i,M,j,1} + EY = E(Z_{i,M,j,1})|T_{i,M,0} = 0)$$

$$+ \int_0^t E(Z_{i,M,j,m+1}|T_{i,M,m} = u)\,\mathrm{d}R_i(u).$$

Let $X_u$ be the uptime in $[0, u]$ for a binary system with availability $a(t)$. From Theorem 3.6 of Aven and Jensen (1999) we have

$$EX_u = \int_0^u a(t)\,\mathrm{d}t.$$

Applying this, we get, from Equation (6.32), for $i = 1, \ldots, n$, $k \in \{1, \ldots, M-1\}$ and $m = 1, 2, \ldots$

$$E(Z_{i,k,j,m}|T_{i,k,m} = u) = E(Y^1_{i,k,j,m}|T_{i,k,m} = u) - E(Y^0_{i,k,j,m}|T_{i,k,m} = u)$$

$$= \int_0^{t-u}\int_0^{t-u-z} [p^j_\phi((\mathbf{0}, (1 - \bar{F}_i^k(z+v)/\bar{F}_i^k(z))_{k-1},$$

$$(\bar{F}_i^k(z+v)/\bar{F}_i^k(z))_k)_i, \boldsymbol{a}(u+z+v)) - p^j_\phi((\mathbf{0}, 1_{k-1}, 0_k)_i,$$

$$\boldsymbol{a}(u+z+v))]\,\mathrm{d}v f_i^k(z)\,\mathrm{d}z.$$

By pivot decomposition this reduces to

$$\begin{aligned}
&\int_0^{t-u}\int_0^{t-u-z}\frac{\bar{F}_i^k(z+v)}{\bar{F}_i^k(z)}I_B^{(i,k,j)}(u+z+v)\,\mathrm{d}v f_i^k(z)\,\mathrm{d}z\\
&=\int_0^{t-u}\int_0^{t-u-v}\frac{\bar{F}_i^k(z+v)}{\bar{F}_i^k(z)}I_B^{(i,k,j)}(u+z+v)f_i^k(z)\,\mathrm{d}z\,\mathrm{d}v\\
&=\int_u^t I_B^{(i,k,j)}(w)\bar{F}_i^k(w-u)\int_0^{w-u}\frac{f_i^k(z)}{\bar{F}_i^k(z)}\,\mathrm{d}z\,\mathrm{d}w\\
&=\int_u^t I_B^{(i,k,j)}(w)\bar{F}_i^k(w-u)(-\ln\bar{F}_i^k(w-u))\,\mathrm{d}w,
\end{aligned}$$

which does not depend on $m$. Similarly, we get

$$\begin{aligned}
&E(Z_{i,M,j,m+1}|T_{i,M,m}=u)\\
&\quad=\int_u^t I_B^{(i,M,j)}(w)\bar{F}_i^M(w-u)(-\ln\bar{F}_i^M(w-u))\,\mathrm{d}w,
\end{aligned}$$

which is also independent of $m$. Inserting these expressions into the expressions for $EY_{i,k,j}(t)$ for $k\in\{1,\ldots,M\}$ completes the proof.

From Natvig (1985b) it follows that, for $k\in\{1,\ldots,M\}$

$$\int_0^\infty \bar{F}_i^k(t)(-\ln\bar{F}_i^k(t))\,\mathrm{d}t=E(T'_{i,k-1,m}-T_{i,k-1,m})\stackrel{d}{=}\mu_i^{k(p)}. \qquad (6.38)$$

Accordingly, this integral equals the expected prolonged time in state $k$ of the $i$th component due to a minimal repair.

Now divide the expressions for $EY_{i,k,j}(t)$ in Theorem 6.4 by $t$ and let $t\to\infty$. Assuming that the first addend in $EY_{i,M,j}(t)$ vanishes, applying a renewal theory argument as in Barlow and Proschan (1975b), we arrive at the following corresponding stationary measures

$$\begin{aligned}
I_N^{(i,j)}&=\lim_{t\to\infty}I_N^{(i,j)}(t)=\frac{[\sum_{k=1}^M I_B^{(i,k,j)}/(\sum_{\ell=0}^M\mu_i^\ell)]\mu_i^{k(p)}}{\sum_{r=1}^n[\sum_{k=1}^M I_B^{(r,k,j)}/(\sum_{\ell=0}^M\mu_r^\ell)]\mu_r^{k(p)}}\\
I_N^{*(i)}&=\sum_{j=1}^M c_j I_N^{(i,j)}.
\end{aligned} \qquad (6.39)$$

Parallel to Theorem 6.2, we arrive at

**Theorem 6.6:** Let $i = 1, \ldots, n,\ j \in \{1, \ldots, M\}$. For a multistate series system, we have

$$I_N^{(i,j)} = \frac{\mu_i^{j(p)} / \sum_{k=j}^{M} \mu_i^k}{\sum_{r=1}^{n} \mu_r^{j(p)} / \sum_{k=j}^{M} \mu_r^k},$$

whereas for a multistate parallel system, we have the dual expression

$$I_N^{(i,j)} = \frac{\mu_i^{j(p)} / \sum_{k=0}^{j-1} \mu_i^k}{\sum_{r=1}^{n} \mu_r^{j(p)} / \sum_{k=0}^{j-1} \mu_r^k}.$$

Hence, also the stationary Natvig measures given by Equation (6.39) for a multistate series system do not depend on component mean times to repair. This generalizes a result in the binary case shown in Natvig and Gåsemyr (2009) and is disappointing. For a multistate parallel system the stationary Natvig measures do depend strongly both on the distributions of component times to jumps downwards and on mean times to repair. This is also true in the binary case, as shown already in Natvig and Gåsemyr (2009).

### 6.4.2 The dual extension of the Natvig measure

As for the Barlow–Proschan measure, we now also take a dual term into account where components which, by being repaired, strongly reduce the expected system time in the worse states are considered very important. In order to formalize this, we introduce, for $i = 1, \ldots, n, m = 1, 2, \ldots$

$T'_{i,M,m}$ = the fictive time of the $m$th jump of the $i$th component into state $M$ after a fictive minimal complete failure of the component at $T_{i,M,m}$; i.e. it is failed to have the same distribution of remaining time in state 0 as it had just before jumping upwards to state $M$.

Define, for $i = 1, \ldots, n, j \in \{1, \ldots, M\}, m = 1, 2, \ldots$

$Y^1_{i,0,j,m}$ = system time in the states $\{0, \ldots, j-1\}$ in the interval $[\min(T_{i,M,m}, t), \min(T'_{i,M,m}, t)]$ just *after* the jump upwards from state 0 to state $M$ of the $i$th component, which, however, immediately undergoes a fictive minimal complete failure.

$Y^0_{i,0,j,m}$ = system time in the states $\{0, \ldots, j-1\}$ in the interval $[\min(T_{i,M,m}, t), \min(T'_{i,M,m}, t)]$ just *after* the jump upwards from state 0 to state $M$ of the $i$th component, assuming that the component stays in the latter state throughout this interval.

Parallel to Equation (6.32) we then introduce the following random variables

$$Z_{i,0,j,m} = Y^1_{i,0,j,m} - Y^0_{i,0,j,m}. \tag{6.40}$$

Thus, $Z_{i,0,j,m}$ can be interpreted as the fictive increase in system time in the states $\{0, \ldots, j-1\}$ in the interval $[\min(T_{i,M,m}, t), \min(T'_{i,M,m}, t)]$ due to a fictive minimal complete failure of the $i$th component when jumping upwards from state 0 to state $M$.

Now, adding up the contributions from the repairs at $T_{i,M,m}$, $m = 1, 2, \ldots,$ and taking the expectation, we get, for $i = 1, \ldots, n$, $j \in \{1, \ldots, M\}$

$$E\left[\sum_{m=1}^{\infty} I(T_{i,0,m} \leq t) Z_{i,0,j,m}\right] \stackrel{d}{=} EY_{i,0,j}(t). \tag{6.41}$$

Parallel to Theorem 6.4, using the argument leading to the last equality in Equation (6.29) we arrive at

**Theorem 6.7:**

$$EY_{i,0,j}(t) = \int_0^t \int_u^t \sum_{k=1}^{M} I_B^{(i,k,j)}(w)\bar{G}_i(w-u)(-\ln \bar{G}_i(w-u))\, \mathrm{d}w\, \mathrm{d}M_i^1(u).$$

We then suggest the following dual generalized Natvig measure, $I^{(i,j)}_{N,D}(t)$, dual generalized weighted Natvig measure, $I^{*(i)}_{N,D}(t)$, extended generalized Natvig measure, $\bar{I}^{(i,j)}_N(t)$, and finally the extended generalized weighted Natvig measure, $\bar{I}^{*(i)}_N(t)$, of the importance of the $i$th component in the time interval $[0, t]$ in repairable systems

$$I^{(i,j)}_{N,D}(t) = EY_{i,0,j}(t) / \sum_{r=1}^{n} EY_{r,0,j}(t) \tag{6.42}$$

$$I^{*(i)}_{N,D}(t) = \sum_{j=1}^{M} c_j I^{(i,j)}_{N,D}(t) \tag{6.43}$$

$$\bar{I}_N^{(i,j)}(t) = \sum_{k=0}^{M} EY_{i,k,j}(t) / \sum_{r=1}^{n} \sum_{k=0}^{M} EY_{r,k,j}(t) \tag{6.44}$$

$$\bar{I}_N^{*(i)}(t) = \sum_{j=1}^{M} c_j \bar{I}_N^{(i,j)}(t), \tag{6.45}$$

tacitly assuming $EY_{i,k,j}(t) < \infty$, $i = 1, \ldots, n$, $k \in \{0, \ldots, M\}$, $j \in \{1, \ldots, M\}$. We obviously have

$$\sum_{i=1}^{n} I_{N,D}^{(i,j)}(t) = 1, \qquad 0 \leq I_{N,D}^{(i,j)}(t) \leq 1 \tag{6.46}$$

$$\sum_{i=1}^{n} I_{N,D}^{*(i)}(t) = 1, \qquad 0 \leq I_{N,D}^{*(i)}(t) \leq 1 \tag{6.47}$$

$$\sum_{i=1}^{n} \bar{I}_N^{(i,j)}(t) = 1, \qquad 0 \leq \bar{I}_N^{(i,j)}(t) \leq 1 \tag{6.48}$$

$$\sum_{i=1}^{n} \bar{I}_N^{*(i)}(t) = 1, \qquad 0 \leq \bar{I}_N^{*(i)}(t) \leq 1. \tag{6.49}$$

Completely parallel to Equation (6.38) we have

$$\int_0^\infty \bar{G}_i(t)(-\ln \bar{G}_i(t))\,\mathrm{d}t = E(T'_{i,M,j} - T_{i,M,j}) \stackrel{d}{=} \mu_i^{0(p)}. \tag{6.50}$$

The corresponding stationary measures are given by

$$I_{N,D}^{(i,j)} = \lim_{t\to\infty} I_{N,D}^{(i,j)}(t) = \frac{[\sum_{k=1}^{M} I_B^{(i,k,j)} / (\sum_{\ell=0}^{M} \mu_i^{\ell})]\mu_i^{0(p)}}{\sum_{r=1}^{n}[\sum_{k=1}^{M} I_B^{(r,k,j)} / (\sum_{\ell=0}^{M} \mu_r^{\ell})]\mu_r^{0(p)}}$$

$$I_{N,D}^{*(i)} = \sum_{j=1}^{M} c_j I_{N,D}^{(i,j)} \tag{6.51}$$

$$\bar{I}_N^{(i,j)} = \lim_{t\to\infty} \bar{I}_N^{(i,j)}(t) = \frac{[\sum_{k=1}^{M} I_B^{(i,k,j)} / (\sum_{\ell=0}^{M} \mu_i^{\ell})](\mu_i^{k(p)} + \mu_i^{0(p)})}{\sum_{r=1}^{n}[\sum_{k=1}^{M} I_B^{(r,k,j)} / (\sum_{\ell=0}^{M} \mu_r^{\ell})](\mu_r^{k(p)} + \mu_r^{0(p)})}$$

$$\bar{I}_N^{*(i)} = \sum_{j=1}^{M} c_j \bar{I}_N^{(i,j)}. \tag{6.52}$$

Note that if $\mu_r^{k(p)}$, $r = 1, \ldots, n, k = 1, \ldots, M$ are all equal, which by Equation (6.38) is the case when all components have the same distribution of the times spent in each of the non complete failure states, then Equation (6.39) reduces to Equation (6.28). Similarly, if $\mu_r^{0(p)}, r = 1, \ldots, n$ are all equal, which by Equation (6.50) is the case when all components have the same repair time distribution, then Equation (6.51) reduces to Equation (6.28). Similarly, if $\mu_r^{k(p)}$, $r = 1, \ldots, n,\ k = 1, \ldots, M$ are all equal and $\mu_r^{0(p)}$, $r = 1, \ldots, n$ are all equal, which by Equations (6.38) and (6.50) is the case when all components have the same distribution of the times spent in each of the non complete failure states and the same repair time distribution, then Equation (6.52) reduces to Equations (6.28) and (6.21).

Parallel to Theorem 6.6, we arrive at

**Theorem 6.8:** Let $i = 1, \ldots, n,\ j \in \{1, \ldots, M\}$. For a multistate series system, we have

$$\bar{I}_N^{(i,j)} = \frac{(\mu_i^{j(p)} + \mu_i^{0(p)}) / \sum_{k=j}^{M} \mu_i^k}{\sum_{r=1}^{n} (\mu_r^{j(p)} + \mu_r^{0(p)}) / \sum_{k=j}^{M} \mu_r^k},$$

whereas for a multistate parallel system, we have the dual expression

$$\bar{I}_N^{(i,j)} = \frac{(\mu_i^{j(p)} + \mu_i^{0(p)}) / \sum_{k=0}^{j-1} \mu_i^k}{\sum_{r=1}^{n} (\mu_r^{j(p)} + \mu_r^{0(p)}) / \sum_{k=0}^{j-1} \mu_r^k}.$$

Hence, the stationary extended Natvig measures for both a multistate series and parallel system do depend on the distributions of component times to jumps downwards and on the distributions of repair times. This generalizes a result in the binary case shown in Natvig and Gåsemyr (2009).

Now consider the special case where the component times to jumps downwards and repair times are Weibull distributed; i.e.

$$\bar{F}_i^k(t) = e^{-(\lambda_i^k t)^{\alpha_i^k}}, \qquad \lambda_i^k > 0,\ \ \alpha_i^k > 0$$

$$\bar{G}_i(t) = e^{-(\lambda_i^0 t)^{\alpha_i^0}}, \qquad \lambda_i^0 > 0,\ \ \alpha_i^0 > 0\,.$$

From Equations (6.38) and (6.50) we get, as shown in Natvig and Gåsemyr (2009), $\mu_i^{k(p)} = \mu_i^k / \alpha_i^k$ for $k \in \{0, \ldots, M\}$. Hence,

Equation (6.52) simplifies to

$$\bar{I}_N^{(i,j)} = \frac{[\sum_{k=1}^{M} I_B^{(i,k,j)}/(\sum_{\ell=0}^{M} \mu_i^{\ell})](\mu_i^k/\alpha_i^k + \mu_i^0/\alpha_i^0)}{\sum_{r=1}^{n}[\sum_{k=1}^{M} I_B^{(r,k,j)}/(\sum_{\ell=0}^{M} \mu_r^{\ell})](\mu_r^k/\alpha_r^k + \mu_r^0/\alpha_r^0)}. \tag{6.53}$$

Now for $k \in \{1, \ldots, M\}$ assume that the shape parameter $\alpha_i^k$ is increasing and $\lambda_i^k$ changing in such a way that $\mu_i^k$ is constant. Hence, according to Equation (6.18), the availability $a_i^k$ is unchanged. Then $\bar{I}_N^{(i,j)}$ is decreasing in $\alpha_i^k$. This is natural since a large $\alpha_i^k > 1$ corresponds to a strongly increasing failure rate and the effect of a minimal repair is small. Hence, according to $\bar{I}_N^{(i,j)}$, the $i$th component is of less importance. If, on the other hand, $\alpha_i^k < 1$ is small, we have a strongly decreasing failure rate and the effect of a minimal repair is large. Hence, according to $\bar{I}_N^{(i,j)}$, the $i$th component is of higher importance. A completely parallel argument is valid for $\alpha_i^0$.

By specializing $\alpha_i^k = \alpha$, $i = 1, \ldots, n$, $k \in \{0, \ldots, M\}$, we get

$$\bar{I}_N^{(i,j)} = \frac{[\sum_{k=1}^{M} I_B^{(i,k,j)}/(\sum_{\ell=0}^{M} \mu_i^{\ell})](\mu_i^k + \mu_i^0)}{\sum_{r=1}^{n}[\sum_{k=1}^{M} I_B^{(r,k,j)}/(\sum_{\ell=0}^{M} \mu_r^{\ell})](\mu_r^k + \mu_r^0)}. \tag{6.54}$$

The following result is now straightforward from Theorem 6.8.

**Theorem 6.9:** Assume component times to jumps downwards and repair times are Weibull distributed with identical shape parameters. Let $i = 1, \ldots, n$, $j \in \{1, \ldots, M\}$. For a multistate series system, we have

$$\bar{I}_N^{(i,j)} = \frac{(\mu_i^j + \mu_i^0)/\sum_{k=j}^{M} \mu_i^k}{\sum_{r=1}^{n}(\mu_r^j + \mu_r^0)/\sum_{k=j}^{M} \mu_r^k},$$

whereas for a multistate parallel system, we have the dual expression

$$\bar{I}_N^{(i,j)} = \frac{(\mu_i^j + \mu_i^0)/\sum_{k=0}^{j-1} \mu_i^k}{\sum_{r=1}^{n}(\mu_r^j + \mu_r^0)/\sum_{k=0}^{j-1} \mu_r^k}.$$

According to this theorem, for a multistate series system the importance of the $i$th component is increasing in $\mu_i^0$, decreasing in $\sum_{k=j+1}^{M} \mu_i^k$ and decreasing in $\mu_i^j$ if $\mu_i^0 > \sum_{k=j+1}^{M} \mu_i^k$; i.e. the poorer the more important. For a multistate parallel system the importance of the $i$th component is increasing in $\mu_i^j$, decreasing in $\sum_{k=1}^{j-1} \mu_i^k$ and decreasing in $\mu_i^0$ if $\mu_i^j > \sum_{k=1}^{j-1} \mu_i^k$; i.e. the better the more important.

This generalizes results shown in Natvig and Gåsemyr (2009) and seems perfectly sensible.

Furthermore, generalizing Theorem 10 in Natvig and Gåsemyr (2009), we have the following more satisfactory theorem than Theorem 6.3.

**Theorem 6.10:** Assume component times to jumps downwards and repair times are Weibull distributed with identical shape parameters. Let the $i$th component be in series (parallel) with the rest of the system. Let, for $j \in \{1, \ldots, M\}$ and for an arbitrary component $k \neq i$ $\sum_{\ell=j}^{M} \mu_i^\ell/(\mu_i^j + \mu_i^0) \leq \mu_k^M/(\mu_k^r + \mu_k^0)(\sum_{\ell=0}^{j-1} \mu_i^\ell/(\mu_i^j + \mu_i^0) \leq \mu_k^0/(\mu_k^r + \mu_k^0))$ for $r = 1, \ldots, M$. Then $\bar{I}_N^{(i,j)} \geq \bar{I}_N^{(k,j)}$. Furthermore, for the numerator of the measure we have, respectively, when the $i$th component is in series and parallel with the rest of the system

$$\frac{\sum_{r=1}^{M} I_B^{(i,r,j)}(\mu_i^r + \mu_i^0)}{\sum_{\ell=0}^{M} \mu_i^\ell} \geq \frac{\sum_{r=1}^{M} I_B^{(k,r,j)}(\mu_k^r + \mu_k^0)}{\sum_{\ell=0}^{M} \mu_k^\ell} + \frac{p_\phi^j((\boldsymbol{e}^0)_k, \boldsymbol{a})(\mu_i^j + \mu_i^0)}{\sum_{\ell=j}^{M} \mu_i^\ell}$$

$$\frac{\sum_{r=1}^{M} I_B^{(i,r,j)}(\mu_i^r + \mu_i^0)}{\sum_{\ell=0}^{M} \mu_i^\ell} \geq \frac{\sum_{r=1}^{M} I_B^{(k,r,j)}(\mu_k^r + \mu_k^0)}{\sum_{\ell=0}^{M} \mu_k^\ell} + \frac{(1 - p_\phi^j((\boldsymbol{e}^M)_k, \boldsymbol{a}))(\mu_i^j + \mu_i^0)}{\sum_{\ell=0}^{j-1} \mu_i^\ell}.$$

**Proof:** When the $i$th component is in series with the rest of the system we have, from Equation (6.54), as in the proof of Theorem 6.3

$$\frac{\sum_{r=1}^{M} I_B^{(i,r,j)}(\mu_i^r + \mu_i^0)}{\sum_{\ell=0}^{M} \mu_i^\ell} = \frac{I_B^{(i,j,j)}(\mu_i^j + \mu_i^0)}{\sum_{\ell=0}^{M} \mu_i^\ell}$$
$$= \frac{\sum_{r=1}^{M} I_B^{(k,r,j)} \sum_{\ell=r}^{M} \mu_k^\ell}{(\sum_{\ell=0}^{M} \mu_k^\ell)(\sum_{\ell=j}^{M} \mu_i^\ell)/(\mu_i^j + \mu_i^0)} + \frac{p_\phi^j((\boldsymbol{e}^0)_k, \boldsymbol{a})(\mu_i^j + \mu_i^0)}{\sum_{\ell=j}^{M} \mu_i^\ell}.$$

Applying the assumption $\sum_{\ell=j}^{M} \mu_i^\ell/(\mu_i^j + \mu_i^0) \leq \mu_k^M/(\mu_k^r + \mu_k^0)$ for $r = 1, \ldots, M$, the results follow. When the $i$th component is in parallel with the rest of the system we similarly have, as in the proof of Theorem 6.3

$$\frac{\sum_{r=1}^{M} I_B^{(i,r,j)}(\mu_i^r + \mu_i^0)}{\sum_{\ell=0}^{M} \mu_i^\ell} = \frac{I_B^{(i,j,j)}(\mu_i^j + \mu_i^0)}{\sum_{\ell=0}^{M} \mu_i^\ell}$$

$$= \frac{\sum_{r=1}^{M} I_B^{(k,r,j)}(\mu_k^r + \mu_k^0) \sum_{\ell=0}^{r-1} \mu_k^\ell / (\mu_k^r + \mu_k^0)}{(\sum_{\ell=0}^{M} \mu_k^\ell)(\sum_{\ell=0}^{j-1} \mu_i^\ell)/(\mu_i^j + \mu_i^0)}$$

$$+ \frac{(1 - p_\phi^j((\boldsymbol{e}^M)_k, \boldsymbol{a}))(\mu_i^j + \mu_i^0)}{\sum_{\ell=0}^{j-1} \mu_i^\ell}.$$

Applying the assumption $\sum_{\ell=0}^{j-1} \mu_i^\ell/(\mu_i^j + \mu_i^0) \leq \mu_k^0/(\mu_k^r + \mu_k^0)$ for $r = 1, \ldots, M$, the results follow.

According to this theorem, for a multistate system where the $i$th component is in series with the rest of the system, the left-hand side of the inequality of the assumption is decreasing in $\mu_i^0$, increasing in $\sum_{\ell=j+1}^{M} \mu_i^\ell$ and increasing in $\mu_i^j$ if $\mu_i^0 > \sum_{\ell=j+1}^{M} \mu_i^\ell$; i.e. the poorer the smaller. The corresponding right-hand side of this inequality is increasing in $\mu_k^M$ and decreasing in $\mu_k^r, r = 1, \ldots, M - 1$ and $\mu_k^0$; i.e. the better the larger. For a multistate system where the $i$th component is in parallel with the rest of the system, the left-hand side of the inequality of the assumption is decreasing in $\mu_i^j$, increasing in $\sum_{\ell=1}^{j-1} \mu_i^\ell$ and increasing in $\mu_i^0$ if $\mu_i^j > \sum_{\ell=1}^{j-1} \mu_i^\ell$; i.e. the better the smaller. The corresponding right-hand side of this inequality is increasing in $\mu_k^0$ and decreasing in $\mu_k^r, r = 1, \ldots, M$; i.e. the poorer the larger. All this seems sensible.

## 6.5 Concluding remarks

In this chapter we have first presented new measures of component importance in nonrepairable multistate systems. Reasonable measures of component importance for repairable systems represent a challenge. In this case Theorem 6.2 and Theorem 6.3 and its dual extension covering the stationary Barlow–Proschan measure are not satisfactory.

Theorem 6.6 covering the stationary Natvig measure for multistate repairable systems is not satisfactory since, for a multistate series system, the measure does not depend on component mean times to repair. However, Theorem 6.8 covering its dual extension seems very sensible. For jumps downwards and repair times which are Weibull distributed, the latter measure is given by Equation (6.53), which has a reasonable performance as a function of the shape parameters. When all shape parameters are equal according to Theorems 6.9 and 6.10, again this measure seems to be sensible.

# 7

# Measures of component importance – a numerical study

## 7.1 Introduction

In Natvig (2011) dynamic and stationary measures of importance of a component in a repairable multistate system were introduced. This is a part of the material presented in the previous chapter. According to the Barlow–Proschan type measures, a component is important if there is a high probability that a change in the component state causes a change in whether or not the system state is above a given state. On the other hand, the Natvig type measures focus on how a change in the component state affects the expected system uptime and downtime relative to the given system state. Extending the work in Natvig *et al.* (2009) from the binary to the multistate case, in this chapter a numerical study of these measures is given for two three-component systems, a bridge system and also applied to an offshore oil and gas production system. In the multistate case the importance of a component is calculated separately for each component state. Thus it may happen that a component is very important at one state and less important, or even irrelevant, at another. Unified measures combining the importance for all component states can be obtained by adding up the importance measures for each individual state. According to these unified measures a component can be important relative to a given system state but not to another. It can be seen that if the distributions of

*Multistate Systems Reliability Theory with Applications* Bent Natvig

the total component times spent in the non complete failure states for the multistate system and the component lifetimes for the binary system are identical, the Barlow–Proschan measure to the lowest system state simply reduces to the binary version of the measure. The extended Natvig measure, however, does not have this property. This indicates that the latter measure captures more information about the system.

In the examples given in the following three sections, covering the paper by Natvig *et al.* (2010b), we will consider a multistate description of both the components and the system. For the components we let $S_i = \{0, 1, 2\}, i = 1, \ldots, n$. We then regard the system as a flow network and let the system state be the amount of flow that can be transported through the network. In order to determine this we start out by identifying the binary minimal cut sets of the network $K_\ell, \ell = 1, \ldots, m$; i.e. the minimal sets of components the removal of which will break the connection between the endpoints of the network. We then apply the well-known max-flow–min-cut theorem, (see Ford and Fulkerson, 1956) and get

$$\phi(\boldsymbol{X}(t)) = \min_{1 \leq \ell \leq m} \sum_{i \in K_\ell} X_i(t). \tag{7.1}$$

For system A treated in Section 7.2 and the offshore oil and gas production system treated in Section 7.4, there is at least one component in series with the rest of the system and we see that we also have $S = \{0, 1, 2\}$. This is in accordance with our general set up. However, for system B treated in Section 7.2 and the bridge structure treated in Section 7.3, there are no components in series with the rest of the system and we end up with $S = \{0, 1, 2, 3, 4\}$. Since our simulation program is written for handling systems that can be considered flow networks, this does not cause any problems. Furthermore, in order to be able to compare with the binary description treated in Natvig *et al.* (2009) we let the distributions of the total component times spent in the non complete failure states for the multistate case be equal to the component lifetime distributions in the binary case except for the exponential distribution case treated in Section 7.4. For simplicity we also assume that the distributions of the times spent in each of the non complete failure states are the same for each component. Finally, we assume that the repair time distributions are the same in the binary and multistate cases.

## 7.2 Component importance in two three-component systems

In this section we will simulate the component importance in two systems with three components. Figure 7.1 shows the systems we will be looking at.

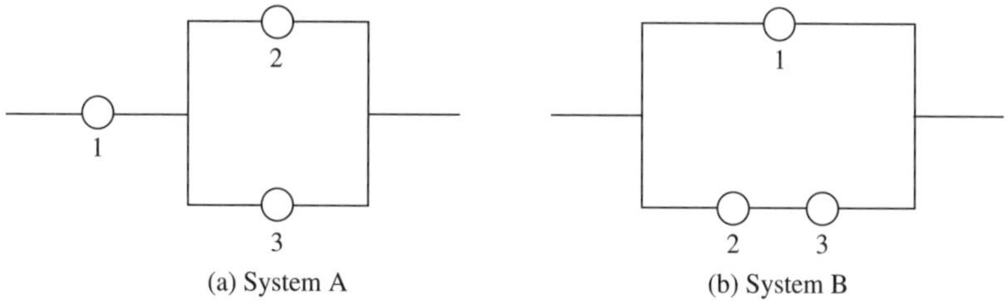


*Figure 7.1 Systems of three components.*

The times spent in each of the non complete failure states and the repair times are assumed to be gamma distributed. We will first see how an increasing variance in the distributions of the times spent in each of the non complete failure states of one of the components influences the component importances. The effect of a decreasing mean time to repair of one of the components will be investigated next.

First let the components have the following distributions of times spent in state $k$ before jumping to state $k-1$ for $k = 1, 2$ and of repair time:

Component 2: $\bar{F}_2^k(t) \sim \text{gamma}(2/c, 3c)$, $k = 1, 2$, $\bar{G}_2(t) \sim \text{gamma}(4, 1/2)$,

Component $i$: $\bar{F}_i^k(t) \sim \text{gamma}(4, 1)$, $k = 1, 2$, $\bar{G}_i(t) \sim \text{gamma}(4, 1/2)$, $i = 1, 3$,

where $c$ is a positive number. The mean times spent in a non complete failure state are $\mu_2^k = 6$ for component 2 and $\mu_i^k = 4$, $i = 1, 3$. All components have mean time to repair equal to $\mu_i^0 = 2$, $i = 1, 2, 3$. The variance associated with the times spent in a non complete failure state of component 2 is $18c$, while the variance associated with the corresponding times of components 1 and 3 is 4. The variances in the repair time distributions are 1 for all components. Tables 7.1 and 7.2 display the results from the simulations for all three versions of the Natvig measure and for $I_B^{(i,j)}(t)$ and $I_{B-P}^{(i,j)}(t)$ for system A for, respectively, system level $j = 1, 2$. Tables 7.3–7.6 do the same for system B for, respectively, system levels 1, 2, 3 and 4.

We first note that for both systems and all system levels, $I_{N,D}^{(i,j)}(t)$ and $I_{B-P}^{(i,j)}(t)$ are practically equal. Since stationarity is reached and the repair time distributions are the same for all three components, this is in accordance with results given in Section 6.4. Furthermore, component 2's importance is increasing in $c$ for both systems and all system levels both for the $I_N^{(i,j)}(t)$ and the extended measure. Hence, according to these measures, the increased uncertainty associated with an increasing variance leads to increased importance of a component.

Table 7.1 Simulations of system A with varying variance in the distributions of the times in a non complete failure state of component 2. Components 1 and 3 have identical distributions. System level is 1. The time horizon is $t = 20\,000$.

| $c$ | $i$ | $I_N^{(i,1)}(t)$ | $I_{N,D}^{(i,1)}(t)$ | $\bar{I}_N^{(i,1)}(t)$ | $I_B^{(i,1)}(t)$ | $I_{B-P}^{(i,1)}(t)$ |
|---|---|---|---|---|---|---|
| | 1 | 0.773 | 0.810 | 0.785 | 0.780 | 0.810 |
| 1/2 | 2 | 0.136 | 0.095 | 0.123 | 0.128 | 0.095 |
| | 3 | 0.091 | 0.095 | 0.092 | 0.092 | 0.095 |
| | 1 | 0.730 | 0.809 | 0.754 | 0.780 | 0.810 |
| 1 | 2 | 0.185 | 0.095 | 0.157 | 0.128 | 0.095 |
| | 3 | 0.086 | 0.095 | 0.089 | 0.092 | 0.095 |
| | 1 | 0.676 | 0.810 | 0.716 | 0.780 | 0.810 |
| 2 | 2 | 0.244 | 0.095 | 0.200 | 0.128 | 0.095 |
| | 3 | 0.079 | 0.095 | 0.084 | 0.092 | 0.095 |
| | 1 | 0.617 | 0.809 | 0.670 | 0.780 | 0.810 |
| 4 | 2 | 0.311 | 0.095 | 0.251 | 0.128 | 0.095 |
| | 3 | 0.072 | 0.095 | 0.079 | 0.092 | 0.095 |

Table 7.2 Simulations of system A with varying variance in the distributions of the times in a non complete failure state of component 2. Components 1 and 3 have identical distributions. System level is 2. The time horizon is $t = 20\,000$.

| $c$ | $i$ | $I_N^{(i,2)}(t)$ | $I_{N,D}^{(i,2)}(t)$ | $\bar{I}_N^{(i,2)}(t)$ | $I_B^{(i,2)}(t)$ | $I_{B-P}^{(i,2)}(t)$ |
|---|---|---|---|---|---|---|
| | 1 | 0.631 | 0.675 | 0.645 | 0.639 | 0.675 |
| 1/2 | 2 | 0.196 | 0.140 | 0.178 | 0.185 | 0.139 |
| | 3 | 0.174 | 0.186 | 0.178 | 0.176 | 0.186 |
| | 1 | 0.581 | 0.675 | 0.609 | 0.639 | 0.675 |
| 1 | 2 | 0.259 | 0.140 | 0.223 | 0.185 | 0.139 |
| | 3 | 0.160 | 0.186 | 0.168 | 0.176 | 0.186 |
| | 1 | 0.523 | 0.674 | 0.566 | 0.639 | 0.675 |
| 2 | 2 | 0.332 | 0.139 | 0.279 | 0.185 | 0.139 |
| | 3 | 0.144 | 0.186 | 0.156 | 0.176 | 0.186 |
| | 1 | 0.462 | 0.674 | 0.516 | 0.639 | 0.674 |
| 4 | 2 | 0.410 | 0.140 | 0.341 | 0.185 | 0.140 |
| | 3 | 0.128 | 0.186 | 0.143 | 0.176 | 0.186 |

Table 7.3 Simulations of system B with varying variance in the distributions of the times in a non complete failure state of component 2. Components 1 and 3 have identical distributions. System level is 1. The time horizon is $t = 20\,000$.

| $c$ | $i$ | $I_N^{(i,1)}(t)$ | $I_{N,D}^{(i,1)}(t)$ | $\bar{I}_N^{(i,1)}(t)$ | $I_B^{(i,1)}(t)$ | $I_{B-P}^{(i,1)}(t)$ |
|---|---|---|---|---|---|---|
| | 1 | 0.479 | 0.524 | 0.493 | 0.487 | 0.524 |
| 1/2 | 2 | 0.261 | 0.191 | 0.239 | 0.248 | 0.190 |
| | 3 | 0.261 | 0.286 | 0.269 | 0.266 | 0.286 |
| | 1 | 0.429 | 0.523 | 0.457 | 0.486 | 0.524 |
| 1 | 2 | 0.337 | 0.191 | 0.294 | 0.248 | 0.191 |
| | 3 | 0.234 | 0.286 | 0.249 | 0.266 | 0.286 |
| | 1 | 0.376 | 0.524 | 0.415 | 0.487 | 0.524 |
| 2 | 2 | 0.419 | 0.190 | 0.359 | 0.248 | 0.190 |
| | 3 | 0.205 | 0.286 | 0.226 | 0.265 | 0.285 |
| | 1 | 0.322 | 0.524 | 0.369 | 0.486 | 0.523 |
| 4 | 2 | 0.502 | 0.190 | 0.429 | 0.248 | 0.191 |
| | 3 | 0.176 | 0.286 | 0.202 | 0.266 | 0.286 |

Table 7.4 Simulations of system B with varying variance in the distributions of the times in a non complete failure state of component 2. Components 1 and 3 have identical distributions. System level is 2. The time horizon is $t = 20\,000$.

| $c$ | $i$ | $I_N^{(i,2)}(t)$ | $I_{N,D}^{(i,2)}(t)$ | $\bar{I}_N^{(i,2)}(t)$ | $I_B^{(i,2)}(t)$ | $I_{B-P}^{(i,2)}(t)$ |
|---|---|---|---|---|---|---|
| | 1 | 0.491 | 0.536 | 0.505 | 0.500 | 0.537 |
| 1/2 | 2 | 0.254 | 0.186 | 0.233 | 0.242 | 0.185 |
| | 3 | 0.255 | 0.279 | 0.262 | 0.259 | 0.278 |
| | 1 | 0.442 | 0.537 | 0.470 | 0.500 | 0.537 |
| 1 | 2 | 0.329 | 0.186 | 0.287 | 0.241 | 0.185 |
| | 3 | 0.229 | 0.277 | 0.243 | 0.259 | 0.278 |
| | 1 | 0.388 | 0.537 | 0.428 | 0.500 | 0.537 |
| 2 | 2 | 0.411 | 0.185 | 0.351 | 0.242 | 0.185 |
| | 3 | 0.201 | 0.278 | 0.221 | 0.259 | 0.278 |
| | 1 | 0.334 | 0.537 | 0.382 | 0.500 | 0.537 |
| 4 | 2 | 0.494 | 0.185 | 0.420 | 0.241 | 0.185 |
| | 3 | 0.173 | 0.278 | 0.197 | 0.259 | 0.278 |

Table 7.5 Simulations of system B with varying variance in the distributions of the times in a non complete failure state of component 2. Components 1 and 3 have identical distributions. System level is 3. The time horizon is $t = 20\,000$.

| $c$ | $i$ | $I_N^{(i,3)}(t)$ | $I_{N,D}^{(i,3)}(t)$ | $\bar{I}_N^{(i,3)}(t)$ | $I_B^{(i,3)}(t)$ | $I_{B-P}^{(i,3)}(t)$ |
|---|---|---|---|---|---|---|
| | 1 | 0.400 | 0.445 | 0.414 | 0.408 | 0.444 |
| 1/2 | 2 | 0.300 | 0.222 | 0.276 | 0.286 | 0.222 |
| | 3 | 0.300 | 0.333 | 0.310 | 0.306 | 0.333 |
| | 1 | 0.354 | 0.444 | 0.379 | 0.408 | 0.445 |
| 1 | 2 | 0.381 | 0.222 | 0.336 | 0.286 | 0.222 |
| | 3 | 0.265 | 0.334 | 0.285 | 0.306 | 0.333 |
| | 1 | 0.305 | 0.444 | 0.340 | 0.408 | 0.444 |
| 2 | 2 | 0.467 | 0.222 | 0.405 | 0.286 | 0.222 |
| | 3 | 0.228 | 0.334 | 0.255 | 0.306 | 0.333 |
| | 1 | 0.257 | 0.445 | 0.299 | 0.408 | 0.444 |
| 4 | 2 | 0.551 | 0.222 | 0.477 | 0.286 | 0.222 |
| | 3 | 0.193 | 0.333 | 0.224 | 0.306 | 0.333 |

Table 7.6 Simulations of system B with varying variance in the distributions of the times in a non complete failure state of component 2. Components 1 and 3 have identical distributions. System level is 4. The time horizon is $t = 20\,000$.

| $c$ | $i$ | $I_N^{(i,4)}(t)$ | $I_{N,D}^{(i,4)}(t)$ | $\bar{I}_N^{(i,4)}(t)$ | $I_B^{(i,4)}(t)$ | $I_{B-P}^{(i,4)}(t)$ |
|---|---|---|---|---|---|---|
| | 1 | 0.334 | 0.375 | 0.347 | 0.341 | 0.375 |
| 1/2 | 2 | 0.333 | 0.250 | 0.307 | 0.318 | 0.250 |
| | 3 | 0.333 | 0.375 | 0.346 | 0.341 | 0.375 |
| | 1 | 0.291 | 0.375 | 0.315 | 0.341 | 0.375 |
| 1 | 2 | 0.418 | 0.251 | 0.371 | 0.318 | 0.250 |
| | 3 | 0.291 | 0.375 | 0.314 | 0.341 | 0.375 |
| | 1 | 0.247 | 0.375 | 0.279 | 0.341 | 0.375 |
| 2 | 2 | 0.506 | 0.250 | 0.442 | 0.318 | 0.250 |
| | 3 | 0.247 | 0.375 | 0.279 | 0.341 | 0.375 |
| | 1 | 0.205 | 0.375 | 0.242 | 0.341 | 0.375 |
| 4 | 2 | 0.589 | 0.250 | 0.516 | 0.318 | 0.250 |
| | 3 | 0.206 | 0.375 | 0.242 | 0.341 | 0.375 |

We will next look at how the extended measure ranks the components. For system A and system level 1, components 2 and 3 are equally important according to $I_{B-P}^{(i,1)}(t)$ irrespective of $c$. Since the variance of the distribution of the times in a non complete failure state of component 2 is greater than the corresponding one of component 3 for all $c$, the former component is more important than the latter according to the extended measure. Except for $c = 1/2$ this is also true for system level 2. However, for both system levels, component 1 is not challenged as the most important for this measure being in series with the rest of the system and by far the most important according to $I_B^{(i,j)}(t)$.

For system B and all system levels, component 3 is more important than component 2 according to $I_{B-P}^{(i,j)}(t)$ irrespective of $c$. This is also true for the extended measure for $c = 1/2$. However, as the variance of the distribution of the times in a non complete failure state of component 2 increases, this component gets increasingly more important according to the extended measure, and finally the most important one for all system levels.

We will now turn our attention to the case where one of the components experiences a decreasing mean time to repair (MTTR). First we will assume that this is the case for component 1, and that components 2 and 3 have identical distributions of times spent in each of the non complete failure states and of repair times. Then the roles of components 1 and 2 are interchanged. More specifically, let the components have the following distributions of times spent in state $k$ before jumping to state $k-1$ for $k = 1, 2$ and of repair time:

$$\text{Component 1: } \bar{F}_1^k(t) \sim \text{gamma}(12, 1), k = 1, 2, \bar{G}_1(t) \sim \text{gamma}(\tfrac{24}{c^2}, c),$$
$$\text{Component } i\text{: } \bar{F}_i^k(t) \sim \text{gamma}(12, 1), k = 1, 2, \bar{G}_i(t) \sim \text{gamma}(4, \tfrac{1}{2}),$$
$$i = 2, 3,$$

where $c$ is a positive number. Component 1's mean time to repair is $\mu_1^0 = 24/c$, while components 2 and 3 have constant mean times to repair equal to $\mu_i^0 = 2,\ i = 2, 3$. The mean times spent in the non complete failure states are equal for all components, thus $\mu_i^k = 12,\ i = 1, 2, 3$.

The variances are constant in all distributions. In the distributions of times spent in each of the non complete failure states, the variances are 12 for all components. The variance in the repair time distribution of component 1 is 24, whereas the variance in the repair time distributions of components 2 and 3 equals 1.

Tables 7.7 and 7.8 display the results from the simulations for component 1 for all three versions of the Natvig measure and for $I_B^{(1,j)}(t)$ and $I_{B-P}^{(1,j)}(t)$ for system A for, respectively, system levels 1 and 2.

Table 7.7 Simulations of system A with decreasing MTTR of component 1. Components 2 and 3 have identical distributions. System level is 1. The time horizon is $t = 20\,000$.

| $c$ | $I_N^{(1,1)}(t)$ | $I_{N,D}^{(1,1)}(t)$ | $\bar{I}_N^{(1,1)}(t)$ | $I_B^{(1,1)}(t)$ | $I_{B-P}^{(1,1)}(t)$ |
|---|---|---|---|---|---|
| 1 | 0.875 | 0.971 | 0.928 | 0.928 | 0.875 |
| 2 | 0.875 | 0.971 | 0.929 | 0.906 | 0.875 |
| 3 | 0.875 | 0.972 | 0.930 | 0.896 | 0.875 |
| 4 | 0.876 | 0.972 | 0.931 | 0.890 | 0.875 |
| 6 | 0.875 | 0.972 | 0.930 | 0.883 | 0.875 |

Table 7.8 Simulations of system A with decreasing MTTR of component 1. Components 2 and 3 have identical distributions. System level is 2. The time horizon is $t = 20\,000$.

| $c$ | $I_N^{(1,2)}(t)$ | $I_{N,D}^{(1,2)}(t)$ | $\bar{I}_N^{(1,2)}(t)$ | $I_B^{(1,2)}(t)$ | $I_{B-P}^{(1,2)}(t)$ |
|---|---|---|---|---|---|
| 1 | 0.650 | 0.898 | 0.775 | 0.774 | 0.650 |
| 2 | 0.651 | 0.900 | 0.778 | 0.720 | 0.651 |
| 3 | 0.650 | 0.902 | 0.779 | 0.696 | 0.650 |
| 4 | 0.651 | 0.903 | 0.781 | 0.682 | 0.651 |
| 6 | 0.650 | 0.903 | 0.780 | 0.667 | 0.650 |

Table 7.9 Simulations of system B with decreasing MTTR of component 1. Components 2 and 3 have identical distributions. System level is 1. The time horizon is $t = 20\,000$.

| $c$ | $I_N^{(1,1)}(t)$ | $I_{N,D}^{(1,1)}(t)$ | $\bar{I}_N^{(1,1)}(t)$ | $I_B^{(1,1)}(t)$ | $I_{B-P}^{(1,1)}(t)$ |
|---|---|---|---|---|---|
| 1 | 0.080 | 0.290 | 0.138 | 0.138 | 0.080 |
| 2 | 0.148 | 0.457 | 0.246 | 0.194 | 0.148 |
| 3 | 0.207 | 0.562 | 0.331 | 0.243 | 0.207 |
| 4 | 0.258 | 0.634 | 0.399 | 0.286 | 0.258 |
| 6 | 0.342 | 0.722 | 0.498 | 0.359 | 0.342 |

Tables 7.9–7.12 do the same for system B for, respectively, system levels 1, 2, 3 and 4. Since components 2 and 3 have interchangeable positions both in system A and B, and identical distributions of times spent in each of the non complete failure states and of repair times, they have the same importance for each of the five measures. These

Table 7.10 Simulations of system B with decreasing MTTR of component 1. Components 2 and 3 have identical distributions. System level is 2. The time horizon is $t = 20\,000$.

| $c$ | $I_N^{(1,2)}(t)$ | $I_{N,D}^{(1,2)}(t)$ | $\bar{I}_N^{(1,2)}(t)$ | $I_B^{(1,2)}(t)$ | $I_{B-P}^{(1,2)}(t)$ |
|---|---|---|---|---|---|
| 1 | 0.315 | 0.685 | 0.460 | 0.460 | 0.316 |
| 2 | 0.380 | 0.749 | 0.536 | 0.460 | 0.381 |
| 3 | 0.409 | 0.774 | 0.568 | 0.460 | 0.409 |
| 4 | 0.425 | 0.787 | 0.586 | 0.460 | 0.425 |
| 6 | 0.442 | 0.798 | 0.602 | 0.460 | 0.442 |

Table 7.11 Simulations of system B with decreasing MTTR of component 1. Components 2 and 3 have identical distributions. System level is 3. The time horizon is $t = 20\,000$.

| $c$ | $I_N^{(1,3)}(t)$ | $I_{N,D}^{(1,3)}(t)$ | $\bar{I}_N^{(1,3)}(t)$ | $I_B^{(1,3)}(t)$ | $I_{B-P}^{(1,3)}(t)$ |
|---|---|---|---|---|---|
| 1 | 0.401 | 0.758 | 0.552 | 0.552 | 0.400 |
| 2 | 0.400 | 0.764 | 0.556 | 0.480 | 0.400 |
| 3 | 0.400 | 0.767 | 0.559 | 0.451 | 0.400 |
| 4 | 0.400 | 0.770 | 0.561 | 0.435 | 0.400 |
| 6 | 0.401 | 0.769 | 0.561 | 0.418 | 0.400 |

Table 7.12 Simulations of system B with decreasing MTTR of component 1. Components 2 and 3 have identical distributions. System level is 4. The time horizon is $t = 20\,000$.

| $c$ | $I_N^{(1,4)}(t)$ | $I_{N,D}^{(1,4)}(t)$ | $\bar{I}_N^{(1,4)}(t)$ | $I_B^{(1,4)}(t)$ | $I_{B-P}^{(1,4)}(t)$ |
|---|---|---|---|---|---|
| 1 | 0.334 | 0.702 | 0.480 | 0.480 | 0.334 |
| 2 | 0.333 | 0.709 | 0.485 | 0.409 | 0.333 |
| 3 | 0.333 | 0.713 | 0.487 | 0.381 | 0.334 |
| 4 | 0.334 | 0.715 | 0.489 | 0.366 | 0.333 |
| 6 | 0.332 | 0.713 | 0.488 | 0.349 | 0.333 |

importances are easily found given the ones for component 1. Now we note, for both systems and all system levels, that $I_N^{(1,j)}(t)$ and $I_{B-P}^{(1,j)}(t)$ are practically equal. Since stationarity is reached and the distributions of times spent in each of the non complete failure states are the same for all three components, this is in accordance with results given in Section 6.4.

For system A for both system levels all measures are practically constant in $c$ for component 1 except $I_B^{(1,j)}(t)$ which is decreasing in $c$. The latter fact follows from Equations (6.20) and (6.21) since the component is in series with the rest of the system and its stationary availability, $a_1^j$, increases due to the decreasing mean time to repair, $\mu_1^0$. For $I_{B-P}^{(1,j)}(t)$ it follows from Equation (6.28) that the increase in the asymptotic failure rate $1/(\sum_{\ell=0}^{M} \mu_1^{\ell})$ as $\mu_1^0$ decreases, compensates for the decrease in $I_B^{(1,j)}(t)$.

For system B, component 1 is in parallel with the rest of the system and the increase in $a_1^1$ as $\mu_1^0$ decreases leads to an increasing $I_B^{(1,1)}(t)$. Since $1/(\sum_{\ell=0}^{M} \mu_1^{\ell})$ increases as well, $I_{B-P}^{(1,j)}(t)$ also increases for $j = 1, 2$. As for system A, $I_N^{(1,j)}(t)$ behaves much like $I_{B-P}^{(1,j)}(t)$.

If one is comparing the results in Tables 7.1, 7.3, 7.7 and 7.9 covering system level 1 to the corresponding results for the binary case given in Natvig *et al*. (2009), the Birnbaum measure, the Barlow–Proschan measure and the dual Natvig measure are mostly identical. The reason is that these measures are not affected by the fictive minimal repairs of the components when jumping downwards from one of the non complete failure states. In particular, the extended Natvig measure, however, is different. This indicates that the latter measure captures more information about the system.

Now, we consider the case where components 1 and 3 are assumed to have identical distributions of times spent in each of the non complete failure states and of repair times. More specifically, let the components have the following distributions of times spent in state $k$ before jumping to state $k-1$ for $k = 1, 2$ and of repair time:

$$\text{Component 2: } \bar{F}_2^k(t) \sim \text{gamma}(4, 1), k = 1, 2, \; \bar{G}_2(t) \sim \text{gamma}(\tfrac{6}{c^2}, \tfrac{c}{2}),$$
$$\text{Component } i\text{: } \bar{F}_i^k(t) \sim \text{gamma}(4, 1), k = 1, 2, \; \bar{G}_i(t) \sim \text{gamma}(6, \tfrac{1}{2}), \quad i = 1, 3,$$

where $c$ is once again a positive number. In this example all components have identical distributions of times spent in each of the non complete failure states with mean times $\mu_i^k = 4, i = 1, 2, 3$. Moreover, the variances associated with these distributions are all equal to 4. The repair time distributions are identical for components 1 and 3 as well. The mean time to repair of these components is $\mu_i^0 = 3, \; i = 1, 3$, while the mean time to repair of component 2 is $\mu_2^0 = 3/c$. The variances in the repair time distributions are 3/2 for all components.

Tables 7.13 and 7.14 display the results from the simulations for all three versions of the Natvig measure and for $I_B^{(i,j)}(t)$ and $I_{B-P}^{(i,j)}(t)$ for

Table 7.13 Simulations of system A with decreasing MTTR of component 2. Components 1 and 3 have identical distributions. System level is 1. The time horizon is $t = 20\,000$.

| $c$ | $i$ | $I_N^{(i,1)}(t)$ | $I_{N,D}^{(i,1)}(t)$ | $\bar{I}_N^{(i,1)}(t)$ | $I_B^{(i,1)}(t)$ | $I_{B-P}^{(i,1)}(t)$ |
|---|---|---|---|---|---|---|
| | 1 | 0.654 | 0.656 | 0.655 | 0.634 | 0.654 |
| 1/2 | 2 | 0.115 | 0.113 | 0.114 | 0.142 | 0.115 |
| | 3 | 0.231 | 0.231 | 0.231 | 0.224 | 0.231 |
| | 1 | 0.681 | 0.683 | 0.682 | 0.673 | 0.682 |
| 3/4 | 2 | 0.137 | 0.135 | 0.136 | 0.147 | 0.136 |
| | 3 | 0.182 | 0.182 | 0.182 | 0.180 | 0.182 |
| | 1 | 0.701 | 0.700 | 0.700 | 0.700 | 0.700 |
| 1 | 2 | 0.150 | 0.150 | 0.150 | 0.150 | 0.150 |
| | 3 | 0.150 | 0.150 | 0.150 | 0.150 | 0.150 |
| | 1 | 0.722 | 0.720 | 0.721 | 0.733 | 0.722 |
| 3/2 | 2 | 0.167 | 0.170 | 0.168 | 0.154 | 0.167 |
| | 3 | 0.111 | 0.110 | 0.111 | 0.113 | 0.111 |
| | 1 | 0.735 | 0.731 | 0.734 | 0.753 | 0.735 |
| 2 | 2 | 0.176 | 0.181 | 0.178 | 0.156 | 0.176 |
| | 3 | 0.088 | 0.088 | 0.088 | 0.090 | 0.088 |
| | 1 | 0.744 | 0.739 | 0.742 | 0.767 | 0.744 |
| 5/2 | 2 | 0.183 | 0.188 | 0.185 | 0.158 | 0.183 |
| | 3 | 0.073 | 0.073 | 0.073 | 0.075 | 0.073 |

system A for, respectively, system levels 1 and 2. Tables 7.15–7.18 do the same for system B for, respectively, system levels 1, 2, 3 and 4.

As for the previous case $I_N^{(i,j)}(t)$ and $I_{B-P}^{(i,j)}(t)$ are practically equal for both systems and all system levels, since the distributions of times spent in each of the non complete failure states are the same for all three components. When $c = 1$, all components have identical distributions. Hence, in accordance with results given in Section 6.4, all measures give the same results in each system for each system level in this case. Irrespective of $c$ the results are very similar for the three Natvig importance measures and $I_{B-P}^{(i,j)}(t)$ in each system for each system level. For system A and both system levels and all measures, the importance of component 3 is decreasing in $c$, while the other components get increasingly more important except for $I_B^{(2,2)}(t)$ which is decreasing in $c$. For system B for all measures and system levels 1 and 2, the importance of component 1 is decreasing in $c$, while the other components get increasingly more important again except for $I_B^{(2,2)}(t)$ which is decreasing in $c$. For the three Natvig importance measures and $I_{B-P}^{(i,j)}(t)$ and system levels 3 and 4, the

Table 7.14 Simulations of system A with decreasing MTTR of component 2. Components 1 and 3 have identical distributions. System level is 2. The time horizon is $t = 20\,000$.

| $c$ | $i$ | $I_N^{(i,2)}(t)$ | $I_{N,D}^{(i,2)}(t)$ | $\bar{I}_N^{(i,2)}(t)$ | $I_B^{(i,2)}(t)$ | $I_{B-P}^{(i,2)}(t)$ |
|---|---|---|---|---|---|---|
| | 1 | 0.595 | 0.598 | 0.596 | 0.570 | 0.596 |
| 1/2 | 2 | 0.167 | 0.162 | 0.165 | 0.203 | 0.166 |
| | 3 | 0.238 | 0.239 | 0.239 | 0.228 | 0.238 |
| | 1 | 0.605 | 0.607 | 0.606 | 0.596 | 0.605 |
| 3/4 | 2 | 0.184 | 0.182 | 0.184 | 0.198 | 0.184 |
| | 3 | 0.210 | 0.211 | 0.211 | 0.207 | 0.210 |
| | 1 | 0.611 | 0.611 | 0.611 | 0.611 | 0.611 |
| 1 | 2 | 0.194 | 0.194 | 0.194 | 0.194 | 0.194 |
| | 3 | 0.195 | 0.195 | 0.195 | 0.195 | 0.195 |
| | 1 | 0.618 | 0.616 | 0.617 | 0.630 | 0.618 |
| 3/2 | 2 | 0.206 | 0.208 | 0.207 | 0.191 | 0.206 |
| | 3 | 0.176 | 0.175 | 0.176 | 0.180 | 0.176 |
| | 1 | 0.622 | 0.617 | 0.620 | 0.640 | 0.622 |
| 2 | 2 | 0.212 | 0.217 | 0.214 | 0.188 | 0.212 |
| | 3 | 0.166 | 0.166 | 0.166 | 0.171 | 0.166 |
| | 1 | 0.623 | 0.619 | 0.621 | 0.647 | 0.624 |
| 5/2 | 2 | 0.217 | 0.222 | 0.219 | 0.187 | 0.216 |
| | 3 | 0.160 | 0.159 | 0.160 | 0.166 | 0.160 |

importance of all components is almost constant in $c$. In particular, for system level 4, according to these measures all components are almost equally important. On the other hand, for $I_B^{(1,j)}(t)$ for system levels 3 and 4, the importance of component 1 is increasing in $c$.

We now turn to the ranking of the components according to the extended measure. We see that component 1 is ranked on top in both systems and for all system levels and almost all values of $c$, so we focus on the ranking of components 2 and 3. For both systems and all system levels, component 3 is ranked before component 2 for $c = 1/2, 3/4$, whereas the ranking is opposite for $c = 3/2, 2, 5/2$. To explain this, we consider each system separately.

We first consider system A where the ranking is the same as for $I_B^{(i,j)}(t)$. When $c < 1$, component 2 has lower stationary availabilities than component 3. Since these components are connected in parallel, according to $I_B^{(i,j)}(t)$ component 3 is ranked before component 2. As soon as $c$

Table 7.15 Simulations of system B with decreasing MTTR of component 2. Components 1 and 3 have identical distributions. System level is 1. The time horizon is $t = 20\,000$.

| $c$ | $i$ | $I_N^{(i,1)}(t)$ | $I_{N,D}^{(i,1)}(t)$ | $\bar{I}_N^{(i,1)}(t)$ | $I_B^{(i,1)}(t)$ | $I_{B-P}^{(i,1)}(t)$ |
|---|---|---|---|---|---|---|
| | 1 | 0.652 | 0.654 | 0.653 | 0.622 | 0.652 |
| 1/2 | 2 | 0.174 | 0.170 | 0.172 | 0.211 | 0.174 |
| | 3 | 0.174 | 0.175 | 0.175 | 0.166 | 0.174 |
| | 1 | 0.586 | 0.588 | 0.587 | 0.575 | 0.586 |
| 3/4 | 2 | 0.207 | 0.205 | 0.206 | 0.222 | 0.207 |
| | 3 | 0.207 | 0.207 | 0.207 | 0.203 | 0.207 |
| | 1 | 0.543 | 0.543 | 0.543 | 0.543 | 0.543 |
| 1 | 2 | 0.229 | 0.228 | 0.229 | 0.228 | 0.228 |
| | 3 | 0.228 | 0.228 | 0.228 | 0.228 | 0.228 |
| | 1 | 0.489 | 0.486 | 0.488 | 0.501 | 0.489 |
| 3/2 | 2 | 0.256 | 0.260 | 0.257 | 0.238 | 0.256 |
| | 3 | 0.255 | 0.254 | 0.254 | 0.261 | 0.255 |
| | 1 | 0.458 | 0.455 | 0.457 | 0.476 | 0.458 |
| 2 | 2 | 0.271 | 0.277 | 0.273 | 0.243 | 0.271 |
| | 3 | 0.271 | 0.268 | 0.270 | 0.281 | 0.271 |
| | 1 | 0.437 | 0.432 | 0.435 | 0.458 | 0.437 |
| 5/2 | 2 | 0.282 | 0.288 | 0.284 | 0.247 | 0.282 |
| | 3 | 0.282 | 0.279 | 0.281 | 0.295 | 0.282 |

gets larger than 1, the roles of component 2 and 3 change. Now component 2 has the higher stationary availabilities of the two, thus, according to $I_B^{(i,j)}(t)$, this component is ranked before component 3. We then turn to system B. Here the ranking of the Natvig measures does not follow $I_B^{(i,j)}(t)$, rather more $I_{B-P}^{(i,j)}(t)$ since the factor $1/\sum_{\ell=0}^{M}\mu_2^{\ell}$ is increasing in $c$ due to $\mu_2^0$ decreasing in $c$.

## 7.3 Component importance in the bridge system

In this section we will investigate the bridge system depicted in Figure 7.2. As in the previous section the times spent in each of the non complete failure states and the repair times are assumed to be gamma distributed.

We will again first see how an increasing variance in the distribution of the times spent in each of the non complete failure states of one of the components influences the component importances. More specifically, let

Table 7.16 Simulations of system B with decreasing MTTR of component 2. Components 1 and 3 have identical distributions. System level is 2. The time horizon is $t = 20\,000$.

| $c$ | $i$ | $I_N^{(i,2)}(t)$ | $I_{N,D}^{(i,2)}(t)$ | $\bar{I}_N^{(i,2)}(t)$ | $I_B^{(i,2)}(t)$ | $I_{B-P}^{(i,2)}(t)$ |
|---|---|---|---|---|---|---|
| | 1 | 0.610 | 0.614 | 0.612 | 0.580 | 0.610 |
| 1/2 | 2 | 0.195 | 0.191 | 0.193 | 0.236 | 0.195 |
| | 3 | 0.195 | 0.196 | 0.195 | 0.185 | 0.195 |
| | 1 | 0.569 | 0.570 | 0.569 | 0.557 | 0.568 |
| 3/4 | 2 | 0.216 | 0.213 | 0.215 | 0.231 | 0.216 |
| | 3 | 0.216 | 0.216 | 0.216 | 0.212 | 0.216 |
| | 1 | 0.544 | 0.544 | 0.544 | 0.544 | 0.544 |
| 1 | 2 | 0.228 | 0.228 | 0.228 | 0.228 | 0.228 |
| | 3 | 0.228 | 0.228 | 0.228 | 0.228 | 0.228 |
| | 1 | 0.517 | 0.514 | 0.516 | 0.528 | 0.517 |
| 3/2 | 2 | 0.241 | 0.245 | 0.243 | 0.225 | 0.242 |
| | 3 | 0.242 | 0.241 | 0.241 | 0.247 | 0.242 |
| | 1 | 0.501 | 0.497 | 0.500 | 0.519 | 0.501 |
| 2 | 2 | 0.249 | 0.255 | 0.252 | 0.223 | 0.249 |
| | 3 | 0.249 | 0.248 | 0.249 | 0.258 | 0.249 |
| | 1 | 0.492 | 0.487 | 0.490 | 0.513 | 0.492 |
| 5/2 | 2 | 0.254 | 0.260 | 0.256 | 0.222 | 0.254 |
| | 3 | 0.254 | 0.253 | 0.254 | 0.265 | 0.254 |

the components have the following distributions of times spent in state $k$ before jumping to state $k-1$ for $k = 1, 2$ and of repair time:

Component 1: $\bar{F}_1^k(t) \sim \text{gamma}(6/c, c), k = 1, 2, \bar{G}_1(t) \sim \text{gamma}(2, 1)$,
Component $i$: $\bar{F}_i^k(t) \sim \text{gamma}(3/2, 2), k = 1, 2, \bar{G}_i(t) \sim \text{gamma}(2, 1)$, $i = 2, 3, 4$,
Component 5: $\bar{F}_5^k(t) \sim \text{gamma}(6, 1), k = 1, 2, \bar{G}_5(t) \sim \text{gamma}(2, 1)$,

where $c$ is a positive number. The mean times spent in each of the non complete failure states of components 1 and 5 are $\mu_i^k = 6, i = 1, 5$ which is twice the size of the corresponding means of the other components. Thus, $\mu_i^k = 3, i = 2, 3, 4$. All the components have the same repair time distributions with expectation $\mu_i^0 = 2, i = 1, \ldots, 5$. The variances in the distributions of times spent in each of the non complete failure states are 6 for all components except for the component 1, which has a variance of $6c$. In the repair time distributions the variances are 2.

The results of the simulations are shown in Tables 7.19–7.22 for all three versions of the Natvig measure and for $I_B^{(i,j)}(t)$ and $I_{B-P}^{(i,j)}(t)$

Table 7.17 Simulations of system B with decreasing MTTR of component 2. Components 1 and 3 have identical distributions. System level is 3. The time horizon is $t = 20\,000$.

| $c$ | $i$ | $I_N^{(i,3)}(t)$ | $I_{N,D}^{(i,3)}(t)$ | $\bar{I}_N^{(i,3)}(t)$ | $I_B^{(i,3)}(t)$ | $I_{B-P}^{(i,3)}(t)$ |
|---|---|---|---|---|---|---|
| | 1 | 0.400 | 0.404 | 0.402 | 0.370 | 0.400 |
| 1/2 | 2 | 0.299 | 0.294 | 0.297 | 0.353 | 0.300 |
| | 3 | 0.300 | 0.303 | 0.301 | 0.277 | 0.300 |
| | 1 | 0.400 | 0.402 | 0.401 | 0.389 | 0.400 |
| 3/4 | 2 | 0.300 | 0.297 | 0.299 | 0.319 | 0.300 |
| | 3 | 0.300 | 0.301 | 0.300 | 0.292 | 0.300 |
| | 1 | 0.400 | 0.400 | 0.400 | 0.400 | 0.400 |
| 1 | 2 | 0.300 | 0.300 | 0.300 | 0.300 | 0.300 |
| | 3 | 0.300 | 0.300 | 0.300 | 0.300 | 0.300 |
| | 1 | 0.400 | 0.397 | 0.399 | 0.411 | 0.400 |
| 3/2 | 2 | 0.300 | 0.304 | 0.302 | 0.281 | 0.300 |
| | 3 | 0.300 | 0.298 | 0.299 | 0.308 | 0.300 |
| | 1 | 0.400 | 0.397 | 0.399 | 0.417 | 0.400 |
| 2 | 2 | 0.300 | 0.306 | 0.302 | 0.270 | 0.300 |
| | 3 | 0.300 | 0.298 | 0.299 | 0.313 | 0.300 |
| | 1 | 0.400 | 0.396 | 0.399 | 0.421 | 0.400 |
| 5/2 | 2 | 0.300 | 0.307 | 0.303 | 0.264 | 0.300 |
| | 3 | 0.300 | 0.297 | 0.299 | 0.315 | 0.300 |

for, respectively, system levels 1, 2, 3 and 4. As for the case presented in Tables 7.1–7.6 all components have the same repair time distribution. Hence, in accordance with results given in Section 6.4, $I_{N,D}^{(i,j)}(t)$ and $I_{B-P}^{(i,j)}(t)$ are practically equal. Parallel to the case mentioned above, component 1's importance is increasing in $c$ for all system levels both for the $I_N^{(1,j)}(t)$ and the extended measure. Hence, again according to these measures, the increased uncertainty associated with an increasing variance leads to increased importance of a component.

The ranks of the extended measure shown in Table 7.23 change a lot as for system B in the case mentioned above. As $c$ increases, and component 1 becomes more uncertain, it also becomes more important. The other components, however, are ranked in the order $2 \approx 4, 5, 3$. When $c = 3/2$, components 1 and 5 have swapped places in the ranking. Component 1 becomes the most important component when $c = 6$ for system levels 1 and 2, but not for system levels 3 and 4.

We will now look at how a decreasing mean time to repair of one of the components influences the importance measures. More specifically, let

Table 7.18 Simulations of system B with decreasing MTTR of component 2. Components 1 and 3 have identical distributions. System level is 4. The time horizon is $t = 20\,000$.

| $c$ | $i$ | $I_N^{(i,4)}(t)$ | $I_{N,D}^{(i,4)}(t)$ | $\bar{I}_N^{(i,4)}(t)$ | $I_B^{(i,4)}(t)$ | $I_{B-P}^{(i,4)}(t)$ |
|---|---|---|---|---|---|---|
| | 1 | 0.334 | 0.336 | 0.334 | 0.306 | 0.333 |
| 1/2 | 2 | 0.333 | 0.328 | 0.331 | 0.389 | 0.333 |
| | 3 | 0.333 | 0.336 | 0.334 | 0.305 | 0.333 |
| | 1 | 0.333 | 0.334 | 0.333 | 0.323 | 0.333 |
| 3/4 | 2 | 0.333 | 0.330 | 0.332 | 0.353 | 0.333 |
| | 3 | 0.334 | 0.336 | 0.335 | 0.324 | 0.334 |
| | 1 | 0.333 | 0.333 | 0.333 | 0.333 | 0.333 |
| 1 | 2 | 0.334 | 0.334 | 0.334 | 0.334 | 0.334 |
| | 3 | 0.333 | 0.333 | 0.333 | 0.333 | 0.333 |
| | 1 | 0.333 | 0.331 | 0.333 | 0.344 | 0.333 |
| 3/2 | 2 | 0.334 | 0.338 | 0.335 | 0.313 | 0.334 |
| | 3 | 0.333 | 0.331 | 0.332 | 0.343 | 0.333 |
| | 1 | 0.333 | 0.330 | 0.332 | 0.349 | 0.334 |
| 2 | 2 | 0.333 | 0.341 | 0.336 | 0.302 | 0.333 |
| | 3 | 0.333 | 0.330 | 0.332 | 0.349 | 0.333 |
| | 1 | 0.333 | 0.329 | 0.331 | 0.352 | 0.333 |
| 5/2 | 2 | 0.334 | 0.341 | 0.337 | 0.296 | 0.334 |
| | 3 | 0.333 | 0.329 | 0.332 | 0.352 | 0.333 |

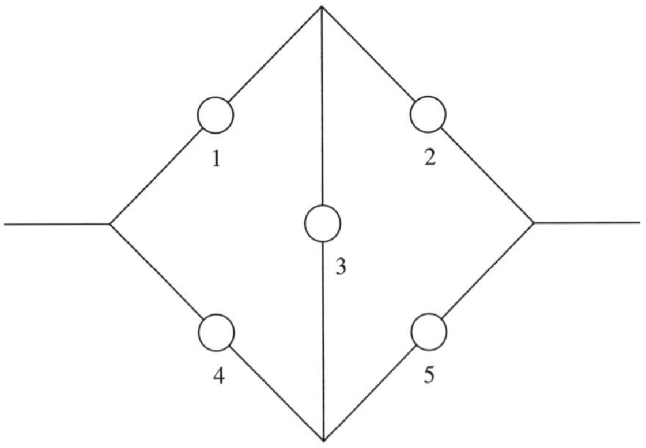


*Figure 7.2 The bridge system.*

Table 7.19 Simulations of the bridge system with increasing variance in the distributions of the times in a non complete failure state of component 1. System level is 1. The time horizon is $t = 20\,000$.

| $c$ | $i$ | $I_N^{(i,1)}(t)$ | $I_{N,D}^{(i,1)}(t)$ | $\bar{I}_N^{(i,1)}(t)$ | $I_B^{(i,1)}(t)$ | $I_{B-P}^{(i,1)}(t)$ |
|---|---|---|---|---|---|---|
| | 1 | 0.147 | 0.202 | 0.168 | 0.271 | 0.202 |
| | 2 | 0.277 | 0.257 | 0.269 | 0.197 | 0.257 |
| 1/2 | 3 | 0.088 | 0.082 | 0.086 | 0.063 | 0.082 |
| | 4 | 0.277 | 0.257 | 0.269 | 0.197 | 0.257 |
| | 5 | 0.211 | 0.202 | 0.208 | 0.271 | 0.202 |
| | 1 | 0.199 | 0.202 | 0.200 | 0.271 | 0.202 |
| | 2 | 0.259 | 0.257 | 0.259 | 0.197 | 0.257 |
| 1 | 3 | 0.083 | 0.082 | 0.083 | 0.063 | 0.082 |
| | 4 | 0.260 | 0.257 | 0.259 | 0.197 | 0.257 |
| | 5 | 0.198 | 0.202 | 0.200 | 0.272 | 0.202 |
| | 1 | 0.235 | 0.202 | 0.223 | 0.272 | 0.202 |
| | 2 | 0.248 | 0.257 | 0.252 | 0.197 | 0.257 |
| 3/2 | 3 | 0.079 | 0.082 | 0.080 | 0.063 | 0.082 |
| | 4 | 0.248 | 0.256 | 0.251 | 0.197 | 0.257 |
| | 5 | 0.189 | 0.202 | 0.194 | 0.271 | 0.202 |
| | 1 | 0.262 | 0.202 | 0.241 | 0.271 | 0.202 |
| | 2 | 0.239 | 0.257 | 0.245 | 0.197 | 0.257 |
| 2 | 3 | 0.076 | 0.082 | 0.078 | 0.063 | 0.082 |
| | 4 | 0.239 | 0.257 | 0.245 | 0.197 | 0.257 |
| | 5 | 0.183 | 0.203 | 0.190 | 0.272 | 0.202 |
| | 1 | 0.385 | 0.202 | 0.329 | 0.271 | 0.202 |
| | 2 | 0.199 | 0.257 | 0.217 | 0.197 | 0.257 |
| 6 | 3 | 0.064 | 0.082 | 0.069 | 0.063 | 0.082 |
| | 4 | 0.199 | 0.257 | 0.217 | 0.197 | 0.257 |
| | 5 | 0.152 | 0.202 | 0.168 | 0.272 | 0.202 |

the components have the following distributions of times spent in state $k$ before jumping to state $k-1$ for $k = 1, 2$ and of repair time:

Component 1: $\bar{F}_1^k(t) \sim \text{gamma}(4, 1), k = 1, 2,\ \bar{G}_1(t) \sim \text{gamma}(6/c^2, c/2)$,

Component $i$: $\bar{F}_i^k(t) \sim \text{gamma}(4, 1), k = 1, 2, \bar{G}_i(t) \sim \text{gamma}(6, 1/2)$, $i = 2, 3, 4, 5$,

where $c$ as above is a positive number. In this example the distributions of the times spent in each of the non complete failure states are equal for all components with mean times $\mu_i^k = 4$, $i = 1, \ldots, 5$. Moreover, the

Table 7.20 Simulations of the bridge system with increasing variance in the distributions of the times in a non complete failure state of component 1. System level is 2. The time horizon is $t = 20\,000$.

| $c$ | $i$ | $I_N^{(i,2)}(t)$ | $I_{N,D}^{(i,2)}(t)$ | $\bar{I}_N^{(i,2)}(t)$ | $I_B^{(i,2)}(t)$ | $I_{B-P}^{(i,2)}(t)$ |
|---|---|---|---|---|---|---|
| | 1 | 0.121 | 0.169 | 0.139 | 0.236 | 0.169 |
| | 2 | 0.315 | 0.296 | 0.308 | 0.237 | 0.297 |
| 1/2 | 3 | 0.074 | 0.070 | 0.072 | 0.056 | 0.070 |
| | 4 | 0.316 | 0.297 | 0.308 | 0.237 | 0.297 |
| | 5 | 0.174 | 0.169 | 0.172 | 0.236 | 0.169 |
| | 1 | 0.165 | 0.169 | 0.167 | 0.236 | 0.169 |
| | 2 | 0.299 | 0.296 | 0.298 | 0.237 | 0.296 |
| 1 | 3 | 0.070 | 0.070 | 0.070 | 0.056 | 0.070 |
| | 4 | 0.300 | 0.296 | 0.298 | 0.237 | 0.296 |
| | 5 | 0.165 | 0.169 | 0.167 | 0.236 | 0.169 |
| | 1 | 0.197 | 0.168 | 0.187 | 0.236 | 0.169 |
| | 2 | 0.288 | 0.297 | 0.291 | 0.237 | 0.296 |
| 3/2 | 3 | 0.068 | 0.070 | 0.069 | 0.056 | 0.070 |
| | 4 | 0.288 | 0.297 | 0.291 | 0.237 | 0.297 |
| | 5 | 0.159 | 0.168 | 0.162 | 0.236 | 0.169 |
| | 1 | 0.222 | 0.169 | 0.203 | 0.236 | 0.169 |
| | 2 | 0.279 | 0.296 | 0.285 | 0.237 | 0.296 |
| 2 | 3 | 0.065 | 0.069 | 0.067 | 0.056 | 0.070 |
| | 4 | 0.279 | 0.297 | 0.285 | 0.237 | 0.297 |
| | 5 | 0.154 | 0.169 | 0.159 | 0.236 | 0.169 |
| | 1 | 0.334 | 0.169 | 0.282 | 0.236 | 0.169 |
| | 2 | 0.239 | 0.296 | 0.257 | 0.237 | 0.296 |
| 6 | 3 | 0.056 | 0.070 | 0.060 | 0.056 | 0.070 |
| | 4 | 0.239 | 0.296 | 0.257 | 0.237 | 0.296 |
| | 5 | 0.132 | 0.169 | 0.144 | 0.236 | 0.169 |

variances associated with these distributions are all equal to 4. Components $2, \ldots, 5$ have mean time to repair equal to 3. The mean time to repair of component 1 is $3/c$, which is decreasing in $c$. The variance in the repair time distributions is $3/2$ for all components.

The results of the simulations are shown in Tables 7.24–7.27 for all three versions of the Natvig measure and for $I_B^{(i,j)}(t)$ and $I_{B-P}^{(i,j)}(t)$ for, respectively, system levels 1, 2, 3 and 4. As for the cases presented in Tables 7.7–7.18 all components have the same distributions of the times spent in each of the non complete failure states. Hence, in accordance with results given in Section 6.4, $I_N^{(i,j)}(t)$ and $I_{B-P}^{(i,j)}(t)$ are practically equal. When $c = 1$, all components have identical distributions. Hence,

Table 7.21 Simulations of the bridge system with increasing variance in the distributions of the times in a non complete failure state of component 1. System level is 3. The time horizon is $t = 20\,000$.

| $c$ | $i$ | $I_N^{(i,3)}(t)$ | $I_{N,D}^{(i,3)}(t)$ | $\bar{I}_N^{(i,3)}(t)$ | $I_B^{(i,3)}(t)$ | $I_{B-P}^{(i,3)}(t)$ |
|---|---|---|---|---|---|---|
| | 1 | 0.114 | 0.160 | 0.132 | 0.226 | 0.160 |
| | 2 | 0.339 | 0.319 | 0.331 | 0.258 | 0.320 |
| 1/2 | 3 | 0.044 | 0.042 | 0.043 | 0.034 | 0.042 |
| | 4 | 0.339 | 0.320 | 0.331 | 0.258 | 0.320 |
| | 5 | 0.164 | 0.160 | 0.162 | 0.225 | 0.160 |
| | 1 | 0.157 | 0.160 | 0.158 | 0.226 | 0.160 |
| | 2 | 0.322 | 0.319 | 0.321 | 0.258 | 0.319 |
| 1 | 3 | 0.042 | 0.042 | 0.042 | 0.034 | 0.042 |
| | 4 | 0.323 | 0.320 | 0.321 | 0.258 | 0.320 |
| | 5 | 0.157 | 0.160 | 0.158 | 0.225 | 0.160 |
| | 1 | 0.187 | 0.160 | 0.177 | 0.226 | 0.160 |
| | 2 | 0.311 | 0.319 | 0.314 | 0.258 | 0.319 |
| 3/2 | 3 | 0.041 | 0.042 | 0.041 | 0.033 | 0.042 |
| | 4 | 0.311 | 0.320 | 0.314 | 0.258 | 0.319 |
| | 5 | 0.151 | 0.160 | 0.154 | 0.226 | 0.160 |
| | 1 | 0.211 | 0.160 | 0.193 | 0.225 | 0.160 |
| | 2 | 0.302 | 0.319 | 0.308 | 0.258 | 0.319 |
| 2 | 3 | 0.039 | 0.042 | 0.040 | 0.034 | 0.042 |
| | 4 | 0.302 | 0.320 | 0.308 | 0.258 | 0.320 |
| | 5 | 0.146 | 0.159 | 0.151 | 0.225 | 0.160 |
| | 1 | 0.320 | 0.160 | 0.269 | 0.226 | 0.160 |
| | 2 | 0.260 | 0.320 | 0.279 | 0.258 | 0.320 |
| 6 | 3 | 0.034 | 0.042 | 0.036 | 0.034 | 0.042 |
| | 4 | 0.260 | 0.319 | 0.279 | 0.258 | 0.319 |
| | 5 | 0.126 | 0.160 | 0.137 | 0.225 | 0.160 |

in accordance with results given in Section 6.4, all measures give the same results for each system level in this case. As is seen from the tables, the results are very similar for all three Natvig importance measures and $I_{B-P}^{(i,j)}(t)$. The importance, according to these measures, of components 1 and 2 is increasing in $c$, while that for components 3 and 4 is decreasing in $c$ for system levels 1 and 2, whereas these importances are approximately constant for system levels 3 and 4. The resulting ranks, which are identical for all the three Natvig measures and $I_{B-P}^{(i,j)}(t)$, are shown in Table 7.28. As in the previous case, component 3 is always the least important component. For all the importance measures there is a turning point at $c = 1$. For this $c$, all components have identical distributions, leaving components 1,

Table 7.22 Simulations of the bridge system with increasing variance in the distributions of the times in a non complete failure state of component 1. System level is 4. The time horizon is $t = 20\,000$.

| $c$ | $i$ | $I_N^{(i,4)}(t)$ | $I_{N,D}^{(i,4)}(t)$ | $\bar{I}_N^{(i,4)}(t)$ | $I_B^{(i,4)}(t)$ | $I_{B-P}^{(i,4)}(t)$ |
|---|---|---|---|---|---|---|
| | 1 | 0.120 | 0.167 | 0.138 | 0.234 | 0.167 |
| | 2 | 0.354 | 0.333 | 0.346 | 0.266 | 0.333 |
| 1/2 | 3 | 0.000 | 0.000 | 0.000 | 0.000 | 0.000 |
| | 4 | 0.354 | 0.333 | 0.346 | 0.266 | 0.333 |
| | 5 | 0.172 | 0.167 | 0.170 | 0.234 | 0.167 |
| | 1 | 0.163 | 0.166 | 0.164 | 0.233 | 0.167 |
| | 2 | 0.337 | 0.334 | 0.336 | 0.266 | 0.333 |
| 1 | 3 | 0.000 | 0.000 | 0.000 | 0.000 | 0.000 |
| | 4 | 0.337 | 0.333 | 0.335 | 0.267 | 0.333 |
| | 5 | 0.164 | 0.167 | 0.165 | 0.234 | 0.167 |
| | 1 | 0.194 | 0.167 | 0.184 | 0.233 | 0.167 |
| | 2 | 0.324 | 0.333 | 0.327 | 0.266 | 0.333 |
| 3/2 | 3 | 0.000 | 0.000 | 0.000 | 0.000 | 0.000 |
| | 4 | 0.324 | 0.333 | 0.328 | 0.267 | 0.334 |
| | 5 | 0.157 | 0.167 | 0.161 | 0.233 | 0.167 |
| | 1 | 0.220 | 0.167 | 0.201 | 0.234 | 0.167 |
| | 2 | 0.314 | 0.333 | 0.321 | 0.266 | 0.333 |
| 2 | 3 | 0.000 | 0.000 | 0.000 | 0.000 | 0.000 |
| | 4 | 0.314 | 0.333 | 0.321 | 0.267 | 0.333 |
| | 5 | 0.153 | 0.167 | 0.158 | 0.234 | 0.167 |
| | 1 | 0.331 | 0.167 | 0.279 | 0.234 | 0.167 |
| | 2 | 0.269 | 0.333 | 0.290 | 0.267 | 0.333 |
| 6 | 3 | 0.000 | 0.000 | 0.000 | 0.000 | 0.000 |
| | 4 | 0.269 | 0.334 | 0.290 | 0.267 | 0.333 |
| | 5 | 0.131 | 0.167 | 0.142 | 0.233 | 0.167 |

Table 7.23 The ranks of the extended measure of component importance corresponding to the results in Tables 7.19–7.22.

| $c$ | Rank system levels 1, 2 | Rank system levels 3, 4 |
|---|---|---|
| 1/2 | $2 \approx 4 > 5 > 1 > 3$ | $2 \approx 4 > 5 > 1 > 3$ |
| 1 | $2 \approx 4 > 1 \approx 5 > 3$ | $2 \approx 4 > 1 \approx 5 > 3$ |
| 3/2 | $2 \approx 4 > 1 > 5 > 3$ | $2 \approx 4 > 1 > 5 > 3$ |
| 2 | $2 \approx 4 > 1 > 5 > 3$ | $2 \approx 4 > 1 > 5 > 3$ |
| 6 | $1 > 2 \approx 4 > 5 > 3$ | $2 \approx 4 > 1 > 5 > 3$ |

Table 7.24 Simulations of the bridge system with decreasing MTTR of component 1. System level is 1. The time horizon is $t = 20\,000$.

| $c$ | $i$ | $I_N^{(i,1)}(t)$ | $I_{N,D}^{(i,1)}(t)$ | $\bar{I}_N^{(i,1)}(t)$ | $I_B^{(i,1)}(t)$ | $I_{B-P}^{(i,1)}(t)$ |
|---|---|---|---|---|---|---|
| | 1 | 0.173 | 0.169 | 0.172 | 0.211 | 0.174 |
| | 2 | 0.198 | 0.199 | 0.199 | 0.189 | 0.198 |
| 1/2 | 3 | 0.074 | 0.074 | 0.074 | 0.071 | 0.074 |
| | 4 | 0.322 | 0.324 | 0.323 | 0.308 | 0.322 |
| | 5 | 0.232 | 0.233 | 0.233 | 0.222 | 0.232 |
| | 1 | 0.233 | 0.234 | 0.233 | 0.233 | 0.233 |
| | 2 | 0.234 | 0.234 | 0.234 | 0.234 | 0.234 |
| 1 | 3 | 0.066 | 0.066 | 0.066 | 0.066 | 0.066 |
| | 4 | 0.233 | 0.233 | 0.233 | 0.233 | 0.233 |
| | 5 | 0.234 | 0.233 | 0.234 | 0.233 | 0.233 |
| | 1 | 0.264 | 0.267 | 0.265 | 0.246 | 0.264 |
| | 2 | 0.251 | 0.250 | 0.251 | 0.258 | 0.252 |
| 3/2 | 3 | 0.062 | 0.062 | 0.062 | 0.064 | 0.062 |
| | 4 | 0.189 | 0.188 | 0.188 | 0.193 | 0.188 |
| | 5 | 0.234 | 0.233 | 0.234 | 0.240 | 0.234 |
| | 1 | 0.295 | 0.302 | 0.297 | 0.259 | 0.295 |
| | 2 | 0.269 | 0.267 | 0.268 | 0.283 | 0.269 |
| 5/2 | 3 | 0.059 | 0.058 | 0.058 | 0.062 | 0.059 |
| | 4 | 0.143 | 0.141 | 0.142 | 0.150 | 0.143 |
| | 5 | 0.235 | 0.232 | 0.234 | 0.246 | 0.235 |

2, 4 and 5 equally important since they are similarly positioned in the system.

As a conclusion, the results of the present section very much parallel those of the previous one.

## 7.4 Application to an offshore oil and gas production system

We will now look at a West African production site for oil and gas based on a memo (Signoret and Clave, 2007). Oil and gas are pumped up from one production well along with water. These substances are separated in a separation unit. We will assume this unit functions perfectly. After being separated the oil is run through an oil treatment unit, which is also assumed to function perfectly. Then the treated oil is exported through a pumping unit.

Table 7.25 Simulations of the bridge system with decreasing MTTR of component 1. System level 2. The time horizon is $t = 20\,000$.

| $c$ | $i$ | $I_N^{(i,2)}(t)$ | $I_{N,D}^{(i,2)}(t)$ | $\bar{I}_N^{(i,2)}(t)$ | $I_B^{(i,2)}(t)$ | $I_{B-P}^{(i,2)}(t)$ |
|---|---|---|---|---|---|---|
| | 1 | 0.202 | 0.197 | 0.200 | 0.243 | 0.202 |
| | 2 | 0.217 | 0.219 | 0.218 | 0.206 | 0.217 |
| 1/2 | 3 | 0.070 | 0.070 | 0.070 | 0.066 | 0.070 |
| | 4 | 0.272 | 0.274 | 0.272 | 0.258 | 0.272 |
| | 5 | 0.239 | 0.240 | 0.239 | 0.227 | 0.239 |
| | 1 | 0.234 | 0.234 | 0.234 | 0.234 | 0.234 |
| | 2 | 0.235 | 0.234 | 0.235 | 0.234 | 0.234 |
| 1 | 3 | 0.063 | 0.063 | 0.063 | 0.063 | 0.063 |
| | 4 | 0.234 | 0.234 | 0.234 | 0.234 | 0.234 |
| | 5 | 0.234 | 0.234 | 0.234 | 0.234 | 0.234 |
| | 1 | 0.247 | 0.251 | 0.249 | 0.230 | 0.247 |
| | 2 | 0.241 | 0.240 | 0.241 | 0.247 | 0.241 |
| 3/2 | 3 | 0.060 | 0.060 | 0.060 | 0.061 | 0.060 |
| | 4 | 0.219 | 0.218 | 0.219 | 0.224 | 0.219 |
| | 5 | 0.233 | 0.232 | 0.232 | 0.238 | 0.232 |
| | 1 | 0.259 | 0.266 | 0.262 | 0.227 | 0.259 |
| | 2 | 0.247 | 0.245 | 0.247 | 0.258 | 0.247 |
| 5/2 | 3 | 0.058 | 0.057 | 0.057 | 0.060 | 0.057 |
| | 4 | 0.205 | 0.204 | 0.205 | 0.214 | 0.205 |
| | 5 | 0.230 | 0.228 | 0.230 | 0.241 | 0.230 |

The gas is sent through two compressors which compress the gas. When both compressors are functioning, we get the maximum amount of gas. However, to obtain at least *some* gas production, it is sufficient that at least one of the compressors is functioning. If this is the case, the uncompressed gas is burned in a flare, which is assumed to function perfectly. The compressed gas is run through a unit where it is dehydrated. This is called a TEG (Tri-Ethylene Glycol) unit. After being dehydrated, the gas is ready to be exported. Some of the gas is used as fuel for the compressors.

The water is first run through a water treatment unit. This unit cleanses the water so that it legally can be pumped back into the wells to maintain the pressure, or back into the sea. If the water treatment unit fails, the whole production stops. The components in the system also need electricity, which comes from two generators. At least one generator must function in order to produce some oil and gas. If both generators fail, the whole system fails. The generators are powered by compressed and dehydrated gas.

Table 7.26 Simulations of the bridge system with decreasing MTTR of component 1. System level is 3. The time horizon is $t = 20\,000$.

| $c$ | $i$ | $I_N^{(i,3)}(t)$ | $I_{N,D}^{(i,3)}(t)$ | $\bar{I}_N^{(i,3)}(t)$ | $I_B^{(i,3)}(t)$ | $I_{B-P}^{(i,3)}(t)$ |
|---|---|---|---|---|---|---|
| | 1 | 0.242 | 0.237 | 0.240 | 0.289 | 0.242 |
| | 2 | 0.242 | 0.244 | 0.243 | 0.227 | 0.242 |
| 1/2 | 3 | 0.031 | 0.031 | 0.031 | 0.029 | 0.031 |
| | 4 | 0.242 | 0.244 | 0.243 | 0.227 | 0.242 |
| | 5 | 0.242 | 0.244 | 0.243 | 0.227 | 0.242 |
| | 1 | 0.242 | 0.243 | 0.242 | 0.242 | 0.242 |
| | 2 | 0.243 | 0.242 | 0.243 | 0.243 | 0.243 |
| 1 | 3 | 0.031 | 0.031 | 0.031 | 0.031 | 0.031 |
| | 4 | 0.242 | 0.242 | 0.242 | 0.242 | 0.242 |
| | 5 | 0.243 | 0.242 | 0.243 | 0.242 | 0.242 |
| | 1 | 0.242 | 0.246 | 0.244 | 0.225 | 0.243 |
| | 2 | 0.242 | 0.241 | 0.242 | 0.248 | 0.242 |
| 3/2 | 3 | 0.031 | 0.031 | 0.031 | 0.031 | 0.031 |
| | 4 | 0.243 | 0.241 | 0.242 | 0.248 | 0.242 |
| | 5 | 0.242 | 0.241 | 0.242 | 0.248 | 0.242 |
| | 1 | 0.242 | 0.248 | 0.245 | 0.211 | 0.242 |
| | 2 | 0.242 | 0.240 | 0.241 | 0.252 | 0.242 |
| 5/2 | 3 | 0.031 | 0.030 | 0.031 | 0.032 | 0.031 |
| | 4 | 0.242 | 0.240 | 0.241 | 0.252 | 0.242 |
| | 5 | 0.242 | 0.241 | 0.242 | 0.252 | 0.242 |

Thus, the simplified production site considered in the present chapter consists of the following eight relevant components, which are assumed to operate independently:

1. Well: a production well where the oil and gas come from.
2. Water cleanser: a component which cleanses the water which is pumped up from the production well along with the oil and gas.
3. Generator 1: generator providing electricity to the system.
4. Generator 2: the same as Generator 1.
5. Compressor 1: a compressor which compresses the gas.
6. Compressor 2: the same as Compressor 1.
7. TEG: a component where the gas is dehydrated.
8. Oil export pump: an oil export pump.

Table 7.27 Simulations of the bridge system with decreasing MTTR of component 1. System level is 4. The time horizon is $t = 20\,000$.

| $c$ | $i$ | $I_N^{(i,4)}(t)$ | $I_{N,D}^{(i,4)}(t)$ | $\bar{I}_N^{(i,4)}(t)$ | $I_B^{(i,4)}(t)$ | $I_{B-P}^{(i,4)}(t)$ |
|---|---|---|---|---|---|---|
| | 1 | 0.250 | 0.244 | 0.248 | 0.297 | 0.249 |
| | 2 | 0.251 | 0.252 | 0.251 | 0.234 | 0.250 |
| 1/2 | 3 | 0.000 | 0.000 | 0.000 | 0.000 | 0.000 |
| | 4 | 0.250 | 0.252 | 0.250 | 0.234 | 0.250 |
| | 5 | 0.250 | 0.252 | 0.251 | 0.235 | 0.250 |
| | 1 | 0.250 | 0.250 | 0.250 | 0.250 | 0.250 |
| | 2 | 0.250 | 0.250 | 0.250 | 0.250 | 0.250 |
| 1 | 3 | 0.000 | 0.000 | 0.000 | 0.000 | 0.000 |
| | 4 | 0.250 | 0.250 | 0.250 | 0.250 | 0.250 |
| | 5 | 0.250 | 0.250 | 0.250 | 0.250 | 0.250 |
| | 1 | 0.250 | 0.254 | 0.252 | 0.233 | 0.250 |
| | 2 | 0.249 | 0.248 | 0.249 | 0.255 | 0.249 |
| 3/2 | 3 | 0.000 | 0.000 | 0.000 | 0.000 | 0.000 |
| | 4 | 0.250 | 0.249 | 0.250 | 0.256 | 0.250 |
| | 5 | 0.251 | 0.249 | 0.250 | 0.256 | 0.251 |
| | 1 | 0.250 | 0.257 | 0.253 | 0.218 | 0.250 |
| | 2 | 0.250 | 0.248 | 0.249 | 0.260 | 0.250 |
| 5/2 | 3 | 0.000 | 0.000 | 0.000 | 0.000 | 0.000 |
| | 4 | 0.250 | 0.248 | 0.249 | 0.261 | 0.250 |
| | 5 | 0.250 | 0.248 | 0.249 | 0.261 | 0.250 |

Table 7.28 The common ranks of the three Natvig measures of component importance corresponding to the results in Tables 7.24–7.27.

| $c$ | Rank system levels 1, 2 | Rank system levels 3, 4 |
|---|---|---|
| 1/2 | $4 > 5 > 2 > 1 > 3$ | $4 \approx 5 \approx 2 \approx 1 > 3$ |
| 1 | $4 \approx 5 \approx 2 \approx 1 > 3$ | $4 \approx 5 \approx 2 \approx 1 > 3$ |
| 3/2 | $1 > 2 > 5 > 4 > 3$ | $4 \approx 5 \approx 2 \approx 1 > 3$ |
| 5/2 | $1 > 2 > 5 > 4 > 3$ | $4 \approx 5 \approx 2 \approx 1 > 3$ |

The structure of the system is shown in Figure 7.3. The components 1, 2, 7 and 8 are all in series with the rest of the system, while the two generators, 3 and 4, operate in parallel with each other. Similarly, the two compressors, 5 and 6, operate in parallel with each other.

In Signoret and Clave (2007) no explicit definition of the system is given. There are several different possible definitions, and in Natvig *et al.* (2009) we used the following: *The oil and gas production site is said to be*

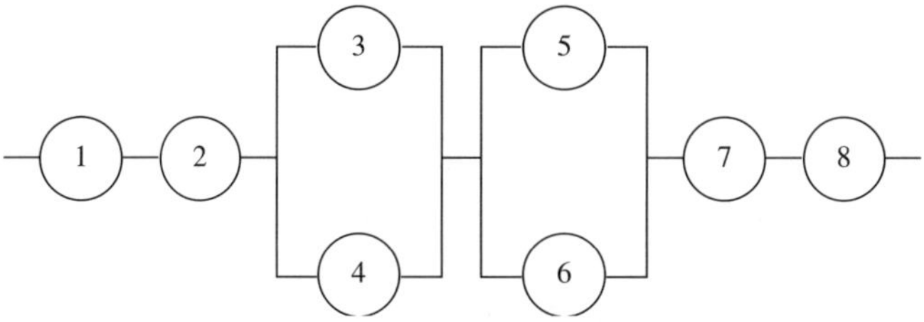


*Figure 7.3 Model of oil and gas production site.*

Table 7.29 Binary failure rates, mean repair times and mean total times spent in the non complete failure states of the components in the oil and gas production site.

| Component | Binary failure rate | $\mu_i^0$ | $\mu_i^1 + \mu_i^2$ |
|---|---|---|---|
| 1 | $2.736 \cdot 10^{-4}$ | 7.000 | 3654.97 |
| 2 | $8.208 \cdot 10^{-3}$ | 0.167 | 121.83 |
| 3 & 4 | $1.776 \cdot 10^{-2}$ | 1.167 | 56.31 |
| 5 & 6 | $1.882 \cdot 10^{-2}$ | 1.083 | 53.11 |
| 7 | $1.368 \cdot 10^{-3}$ | 0.125 | 730.99 |
| 8 | $5.496 \cdot 10^{-4}$ | 0.125 | 1819.51 |

*functioning if it can produce some amount of both oil and gas. Otherwise the system has failed.*

Table 7.29 shows the given binary failure rates, mean repair times and mean total times spent in the non complete failure states of the components in the system. The time unit is *days*. The mean total times spent in the non complete failure states are considerably larger than the mean repair times. For some components (the well, the TEG unit and the oil export pump), the former means are actually several years.

It should be stressed that the memo (Signoret and Clave, 2007) does not discuss what might be reasonable life- and repair time distributions. Hence, the choice of the exponential and gamma distributions in the following may be difficult to justify empirically. Our aim is, however, to illustrate how different choices of such distributions influence the Natvig measures. In order to avoid making this sensitivity analysis too extensive, we have skipped the Weibull distribution.

### 7.4.1 Exponentially distributed repair times and times spent in each of the non complete failure states

In this subsection we assume that the components both have exponentially distributed times spent in each of the non complete failure states and exponentially distributed repair times. The failure rates in the former distributions are twice the inverses of the mean total times spent in the non complete failure states, while the repair rates are the inverses of the mean repair times. Thus, all the parameters needed in the simulations can be derived from Table 7.29. The time horizon $t$ is set to 100 000 days.

In Tables 7.30 and 7.31 we see that $I_N^{(i,j)}(t)$ is almost equal to its extended version $\bar{I}_N^{(i,j)}(t)$ for both system levels. This is because $E[Y_{i,k,j}(t)]$ for $k = 1, \ldots, M$ are very large compared to $E[Y_{i,0,j}(t)]$ for all components. Hence, the contributions of the latter terms in Equation (6.44) are too small to make any difference.

The reason for this is that the repair times of the components are much shorter than the corresponding times spent in each of the non complete failure states. Hence, the fictive prolonged repair times of the components due to the fictive minimal complete failures are much shorter than the fictive prolonged times spent in each of the non complete failure states due to the fictive minimal repairs. In particular, the fictive prolonged repair times will, due to the much longer times spent in each of the non complete failure states, mostly end long before the next real repair. Hence, it is very unlikely that the fictive minimal complete failure periods will overlap.

Table 7.30 Component importance using exponential distributions. System level is 1. The time horizon is $t = 100\,000$.

| Component | $I_N^{(i,1)}(t)$ | $I_{N,D}^{(i,1)}(t)$ | $\bar{I}_N^{(i,1)}(t)$ | $I_B^{(i,1)}(t)$ | $I_{B-P}^{(i,1)}(t)$ |
|---|---|---|---|---|---|
| 1 | 0.244 | 0.438 | 0.244 | 0.247 | 0.024 |
| 2 | 0.250 | 0.318 | 0.250 | 0.247 | 0.731 |
| 3 & 4 | 0.007 | 0.002 | 0.007 | 0.007 | 0.005 |
| 5 & 6 | 0.005 | 0.092 | 0.005 | 0.005 | 0.033 |
| 7 | 0.249 | 0.039 | 0.248 | 0.247 | 0.122 |
| 8 | 0.246 | 0.016 | 0.246 | 0.247 | 0.049 |

Table 7.31 Component importance using exponential distributions. System level is 2. The time horizon is $t = 100\,000$.

| Component | $I_N^{(i,2)}(t)$ | $I_{N,D}^{(i,2)}(t)$ | $\bar{I}_N^{(i,2)}(t)$ | $I_B^{(i,2)}(t)$ | $I_{B-P}^{(i,2)}(t)$ |
|---|---|---|---|---|---|
| 1 | 0.197 | 0.121 | 0.196 | 0.198 | 0.010 |
| 2 | 0.202 | 0.088 | 0.201 | 0.200 | 0.285 |
| 3 & 4 | 0.051 | 0.049 | 0.051 | 0.050 | 0.154 |
| 5 & 6 | 0.051 | 0.338 | 0.053 | 0.052 | 0.166 |
| 7 | 0.200 | 0.011 | 0.199 | 0.199 | 0.048 |
| 8 | 0.197 | 0.005 | 0.196 | 0.198 | 0.019 |

As a conclusion, it is very sensible for this case study that $I_N^{(i,j)}(t)$ is equal to $\bar{I}_N^{(i,j)}(t)$.

We also observe from Tables 7.30 and 7.31 that, for the two almost equal measures, the components 1, 2, 7 and 8 that are in series with the rest of the system have approximately the same importance for both system levels. This can be seen for system level 2 by the following argument. Since $t = 100\,000$ days we have reached stationarity. Furthermore, for the exponential distribution $\mu_i^{k(p)} = \mu_i^k$ for $k = 0, \ldots, M$. If components $i$ and $\ell$ both are in series with the rest of the system, by conditioning on the state of component $\ell$ and applying Equation (6.18), the numerator of Equation (6.39) equals $p_\phi^2((\boldsymbol{e}^2)_i, (\boldsymbol{e}^2)_\ell, \boldsymbol{a})a_i^2 a_\ell^2$. By a parallel argument this is also the numerator of $I_N^{(\ell,2)}$.

Note also that the remaining components that are parts of parallel modules have almost identical importances according to the two almost equal measures and are much less important than the ones in series with the rest of the system for both system levels. According to an argument in Natvig *et al.* (2009) the reason for this in the binary case is that all components 3, 4, 5 and 6 have almost identical availabilities, as is seen from Table 7.29.

The ranks of the component importance for the three versions of the Natvig measure are given in Table 7.32. We suggest applying the ranking based on the measure $\bar{I}_N^{(i,j)}(t)$ which is almost identical to the one of $I_N^{(i,j)}(t)$.

### 7.4.2 Gamma distributed repair times and times spent in each of the non complete failure states

In this subsection we assume instead that the components have gamma distributed times spent in each of the non complete failure states and gamma distributed repair times. More specifically, we assume that for

Table 7.32 The ranks of the component importance for system levels 1 and 2 for the three versions of the Natvig measure according to the results given in Tables 7.30 and 7.31.

| Measure | Rank system level 1 | Rank system level 2 |
|---|---|---|
| $I_N^{(i)}(t)$ | $2 > 7 > 8 > 1 > 3 \approx 4 > 5 \approx 6$ | $2 > 7 > 8 \approx 1 > 3 \approx 4 \approx 5 \approx 6$ |
| $I_{N,D}^{(i)}(t)$ | $1 > 2 > 5 \approx 6 > 7 > 8 > 3 \approx 4$ | $5 \approx 6 > 1 > 2 > 3 \approx 4 > 7 > 8$ |
| $\bar{I}_N^{(i)}(t)$ | $2 > 7 > 8 > 1 > 3 \approx 4 > 5 \approx 6$ | $2 > 7 > 8 \approx 1 > 5 \approx 6 > 3 \approx 4$ |

$i = 1, \ldots, 8$, the times spent in each of the non complete failure states of the $i$th component have the densities:

$$f_i(t) = \frac{1}{(\beta_i)^{\alpha_i/2}\Gamma(\alpha_i/2)} t^{\alpha_i/2-1} \exp(-t/\beta_i),$$

while the repair times of the $i$th component have the densities:

$$g_i(t) = \frac{1}{(\beta_i')^{\alpha_i'}\Gamma(\alpha_i')} t^{\alpha_i'-1} \exp(-t/\beta_i').$$

By choosing different values for the density parameters it is possible to alter the variances in the distributions of the times spent in each of the non complete failure states and still keep the expectations fixed. In order to see the effect of this on the importance measures, we focus on component 1 where we consider five different parameter combinations for the distribution of the times spent in each of the non complete failure states. For all these combinations the expected time spent in each of the non complete failure states is 1827.49 days, but the variance varies between $9.135 \cdot 10^2$ and $5.85 \cdot 10^5$. Table 7.33 lists these parameter combinations. For the remaining gamma densities we use the parameters listed in Tables 7.34 and 7.35. All parameters are chosen such that the expectations in the repair time distributions and in the total times spent in the non complete failure states match the corresponding values given in Table 7.29. We also use the same time horizon $t = 100\,000$ days as in the previous subsection.

Tables 7.36–7.40 display the results obtained from simulations using the parameters listed in Tables 7.33–7.35. As for the case with exponentially distributed repair times and times spent in each of the non complete failure states, $I_N^{(i,j)}(t)$ is almost equal to its extended version $\bar{I}_N^{(i,j)}(t)$ for both system levels.

Table 7.33 Parameter sets for the distribution of times spent in each of the non complete failure states for component 1.

| Set | $\alpha_1$ | $\beta_1$ | Variance |
|---|---|---|---|
| 1 | 7309.940 | 0.500 | $9.135 \cdot 10^2$ |
| 2 | 550.033 | 6.645 | $1.215 \cdot 10^4$ |
| 3 | 101.493 | 36.012 | $6.58 \cdot 10^4$ |
| 4 | 45.687 | 80.000 | $1.462 \cdot 10^5$ |
| 5 | 11.422 | 319.994 | $5.85 \cdot 10^5$ |

Table 7.34 Parameters in the distribution of times spent in each of the non complete failure states for components 2, ..., 8.

| Component | $\alpha_i$ | $\beta_i$ | Variance |
|---|---|---|---|
| 2 | 30.000 | 4.062 | $2.475 \cdot 10^2$ |
| 3 & 4 | 30.000 | 1.877 | $5.285 \cdot 10^1$ |
| 5 & 6 | 10.000 | 5.311 | $1.411 \cdot 10^2$ |
| 7 | 179.958 | 4.062 | $1.485 \cdot 10^3$ |
| 8 | 218.219 | 8.338 | $7.585 \cdot 10^3$ |

Table 7.35 Parameters in the repair time distributions of components 1, ..., 8.

| Component | $\alpha_i'$ | $\beta_i'$ | Variance |
|---|---|---|---|
| 1 | 3.500 | 2.000 | $1.400 \cdot 10^1$ |
| 2 | 0.668 | 0.250 | $4.175 \cdot 10^{-2}$ |
| 3 & 4 | 3.000 | 0.389 | $4.540 \cdot 10^{-1}$ |
| 5 & 6 | 1.500 | 0.722 | $7.819 \cdot 10^{-1}$ |
| 7 | 1.000 | 0.125 | $1.563 \cdot 10^{-2}$ |
| 8 | 1.000 | 0.125 | $1.563 \cdot 10^{-2}$ |

Table 7.36 Component importance using gamma distributions. Variance of times spent in each of the non complete failure states of component 1: $9.135 \cdot 10^2$.

| System level | Component | $I_N^{(i,j)}(t)$ | $I_{N,D}^{(i,j)}(t)$ | $\bar{I}_N^{(i,j)}(t)$ | $I_B^{(i,j)}(t)$ | $I_{B-P}^{(i,j)}(t)$ |
|---|---|---|---|---|---|---|
| | 1 | 0.032 | 0.249 | 0.035 | 0.245 | 0.023 |
| | 2 | 0.519 | 0.414 | 0.518 | 0.245 | 0.694 |
| 1 | 3 & 4 | 0.010 | 0.058 | 0.011 | 0.005 | 0.030 |
| | 5 & 6 | 0.018 | 0.080 | 0.019 | 0.005 | 0.031 |
| | 7 | 0.206 | 0.042 | 0.204 | 0.245 | 0.115 |
| | 8 | 0.186 | 0.016 | 0.183 | 0.245 | 0.046 |
| | 1 | 0.020 | 0.062 | 0.021 | 0.196 | 0.010 |
| | 2 | 0.309 | 0.095 | 0.300 | 0.201 | 0.284 |
| 2 | 3 & 4 | 0.078 | 0.071 | 0.082 | 0.052 | 0.155 |
| | 5 & 6 | 0.139 | 0.244 | 0.144 | 0.052 | 0.165 |
| | 7 | 0.122 | 0.010 | 0.117 | 0.198 | 0.047 |
| | 8 | 0.113 | 0.004 | 0.109 | 0.199 | 0.019 |

Table 7.37 Component importance using gamma distributions. Variance of times spent in each of the non complete failure states of component 1: $1.215 \cdot 10^4$.

| System level | Component | $I_N^{(i,j)}(t)$ | $I_{N,D}^{(i,j)}(t)$ | $\bar{I}_N^{(i,j)}(t)$ | $I_B^{(i,j)}(t)$ | $I_{B-P}^{(i,j)}(t)$ |
|---|---|---|---|---|---|---|
| | 1 | 0.107 | 0.246 | 0.109 | 0.245 | 0.023 |
| | 2 | 0.479 | 0.417 | 0.479 | 0.245 | 0.694 |
| 1 | 3 & 4 | 0.010 | 0.059 | 0.010 | 0.005 | 0.030 |
| | 5 & 6 | 0.017 | 0.081 | 0.018 | 0.005 | 0.031 |
| | 7 | 0.189 | 0.042 | 0.187 | 0.245 | 0.115 |
| | 8 | 0.172 | 0.016 | 0.169 | 0.245 | 0.046 |
| | 1 | 0.066 | 0.059 | 0.065 | 0.197 | 0.009 |
| | 2 | 0.295 | 0.096 | 0.287 | 0.200 | 0.285 |
| 2 | 3 & 4 | 0.075 | 0.174 | 0.079 | 0.052 | 0.155 |
| | 5 & 6 | 0.132 | 0.244 | 0.137 | 0.051 | 0.164 |
| | 7 | 0.117 | 0.009 | 0.113 | 0.198 | 0.047 |
| | 8 | 0.109 | 0.004 | 0.105 | 0.198 | 0.019 |

Table 7.38 Component importance using gamma distributions. Variance of times spent in each of the non complete failure states of component 1: $6.58 \cdot 10^4$.

| System level | Component | $I_N^{(i,j)}(t)$ | $I_{N,D}^{(i,j)}(t)$ | $\bar{I}_N^{(i,j)}(t)$ | $I_B^{(i,j)}(t)$ | $I_{B-P}^{(i,j)}(t)$ |
|---|---|---|---|---|---|---|
| | 1 | 0.219 | 0.241 | 0.220 | 0.245 | 0.023 |
| | 2 | 0.419 | 0.418 | 0.419 | 0.245 | 0.694 |
| 1 | 3 & 4 | 0.008 | 0.058 | 0.009 | 0.005 | 0.030 |
| | 5 & 6 | 0.015 | 0.081 | 0.016 | 0.005 | 0.031 |
| | 7 | 0.166 | 0.044 | 0.164 | 0.245 | 0.115 |
| | 8 | 0.150 | 0.018 | 0.148 | 0.245 | 0.046 |
| | 1 | 0.141 | 0.056 | 0.138 | 0.197 | 0.009 |
| | 2 | 0.272 | 0.095 | 0.265 | 0.201 | 0.285 |
| 2 | 3 & 4 | 0.068 | 0.172 | 0.072 | 0.051 | 0.155 |
| | 5 & 6 | 0.122 | 0.245 | 0.126 | 0.052 | 0.165 |
| | 7 | 0.107 | 0.010 | 0.104 | 0.199 | 0.047 |
| | 8 | 0.100 | 0.004 | 0.097 | 0.198 | 0.019 |

We now observe that for these two almost equal measures, the components 1, 2, 7 and 8 that are in series with the rest of the system have different importances for both system levels as opposed to the exponential case. However, the remaining components that are parts of parallel modules are still much less important for system level 1.

Furthermore, we see that for both system levels the extended component importance of component 1 is increasing with increasing variances,

Table 7.39 Component importance using gamma distributions. Variance of times spent in each of the non complete failure states of component 1: $1.462 \cdot 10^5$.

| System level | Component | $I_N^{(i,j)}(t)$ | $I_{N,D}^{(i,j)}(t)$ | $\bar{I}_N^{(i,j)}(t)$ | $I_B^{(i,j)}(t)$ | $I_{B-P}^{(i,j)}(t)$ |
|---|---|---|---|---|---|---|
| | 1 | 0.298 | 0.250 | 0.297 | 0.245 | 0.023 |
| | 2 | 0.377 | 0.415 | 0.378 | 0.245 | 0.694 |
| 1 | 3 & 4 | 0.008 | 0.058 | 0.008 | 0.005 | 0.030 |
| | 5 & 6 | 0.013 | 0.080 | 0.014 | 0.005 | 0.031 |
| | 7 | 0.150 | 0.043 | 0.149 | 0.245 | 0.115 |
| | 8 | 0.134 | 0.017 | 0.133 | 0.245 | 0.046 |
| | 1 | 0.199 | 0.055 | 0.194 | 0.197 | 0.009 |
| | 2 | 0.254 | 0.099 | 0.248 | 0.201 | 0.285 |
| 2 | 3 & 4 | 0.064 | 0.173 | 0.067 | 0.052 | 0.155 |
| | 5 & 6 | 0.113 | 0.243 | 0.118 | 0.051 | 0.164 |
| | 7 | 0.100 | 0.010 | 0.097 | 0.199 | 0.047 |
| | 8 | 0.093 | 0.004 | 0.090 | 0.198 | 0.019 |

Table 7.40 Component importance using gamma distributions. Variance of times spent in each of the non complete failure states of component 1: $5.85 \cdot 10^5$.

| System level | Component | $I_N^{(i,j)}(t)$ | $I_{N,D}^{(i,j)}(t)$ | $\bar{I}_N^{(i,j)}(t)$ | $I_B^{(i,j)}(t)$ | $I_{B-P}^{(i,j)}(t)$ |
|---|---|---|---|---|---|---|
| | 1 | 0.466 | 0.249 | 0.464 | 0.245 | 0.023 |
| | 2 | 0.286 | 0.416 | 0.287 | 0.245 | 0.694 |
| 1 | 3 & 4 | 0.006 | 0.058 | 0.006 | 0.005 | 0.030 |
| | 5 & 6 | 0.010 | 0.081 | 0.011 | 0.005 | 0.031 |
| | 7 | 0.114 | 0.041 | 0.113 | 0.245 | 0.115 |
| | 8 | 0.103 | 0.016 | 0.102 | 0.245 | 0.046 |
| | 1 | 0.337 | 0.056 | 0.329 | 0.197 | 0.009 |
| | 2 | 0.210 | 0.097 | 0.207 | 0.200 | 0.285 |
| 2 | 3 & 4 | 0.053 | 0.173 | 0.056 | 0.051 | 0.155 |
| | 5 & 6 | 0.094 | 0.244 | 0.098 | 0.051 | 0.164 |
| | 7 | 0.083 | 0.010 | 0.081 | 0.199 | 0.047 |
| | 8 | 0.077 | 0.004 | 0.075 | 0.198 | 0.019 |

and decreasing shape parameters $\alpha_1/2$, all greater than 1, in its distribution of times spent in each of the non complete failure states. Since we have reached stationarity, this observation, based on simulations and assuming gamma distributed repair times and times spent in each of the non complete failure states, is in accordance with the theoretical result concerning the Weibull distribution described following Equation (6.53).

Table 7.41 The ranks of the extended component importance according to the results given in Tables 7.36–7.40.

| Table | Rank system level 1 | Rank system level 2 |
|---|---|---|
| 7.36 | $2 > 7 > 8 > 1 > 5 \approx 6 > 3 \approx 4$ | $2 > 5 \approx 6 > 7 > 8 > 3 \approx 4 > 1$ |
| 7.37 | $2 > 7 > 8 > 1 > 5 \approx 6 > 3 \approx 4$ | $2 > 5 \approx 6 > 7 > 8 > 3 \approx 4 > 1$ |
| 7.38 | $2 > 1 > 7 > 8 > 5 \approx 6 > 3 \approx 4$ | $2 > 1 > 5 \approx 6 > 7 > 8 > 3 \approx 4$ |
| 7.39 | $2 > 1 > 7 > 8 > 5 \approx 6 > 3 \approx 4$ | $2 > 1 > 5 \approx 6 > 7 > 8 > 3 \approx 4$ |
| 7.40 | $1 > 2 > 7 > 8 > 5 \approx 6 > 3 \approx 4$ | $1 > 2 > 5 \approx 6 > 7 > 8 > 3 \approx 4$ |

Table 7.41 displays the ranks of the components according to the extended measure for both system levels. Along with the increased importance for both system levels, according to the extended measure, of component 1 as $\alpha_1/2$ decreases, we observe from this table a corresponding improvement in its rank. All the other components are ranked in the same order for every value of $\alpha_1/2$ both for system level 1 and 2. This is as expected from Equation (6.52), since the ordering is determined by its numerator. For all components except component 1, the numerator depends on the distributions of repair time and the times spent in each of the non complete failure states of this component only through $a_1^j$, $j = 1, 2$, which is kept fixed when varying $\alpha_1/2$. We also see that the components that are in series with the rest of the system for both system levels are ranked according to the shape parameter $\alpha_i/2$, such that components with the smaller shape parameters are more important. Note also that, according to the ranks, components 1, 7 and 8 are more important on system level 1 than 2, while it is the other way round for components 5 and 6.

Comparing the ranks of the extended component importance in Table 7.41 for the gamma distributions to those in Table 7.32 for the exponential distributions, we see that in the former case for both system levels components 5 and 6 in one of the parallel modules are more important than components 3 and 4 in the other parallel module, whereas they are roughly equally important in the latter case. This is due to the fact that in the former case the repair times and the times spent in each of the non complete failure states of components 5 and 6 have a larger variance than the corresponding ones of components 3 and 4. In the case where exponential distributions were used, the corresponding variances were roughly similar. The ranking of the components 2, 7 and 8 that are in series with the rest of the system is common for both system levels in these tables.

## 7.5 Concluding remarks

This chapter is based on ideas, concepts and theoretical results, as treated in Natvig (2011) and reviewed in Chapter 6, for the Natvig measures of component importance for repairable multistate systems, and its extended version. Two three-component systems and the bridge system were analyzed. We saw that an important feature of the Natvig measures is that they reflect the degree of uncertainty in the distributions of the repair time and of the times spent in each of the non complete failure states of the components. Finally, the theory was applied to an offshore oil and gas production system. First the times were assumed to be exponentially distributed and then gamma distributed both in accordance with the data given in the memo by Signoret and Clave (2007). The time horizon was set at 100 000 days so stationarity was reached.

A finding from the simulations of this case study is that the results for the original Natvig measure and its extended version, also taking a dual term into account, are almost identical. This is perfectly sensible since the dual term vanishes because the fictive prolonged repair times are much shorter than the fictive prolonged times spent in each of the non complete failure states. The weaknesses of this system are linked to the times spent in each of the non complete failure states and not the repair times.

Component 1 is the well, which is in series with the rest of the system. For this component we see that the extended component importance in the gamma case is increasing with increasing variances, and decreasing shape parameters, all greater than 1, in the distributions of times spent in each of the non complete failure states. This is in accordance with a theoretical result for the Weibull distribution. Along with this increased importance we also observed a corresponding improvement in its ranking.

Based on the presentation in Natvig (2011) we feel that the Natvig measures of component importance for repairable multistate systems on the one hand represent a theoretical novelty. On the other hand the case study indicates a great potential for applications, especially due to the simulation methods applied, as presented in Huseby and Natvig (2010).

# 8

# Probabilistic modeling of monitoring and maintenance

## 8.1 Introduction and basic marked point process

In Gåsemyr and Natvig (2001) partial monitoring of components with applications to preventive system maintenance was considered for a binary monotone system of binary components. This was extended to an MMS in Gåsemyr and Natvig (2005), and the purpose of the present chapter is to review this work. In the former paper the framework for the probabilistic modeling was a marked point process linked to the observed events only. In this section we start out closer to the spirit of Arjas (1989) by introducing a marked point process with complete monitoring of all components, and hence of the system, as the basic reference framework. In the first four sections we only consider a single operational period of the system without any repair or maintenance of the components, ending when the system deteriorates to a certain state or due to interruption by some sort of censoring.

We define a marked point process $\{(T_\ell, U_\ell); \ell \geq 1\}$, where $T_1 < T_2 < \cdots$ is the ordered sequence of times of component state changes and $U_\ell = (I_\ell, J_\ell, K_\ell)$ is a description of the event, the marked point, at $T_\ell$. $I_\ell$ is the component that changes state at $T_\ell$, changing from $J_\ell$ to $K_\ell$. We denote by $\mathcal{F}_{t^-}$ the pre-$t$ history, $\mathcal{F}_t$ the history including time $t$, and let $\mathbb{F} = \{\mathcal{F}_{t^-}\}_{t \geq 0}$ be the filtration, i.e. the collection of pre-$t$ histories as time runs. $\mathcal{F}_0$ may, for instance, contain information on the initial component states, which normally are supposed to be $\boldsymbol{M}$. We can think of the pre-$t$ marked points as forming a set-valued stochastic process $F_{t^-} = \{(T_\ell, U_\ell); T_\ell < t\}$, called the history process. Replacing $<$ by $\leq$, we obtain, by definition, a process $F_t$. Define $T(t) = \max\{T_\ell : T_\ell < t\}$ and $N(t) = \sum_{\ell=1} I(T_\ell < t)$. As before, $\boldsymbol{X}(t) = (X_1(t), \ldots, X_n(t))$ is the random component state vector at time $t$. We define $T(i,k) = \min\{t : X_i(t) \leq k\}$; $k \in S_i$, $i = 1, \ldots, n$. The random variables $T(t)$, $N(t)$ and $\boldsymbol{X}(s)$, $s < t$ are functions of the history process $F_{t^-}$. Let $f_{t^-}$ be a realization of $F_{t^-}$. The set of possible history processes up to $t$ is denoted by $\mathcal{H}^F_{t^-}$.

The intensity for a transition from state $j$ to $k$ for component $i$ at time $t$, given that $F_{t^-} = f_{t^-}$ is denoted by $\lambda_{i,j,k}(t; f_{t^-})$ and defined to be 0 on $\{X_i(T(t)) \neq j\}$. We assume that these intensities can be modeled as explicit functions in terms of the relevant parameter vector $\boldsymbol{\theta}$ based on physical insight and standard distributional assumptions. A very general way of expressing such intensities, allowing for different types of dependencies between the components, is

$$\lambda_{i,j,k}(t; f_{t^-}) = \sum_{r=1}^{R_{i,j,k}} I(f_{t^-} \in \mathcal{A}_t(i,j,k,r))\mu_{i,j,k,r}(t; \psi_{i,j,k,r}(f_{t^-})), \quad (8.1)$$

where the sets $\mathcal{A}_t(i,j,k,r)$, $k \in S_i$ with $k < j$, $r = 1, \ldots, R_{i,j,k}$ are a partition of the set of possible histories $f_{t^-} \in \mathcal{H}^F_{t^-}$ for which $X_i(T(t)) = j$. $\psi_{i,j,k,r}(f_{t^-})$ is some function of the history process and the $\mu_{i,j,k,r}$s are parametric functions with parameters $\psi_{i,j,k,r}(f_{t^-})$ and some subvector of $\boldsymbol{\theta}$.

If, for $t' > t$, $f_{t'^-} = f_{t^-}$, it is natural to assume that $I(f_{t'^-} \in \mathcal{A}_{t'}(i,j,k,r)) = I(f_{t^-} \in \mathcal{A}_t(i,j,k,r))$. Hence, the component transition process is determined by the same parametric functions as long as no new events take place. We also define the random variable

$$R_t(i,j,k) = \sum_{r=1}^{R_{i,j,k}} rI(F_{t^-} \in \mathcal{A}_t(i,j,k,r)). \quad (8.2)$$

Accordingly, we may write

$$\lambda_{i,j,k}(t; F_{t^-}) = \mu_{i,j,k,R_t(i,j,k)}(t; \psi_{i,j,k,R_t(i,j,k)}(F_{t^-})). \quad (8.3)$$

As an example we may have $\mathcal{A}_t(i,j,r) = \{f_{t^-} : \boldsymbol{x}(T(t)) \in C_{i,j,r}\}$. Here $C_{i,j,r}$, $r = 1, \ldots, R_{i,j}$ is a partition of the set of possible component state vectors $\boldsymbol{x}$ for which $x_i = j$. Another example is

$$\mathcal{A}_t(i,r) = \{f_{t^-} : \sigma_{i,r-1} < t \leq \sigma_{i,r}\}, \quad (8.4)$$

where the $\sigma_{i,r}$s are $\mathcal{F}_{t^-}$-stopping times.

In Section 8.2 we consider a marked point process linked to partial monitoring of some components from certain time points, called inspection times, onwards. Mainly, we assume that the inspection strategy is determined by the observed component history process only, and as in Gåsemyr and Natvig (2001) that it is noninformative on the component states not yet monitored and also has a certain stability. The corresponding likelihood is established by Jacod's likelihood formula.

Information from the observed system history process is incorporated in Section 8.3 and corresponding likelihood formulae arrived at. One source of complexity is that for each change of state of the system, we must take into account the various possible component transitions being consistent with all information linked to the system event. In Section 8.4 it is shown, as in Gåsemyr and Natvig (2001), that this source of complexity can be eliminated by introducing the notion of cause control. Then the inspection strategy is constructed in such a way that the actual component state transition, occurring at the same time as the system state transition, is observed. Maintenance and repair is considered in Section 8.5. In Section 8.6 the theory is applied to the electrical power generation system for two nearby oilrigs with some standby components as presented in Section 1.2. In Section 8.7 we consider how to arrive at the posterior distribution for the parameter vector $\boldsymbol{\theta}$ by a standard simulation procedure, the data augmentation method. The idea is to extend the observed data to the complete component history process.

## 8.2 Partial monitoring of components and the corresponding likelihood formula

We start out following Gåsemyr and Natvig (2001) by considering a monitoring scheme in which a subset $\{1, \ldots, p\}$ of the components is monitored from time 0 onwards, while the components in another subset $\{p+1, \ldots, p+q\}$, where $1 \leq p < p+q \leq n$, are conditionally monitored, i.e. they are monitored from certain time points $\tau_i$ onwards, $i \in \{p+1, \ldots, p+q\}$. These time points are called inspection times and are here assumed to depend on the observed component history process according to a specific strategy determined in advance. More specifically, they will coincide with the times of component changes. We let $\tau_i = 0$ for $i \in \{1, \ldots, p\}$.

In order to account for all information acquired by this partial monitoring scheme, we extend the $U_\ell$s in the marked point process considered in Section 8.1. Now $U_\ell = (I_\ell, J_\ell, K_\ell, \boldsymbol{Y}_\ell)$, where $\boldsymbol{Y}_\ell = (Y_{\ell,p+1}, \ldots, Y_{\ell,p+q})$ with $Y_{\ell,i} = -1$ if the $i$th component is not inspected at time $T_\ell$ and equals

$X_i(T_\ell)$ otherwise, $i \in \{p+1, \ldots, p+q\}$. The filtration generated by this new marked point process is denoted by $\mathbb{E} = \{\mathcal{E}_{t^-}\}_{t \geq 0}$ and the history process by $E_{t^-}$. $E_t$ is defined analogously to $F_t$. Due to the partial monitoring, we have $\mathcal{E}_{t^-} \subseteq \mathcal{F}_{t^-}$. We assume $\mathcal{E}_0 = \mathcal{F}_0$. Moreover, if components are partially monitored according to a nonrandomized inspection strategy, $E_{t^-}$ is a deterministic function of $F_{t^-}$. Technically, the inspection times $\tau_i$ are $\mathcal{E}_{t^-}$-stopping times, i.e. information about $I(\tau_i < t)$ is contained in $E_{t^-}$.

We denote by $\rho^{\mathbb{E}}_{i,j,k,\boldsymbol{y}}(t)$ the intensities for the marked points given $E_{t^-}$, i.e.

$$\begin{aligned}\rho^{\mathbb{E}}_{i,j,k,\boldsymbol{y}}(t) = \lim_{dt \to 0}(1/dt)P[T_{N(t)+1} &\in [t, t+dt), I_{N(t)+1} = i,\\ J_{N(t)+1} = j, K_{N(t)+1} &= k, \boldsymbol{Y}_{N(t)+1} = \boldsymbol{y} | E_{t^-}].\end{aligned} \tag{8.5}$$

Leaving out $\boldsymbol{Y}_\ell$ from $U_\ell$ we denote the corresponding intensities by $\alpha^{\mathbb{E}}_{i,j,k}(t)$. Note that these intensities are 0 on $\{X_i(T(t)) \neq j\}$ and on $\{\tau_i \geq t\}$. When $\mathbb{E} = \mathbb{F}$, we obviously have $\alpha^{\mathbb{E}}_{i,j,k}(t) = \lambda_{i,j,k}(t; F_{t^-})$.

We now formulate the fundamental assumption on the inspection strategy generalizing (5) of Gåsemyr and Natvig (2001). Consider the extended history process $E^+_{t^-} = E_{t^-} \cup V_t$, where $V_t = \{T_\ell, I_\ell, J_\ell, K_\ell\}$ if $T_\ell = t$ for some $\ell$ and $V_t = \emptyset$ otherwise. For $\boldsymbol{y} \in \mathcal{Y}_t$, the set of possible inspection outcomes given $E^+_{t^-}$, we assume

$$\begin{aligned}\lim_{dt \to 0} P[\boldsymbol{Y}_{N(t)+1} = \boldsymbol{y} | E_{t^-}, T_{N(t)+1} &\in [t, t+dt), I_{N(t)+1} = i,\\ J_{N(t)+1} = j, K_{N(t)+1} &= k]\\ = P\Big[\bigcap_{\{i \in C: y_i \neq -1\}} &(X_i(t) = y_i) | E_{t^-}, T_{N(t)+1} = t,\\ &I_{N(t)+1} = i, J_{N(t)+1} = j, K_{N(t)+1} = k\Big].\end{aligned} \tag{8.6}$$

As in Gåsemyr and Natvig (2001), Equation (8.6) expresses the fact that the inspection strategy is noninformative on the component states not yet monitored and also has a certain stability. From Equation (8.5) and Equation (8.6) we have, for $T_{N(t)} < t \leq T_{N(t)+1}$

$$\begin{aligned}\rho^{\mathbb{E}}_{i,j,k,\boldsymbol{y}}(t) = \alpha^{\mathbb{E}}_{i,j,k}(t) P\Big[\bigcap_{\{i \in C: y_i \neq -1\}} &(X_i(t) = y_i) | E_{t^-},\\ T_{N(t)+1} = t, I_{N(t)+1} = i, J_{N(t)+1} &= j, K_{N(t)+1} = k\Big].\end{aligned} \tag{8.7}$$

if $\boldsymbol{y} \in \mathcal{Y}_t$ and 0 otherwise. The likelihood for $\boldsymbol{\theta}$ based on $\mathcal{E}_t$ is given by Jacod's likelihood formula

$$L(\boldsymbol{\theta}; \mathcal{E}_t) = \prod_{\ell=1}^{N(t)} \rho^{\mathbb{E}}_{I_\ell, J_\ell, K_\ell, \boldsymbol{Y}_\ell}(T_\ell) \exp\Big( - \sum_{i,j,k,\boldsymbol{y}} \int_0^t \rho^{\mathbb{E}}_{i,j,k,\boldsymbol{y}}(s)\mathrm{d}s\Big). \tag{8.8}$$

Noting that $\alpha^{\mathbb{E}}_{i,j,k}(t) = \sum_{\boldsymbol{y} \in \mathcal{Y}_t} \rho^{\mathbb{E}}_{i,j,k,\boldsymbol{y}}(t)$ from the additivity of mark-specific intensities, introducing $\alpha^{\mathbb{E}}_{i,j}(t) = \sum_{\{k \in S_i : k < j\}} \alpha^{\mathbb{E}}_{i,j,k}(t)$ and applying Equation (8.7) we get

$$L(\boldsymbol{\theta}; \mathcal{E}_t) = \prod_{\ell=1}^{N(t)} \alpha^{\mathbb{E}}_{I_\ell, J_\ell, K_\ell}(T_\ell) \prod_{\ell=1}^{N(t)} P\Big[ \bigcap_{\{i \in C : y_i \neq -1\}} (X_i(t) = y_i) | E^+_{t^-} \Big]_{t=T_\ell, \boldsymbol{y}=\boldsymbol{Y}_\ell}$$

$$\times \exp\Big(-\Big(\sum_{i=1}^{n} \sum_{\ell=\min\{\ell' : T_{\ell'} = \tau_i\}} \int_{T_\ell}^{\min(t, T_{\ell+1})} \alpha^{\mathbb{E}}_{i, X_i(T^-_\ell)}(s)\mathrm{d}s\Big)\Big). \tag{8.9}$$

## 8.3 Incorporation of information from the observed system history process

Let $T = \inf\{t : \phi(\boldsymbol{X}(t)) \leq m\}$, where $m$, with $M > m \geq 0$, represents a certain system degradation level. Furthermore, let $V > 0$ be a censoring time, either fixed in advance or a random variable, being independent of $\mathbb{F}$ and not depending on our parameter vector $\boldsymbol{\theta}$. The operational period of the system is $[0, t]$, where $t = T \wedge V$. We suppose that $T$ is the only possible event time recorded in the system history process, denoted by $C_t$. Let $A = \{i : \tau_i < t\}$ and $I$ be the component having a transition at $T$.

There are several possibilities for the information content in $C_T$. It seems of special interest to consider the following variants:

1. We observe either (a) $\phi(\boldsymbol{X}(T)) \leq m$ or (b) $\phi(\boldsymbol{X}(T)) = m_r \leq m$, where $0 \leq m_r$.

2. We observe either (a) $X_i(T) = x_i$ for $i \in A^c$, corresponding to an autopsy, or (b) no inspections are performed at $T$ or (c) intermediate cases where some components in $A^c$ are inspected at $T$.

One of these variants must be chosen in advance as part of the model. Let $\boldsymbol{Y} = (Y_i)_{i \in A^c}$ with $Y_i = -1$ if the $i$th component is not inspected at time $T$ and equal to $X_i(T)$ otherwise. The marked point process defined by $C_t$ has a single event time $T$ with a single marked point $\boldsymbol{Y}$, whose cumulative hazard is discontinuous at $T$.

There are different forms of the additional factor in the likelihood, $L(\boldsymbol{\theta}; \mathcal{C}_t|\mathcal{E}_t)$, which is based on the information on the system level. This likelihood factor depends on which of the following randomly occurring events takes place

(i) $t = V$, corresponding to censoring

(ii) $\tau_I < T = t$

(iii) $\tau_I \geq T = t$

In case (i) the likelihood factor does not depend on the different variants in 1. and 2. and is simply of the form

$$L(\boldsymbol{\theta}; \mathcal{C}_t|\mathcal{E}_t) = P[\phi(\boldsymbol{X}(t)) > m|E_t] = \sum_{\boldsymbol{x}^{A^C} \in \mathcal{D}(\boldsymbol{x}^A)} P[\boldsymbol{X}^{A^C}(t) = \boldsymbol{x}^{A^C}|E_t], \tag{8.10}$$

where, for $B \subset C = \{1, \ldots, n\}$

$$\mathcal{D}(\boldsymbol{x}^B) = \{\boldsymbol{x}^{B^C} : \phi(\boldsymbol{x}^B, \boldsymbol{x}^{B^C}) > m\}. \tag{8.11}$$

We now consider the variant 1(a), 2(a) for case (ii). Using Jacod's likelihood formula, one can show that

$$\begin{aligned} &L(\boldsymbol{\theta}; \mathcal{C}_t|\mathcal{E}_t) \\ &\quad = P(T \geq t|E_t)P[T = T_{N(t)+1}, \boldsymbol{X}^{A^C}(t) \\ &\quad = \boldsymbol{y}|E_t, T \geq T_{N(t)+1}]_{t=T_{N(t)+1}, \boldsymbol{y}=\boldsymbol{Y}} \\ &\quad = P[T = t, \boldsymbol{X}^{A^C}(t) = \boldsymbol{y}|E_t]_{t=T, \boldsymbol{y}=\boldsymbol{Y}} = P[\boldsymbol{X}^{A^C}(t) = \boldsymbol{y}|E_t]_{t=T, \boldsymbol{y}=\boldsymbol{Y}}. \end{aligned} \tag{8.12}$$

To arrive at the corresponding result for case (iii) we define

$$\begin{aligned} \tilde{\alpha}^{\mathbb{E}}_{i,j,k}(t) &= \lim_{\mathrm{d}t \to 0} (1/\mathrm{d}t)P[T(i,k) \in [t, t + \mathrm{d}t), X_i(T(i,k)) \\ &= k|E_{t^-}, X_i(t^-) = j]. \end{aligned} \tag{8.13}$$

Again using Jacod's likelihood formula, we get

$$L(\boldsymbol{\theta}; \mathcal{C}_t|\mathcal{E}_t) = P(T \geq t|E_t)$$
$$\times \sum_{\{(i,j):i\in A^c, Y_i=k<j\leq M\}} \lim_{dt\to 0}(1/dt)P[T(i,k)$$
$$= T \in [t, t+dt), X_i(T(i,k)^-) = j, X_i(T(i,k)) = k, \boldsymbol{X}^{A^c-\{i\}}(t)$$
$$= \boldsymbol{y}^{A^c-\{i\}}|E_t, T \geq t]_{t=T(i,k), \boldsymbol{y}^{A^c-\{i\}}=\boldsymbol{Y}^{A^c-\{i\}}}$$
$$= \sum_{\{(i,j):i\in A^c, Y_i=k<j\leq M\}} I(\phi(j_i, \boldsymbol{X}^A(t), \boldsymbol{Y}^{A^c-\{i\}}) > m)P[X_i(t^-) = j|E_t]$$
$$\times \tilde{\alpha}^{\mathbb{E}}_{i,j,k}(t)P[\boldsymbol{X}^{A^c-\{i\}}(t) = \boldsymbol{y}^{A^c-\{i\}}|E_t, X_i(T(i,k)^-) = j,$$
$$X_i(T(i,k)) = k, T(i,k)]_{t=T(i,k), \boldsymbol{y}^{A^c-\{i\}}=\boldsymbol{Y}^{A^c-\{i\}}}. \quad (8.14)$$

It should be noted that for both the cases (ii) and (iii) $T = t$ and hence $\phi(\boldsymbol{X}(T)) = \phi(\boldsymbol{X}^A(t), \boldsymbol{Y}^{A^c})$ is known and hence the variant 1(b) is covered as well.

Analogous to Equation (8.11) we define

$$\mathcal{D}((i,j,k); \boldsymbol{x}^B)$$
$$= \{\boldsymbol{x}^{B^c-\{i\}} : \phi(j_i, \boldsymbol{x}^B, \boldsymbol{x}^{B^c-\{i\}}) > m, \phi(k_i, \boldsymbol{x}^B, \boldsymbol{x}^{B^c-\{i\}}) \leq m\}. \quad (8.15)$$

The likelihood factor for the variant 1(a), 2(b) and case (ii) is, parallel to Equation (8.12), simply of the form

$$L(\boldsymbol{\theta}; \mathcal{C}_t|\mathcal{E}_t) = \sum_{\boldsymbol{x}^{A^c}\in\mathcal{D}((I_{N(t)+1}, J_{N(t)+1}, K_{N(t)+1}); \boldsymbol{X}^{A-\{I_{N(t)+1}\}}(t))} P[\boldsymbol{X}^{A^c}(t) = \boldsymbol{x}^{A^c}|E_t]_{t=T}. \quad (8.16)$$

The corresponding result for case (iii) is, parallel to Equation (8.14), given by

$$L(\boldsymbol{\theta}; \mathcal{C}_t|\mathcal{E}_t) = \sum_{\{(i,j,k):i\in A^c, 0\leq k<j\leq M\}} P[X_i(t^-) = j|E_t]\tilde{\alpha}^{\mathbb{E}}_{i,j,k}(t)$$
$$\times \sum_{\boldsymbol{x}^{A^c-\{i\}}\in\mathcal{D}((i,j,k);\boldsymbol{X}^A(t))} P[\boldsymbol{X}^{A^c-\{i\}}(t) = \boldsymbol{x}^{A^c-\{i\}}|E_t,$$
$$X_i(T(i,k)^-) = j, X_i(T(i,k)) = k, T(i,k)]_{t=T(i,k)}. \quad (8.17)$$

To arrive at the parallel results with 1(a) replaced by 1(b) we just replace the set $\mathcal{D}((i, j, k); \boldsymbol{x}^B)$, given by Equation (8.15), by the set

$$\mathcal{D}_{m_r}((i, j, k); \boldsymbol{x}^B) = \{\boldsymbol{x}^{B^c-\{i\}} : \phi(j_i, \boldsymbol{x}^B, \boldsymbol{x}^{B^c-\{i\}}) > m,$$
$$\phi(k_i, \boldsymbol{x}^B, \boldsymbol{x}^{B^c-\{i\}}) = m_r\}. \tag{8.18}$$

Let us finally consider the variant 1(a), 2(c) for case (ii). We denote by $H$ the nonempty subset of $A^c$ which is inspected at $T$. Similar to Equation (8.16) we arrive at

$$L(\boldsymbol{\theta}, \mathcal{C}_t|\mathcal{E}_t)$$
$$= \sum_{\boldsymbol{x}^{(A\cup H)^c} \in \mathcal{D}((I_{N(t)+1}, J_{N(t)+1}, K_{N(t)+1}), \boldsymbol{X}^{(A\cup H)-\{I_{N(t)+1}\}}(t))} P[\boldsymbol{X}^{(A\cup H)^c}(t) = \boldsymbol{x}^{(A\cup H)^c},$$
$$\boldsymbol{Y} = \boldsymbol{y}|E_t]_{t=T, \boldsymbol{y}=\boldsymbol{Y}}. \tag{8.19}$$

By making an assumption on noninformativity and stability at the system level, parallel to Equation (8.6), we arrive at the following simplified version

$$L(\boldsymbol{\theta}, \mathcal{C}_t|\mathcal{E}_t) = \sum_{\boldsymbol{x}^{(A\cup H)^c} \in \mathcal{D}((I_{N(t)+1}, J_{N(t)+1}, K_{N(t)+1}), \boldsymbol{X}^{(A\cup H)-\{I_{N(t)+1}\}}(t))} P\big[\boldsymbol{X}^{A^c}(t)$$
$$= (\boldsymbol{x}^{(A\cup H)^c}, \boldsymbol{y}^H)|E_t\big]_{t=T, \boldsymbol{y}=\boldsymbol{Y}}. \tag{8.20}$$

The corresponding result for case (iii) can be arrived at similar to Equation (8.17).

## 8.4 Cause control and transition rate control

In this section we discuss ways of constructing inspection strategies that may prevent the likelihood of prohibitive complexity in larger models. First, for the case (iii) in Section 8.3, where an unmonitored component has a transition at $T$, we see from Equation (8.14) and Equation (8.17) that we must take into account the various possible component transitions being consistent with all information linked to the system event. This source of complexity in $L(\boldsymbol{\theta}; \mathcal{C}_t|\mathcal{E}_t)$ is eliminated if the inspection strategy is constructed in such a way that the case (iii) can never take place, i.e. such that the actual component transition occurring at $T$ is observed. Hence, the immediate cause of the system transition at $T$ is known. As in Gåsemyr and Natvig (2001) we call this cause control. Suppose, more generally, that the system transition times $T^v = \inf\{t : \phi(\boldsymbol{X}(t)) \leq m_v\}$,

$v = 1, \dots, r$ are observed, where $M > m_1 > m_2 > \cdots > m_r \geq 0$. $m_r$ is the lowest system degradation level of interest in the operational period. Some, or all, the $T^v$, $v = 1, \dots, r$ may coincide since a component transition can cause a system transition to a much lower level. We now construct an inspection strategy such that the actual component transitions occurring at $T^v$, $v = 1, \dots, r$, are observed. Denote by $R(s) \in \{1, \dots, p+q\}$ the set of components that are monitored at time $s$, where we possibly must choose $p + q = n$. Choose $R(0)$ such that $m_1 + 1 \leq \phi(\boldsymbol{x}^{R(0)}(0), \boldsymbol{x}^{R(0)^c})$ for all $\boldsymbol{x}^{R(0)^c}$. If $p + q = n$, this can be trivially obtained for $\boldsymbol{x}^{R(0)}(0) = \boldsymbol{M}$. Such $R(0)$ ensures that if the system jumps below $m_1 + 1$, one of the monitored components jumps simultaneously. Now consider an arbitrary observed component transition time $s$. The cause control inspection strategy is constructed inductively as follows.

1. By induction, it is known that $m_v + 1 \leq \phi(\boldsymbol{x}^{R(s^-)}(s^-), \boldsymbol{x}^{R(s^-)^c}) \leq m_{v-1}$ for all $\boldsymbol{x}^{R(s^-)^c}$ and for some $v \in \{1, \dots, r\}$, where $m_0 = M$. Due to the transition at $s$ we possibly have $\phi(\boldsymbol{x}^{R(s^-)}(s), \boldsymbol{x}^{R(s^-)^c}) \leq m_v$ for some $\boldsymbol{x}^{R(s^-)^c}$, in which case we proceed to 2 in order to update $R(s^-)$. If not, stop inspecting.

2. At $s$, inspect components iteratively updating $R(s^-)$ to $R(s)$, until either $m_v + 1 \leq \phi(\boldsymbol{x}^{R(s)}(s), \boldsymbol{x}^{R(s)^c})$ or $\phi(\boldsymbol{x}^{R(s)}(s), \boldsymbol{x}^{R(s)^c}) \leq m_v$, for all $\boldsymbol{x}^{R(s)^c}$. In the former case, we obviously stop inspecting. In the latter case, being sure that the system state has moved at least one degradation level of interest down, continue inspecting if necessary until either $m_{v+1} + 1 \leq \phi(\boldsymbol{x}^{R(s)}(s), \boldsymbol{x}^{R(s)^c})$ or $\phi(\boldsymbol{x}^{R(s)}(s), \boldsymbol{x}^{R(s)^c}) \leq m_{v+1}$, for all $\boldsymbol{x}^{R(s)^c}$. Again stop inspecting in the former case since we know that $m_{v+1} + 1 \leq \phi(\boldsymbol{x}^{R(s)}(s), \boldsymbol{x}^{R(s)^c}) \leq m_v$ for all $\boldsymbol{x}^{R(s)^c}$. Continue inspecting in the latter case, being sure that the system state has moved at least another degradation level of interest down.

The whole procedure stops when, for the first time, $\phi(\boldsymbol{x}^{R(t)}(t), \boldsymbol{x}^{R(t)^c}) \leq m_r$ for all $\boldsymbol{x}^{R(t)^c}$ and for some observed component transition at $t$, being sure that the system state has moved below the lowest degradation level of interest. Note that there is no reason for proceeding when there are no more components in $\{p + 1, \dots, n\}$ to inspect. If $m_v$ runs through the states $M, M - 1, \dots, 1, 0$ the cause control strategy implies that the system state transitions are always known from the available information at the component level. In this case, no new information is acquired from observation at the system level except possibly the autopsy data at the end of the operational period.

Another source of complexity is in the likelihood factor $L(\boldsymbol{\theta}; \mathcal{E}_t)$ given by Equation (8.9). Here the possibility of a large number of different orderings of past unobserved events in the history process $F_{t^-}$ may have to be taken into account. This complexity is addressed by introducing what we call transition rate control. The idea is to construct inspection strategies in such a way that the history process $E_{t^-}$ contains all relevant information for the hazard rates of component state transitions, given the full history process $F_{t^-}$. This means that the functions $I(f_{t^-} \in \mathcal{A}_t(i, j, k, r))$ and $\Psi_{i,j,k,r}(f_{t^-})$ being entered in Equation (8.1), defined on $\mathcal{H}^F_{t^-}$, are actually determined by $e_{t^-} \in \mathcal{H}^E_{t^-}$. Hence, these functions are supposed to be $\mathcal{E}_{t^-}$-measurable. Furthermore, this means that $R_t(i, j, k)$ given in Equation (8.2) is $\mathcal{E}_{t^-}$-measurable. In particular, the $\sigma_{i,r}$s in Equation (8.4) are now $\mathcal{E}_{t^-}$-stopping times.

Assuming transition rate control very much simplifies the likelihood factor in Equation (8.9). We will not give the simplified version here. In Section 8.6 we consider the electrical power generation system with some standby components where the assumptions of cause control and transition rate control enter in a natural way. It should be admitted that both these assumptions are very strong. This gives the motivation for the data augmentation simulation method presented in Section 8.7, which is an approach that is less sensitive to the complexities discussed in the present section without assuming cause control or transition rate control.

## 8.5 Maintenance, repair and aggregation of operational periods

We start this section by describing a reasonably flexible standard framework for incorporating maintenance or repair. Let $S^\ell$, $\ell = 1, 2, \ldots$ denote the successive times at which maintenance or repair decisions are made and followed up by corresponding actions. The intervals $(0, S^1], (S^1, S^2], \ldots$ are called operational periods. The times $S^\ell$ may be random, for instance occurring when the system state drops below a certain threshold level, or deterministic corresponding to periodic maintenance. When $S^\ell$ is the occurrence time of a random event, this event may be completely characterized by information at the component level, or contain additional information obtained by observation at the system level. In addition, further information may be collected as a basis for the maintenance or repair decision through an inspection. Any such information, which is not acquired as a consequence of the component monitoring scheme, is summarized in the random vector $\boldsymbol{V}^\ell$ generalizing $\boldsymbol{Y}$ in Section 8.3. Except for this information acquired at the end of the operational period, no information is obtained from observation at the system level.

The maintenance or repair actions may be considered as taking zero operational time, actually taking a negligible amount of time or being performed during an operational break. Alternatively, the actions may be modeled by means of repair time random variables. In this case, the intensities $\lambda_{i,j,k}(t; F_{t}-)$ may be nonzero also for $k > j$, in the repair phase. The levels to which the repair action brings the components may be deterministic or random.

The history process $F_t$ is defined basically as before, although our tacit assumption that no two component transitions occur simultaneously may be less natural when also recording repair processes, and hence some adjustments may have to be made. The observed history processes $E_{t}-$ and $E_t$ are also defined as before.

In the special case when maintenance or repair actions take zero operational time, denote by $\mathbf{Z}^{\ell}$ the random vector of states of the presently monitored components immediately after $S^{\ell}$. The $\sigma$-algebra $\mathcal{B}^{\ell}$ containing all information acquired in $(0, S^{\ell}]$ is defined inductively by $\mathcal{B}^{\ell} = \mathcal{B}^{\ell-1} \vee \sigma\{E_{S^{\ell}}, S^{\ell}, \mathbf{V}^{\ell}, \mathbf{Z}^{\ell}\}$. Furthermore, define $M(t) = \max\{\ell : S^{\ell} < t\}$ and $\mathcal{E}_t = \mathcal{B}^{M(t)} \vee \sigma\{E_t\}$, $\mathcal{B}_t = \mathcal{E}_t \vee \sigma\{I(S^{M(t)+1} > t)\}$ if $S^{M(t)+1} > t$ and $\mathcal{B}_t = \mathcal{B}^{M(t)+1}$ if $S^{M(t)+1} = t$. With these definitions, the likelihood contributions from different operational periods can be linked together as follows

$$L(\boldsymbol{\theta}; \mathcal{B}_t) = \Big[\prod_{\ell=1}^{M(t)} L(\boldsymbol{\theta}; \mathcal{B}^{\ell}|\mathcal{B}^{\ell-1})\Big] L(\boldsymbol{\theta}; \mathcal{B}_t|\mathcal{B}^{M(t)}). \tag{8.21}$$

The $\ell$th operational period may be treated as the unspecified operational period considered in the previous sections. $\mathcal{B}^{\ell-1}$ may be regarded as the initial $\sigma$-algebra analogous to $\mathcal{E}_0 = \mathcal{F}_0$ previously.

Note that the times $S^{\ell}$ ending the operational periods serve two purposes. First, they constitute times of opportunity for interruption of the gradual degradation of the system by a maintenance or repair action, bringing the system back to a higher performance level. Second, they provide opportunities for incorporating information gathered at the system level into the collection of background information that may be used in future maintenance or repair decisions, and also integrated in the probabilistic analysis of future observations in the history process. If both the inspection strategy and the maintenance or repair decisions regime are completely specified in terms of the component history process, the division into operational periods is unnecessary from a probabilistic point of view. The system may then be analyzed according to the set up of the previous sections, except that we must allow for positive repair rates $\lambda_{i,j,k}(t; F_{t}-)$ with $k > j$. We may then also consider models where maintenance or repair actions are performed more or less continuously, e.g. with each $T_\ell$ as an opportunity for such actions.

The entire elimination of the need, from a probabilistic point of view, of the division into operational periods can, for instance, be achieved in the following type of model. The maintenance or repair decision opportunity times $S^\ell$ are of the form $S^\ell = \inf\{t > S^{\ell-1} : \phi(\boldsymbol{X}(t)) \leq m\}$. We assume, as earlier, that no information on the system level is obtained in $(S^{\ell-1}, S^\ell)$. The inspection strategy is chosen to be cause controlling with respect to a transition to a state $\leq m$ for the system. If it is affirmed that $\phi(\boldsymbol{X}(t)) \leq m$, further inspection decisions and a maintenance or repair action may be made, but only based on information available at the component level.

## 8.6 The offshore electrical power generation system

In Section 3.8, for the offshore electrical power generation system as presented in Section 1.2, bounds for the availabilities and unavailabilities to various levels, in a fixed time interval, for the MMS $\phi_1(\boldsymbol{x})$ and $\phi_2(\boldsymbol{x})$ are given based on corresponding information on the components. This approach does not take the dynamics into account, which is the main theme of the present chapter and the following model description.

We start by considering the inspection and repair strategies. For $i \in \{1, 2, 3, 4, 5\}$ we assume that the $i$th component is in unmonitored standby in the intervals $[\gamma_{i,\ell-1}, \tau_{i,\ell})$, is monitored and operative or waiting for repair in the intervals $[\tau_{i,\ell}, \rho_{i,\ell})$ and is monitored and being repaired in the intervals $[\rho_{i,\ell}, \gamma_{i,\ell})$, $\ell = 1, 2, \ldots$, where $\gamma_{i,0} = 0$. $\tau_{i,1}$ corresponds to $\tau_i$ in the previous notation. For $i \in \{1, 3\}$, i.e. for the main generators, we assume that the $i$th component is never in standby, i.e. $\tau_{i,\ell} = \gamma_{i,\ell-1}$, $\ell = 1, 2, \ldots$. Furthermore, noting that the whole set up ensures that the state of component 4 is always known when components 2 and 5 are under inspection, we assume, for $\ell = 1, 2, \ldots$

$$\tau_{4,\ell} = \inf\{t > \gamma_{4,\ell-1} : (X_1(t) \leq 2) \cup (X_3(t) \leq 2)\}$$

$$\tau_{2,\ell} = \inf\{t > \gamma_{2,\ell-1} : [(X_1(t) \leq 2) \cup (X_3(t) \leq 2)] \cap (X_4(t) = 4)\}$$

$$\tau_{5,\ell} = \inf\{t > \gamma_{5,\ell-1} : (X_2(t) \geq 2) \cap (X_3(t) \leq 2) \cap (X_4(t) = 4)\}.$$

Assuming that the repair of the standby generator has lower priority compared to the main generators, and also noting that for $i \in \{2, 4, 5\}$ some intervals may reduce to the empty set, we get, for $\ell = 1, 2, \ldots$

$$\rho_{1,\ell} = \inf\{t > \tau_{1,\ell} : (X_1(t) \leq 2) \cap (X_3(t) = 4)\}$$

$$\rho_{2,\ell} = \inf\{t > \tau_{2,\ell} : X_1(t) = X_3(t) = 4\}$$

$$\rho_{3,\ell} = \inf\{t > \tau_{3,\ell} : (X_3(t) \leq 2) \cap (X_1(t) = 4)\}$$

$$\rho_{4,\ell} = \inf\{t > \tau_{4,\ell} : (X_4(t) \leq 2) \cup (X_1(t) = X_3(t) = 4)\}$$

$$\rho_{5,\ell} = \inf\{t > \tau_{5,\ell} : (X_5(t) \leq 2) \cup (X_3(t) = 4)\}$$

$$\gamma_{i,\ell} = \inf\{t > \rho_{i,\ell} : X_i(t) = 4\} \qquad i \in \{1, 2, 3, 4, 5\}.$$

Since both the inspection strategy and the repair decisions regime are completely specified in terms of the component history process, as mentioned in the previous section the division into operational periods is unnecessary just allowing for positive repair rates $\lambda_{i,j,k}(t; F_{t}-)$ with $k > j$. Considering the structure functions $\phi_2(x)$ and $\phi_2^*(x)$ given respectively by Equations (1.4) and (1.6), we assume that the control unit by a downward state transition in standby cannot cause a system state transition before being activated. By the construction of the inspection strategy, this is also the case for the standby generator and the subsea cables. Hence, the system state transitions are always known from the available information at the component level and we have a cause controlling inspection strategy. To also have transition rate control we must assume that a downward state transition of a component in standby cannot change any component transition rate. It is reasonable to assume that such a downward state transition of a component in standby can only affect the transition rates of the same component. This leaves us with having to assume that the time of the downward state transition to level 2 of a component in standby cannot affect the transition rate from 2 to 0 of the same component. For the cables, $L$, the latter transition rate is the failure rate of a simple cable. If the two simple cables are identical, the assumption is reasonable. For the generator $A_2$ we assume that only a transition from 4 to 0 in standby is possible. Hence, we only have to make this assumption for the control unit. Now applying Equation (8.1) we specify

$$\begin{aligned}\lambda_{i,j,k}(t;F_{t}-) = {} & \sum_{\ell=1} I([\gamma_{i,\ell-1} \leq t < \tau_{i,\ell}))\mu_{i,j,k,1}(t;\, \psi_{i,j,k,1}(E_{t}-)) \\ & + \sum_{\ell=1} I([\tau_{i,\ell} \leq t < \rho_{i,\ell}))\mu_{i,j,k,2}(t;\, \psi_{i,j,k,2}(E_{t}-)) \\ & + \sum_{\ell=1} I([\rho_{i,\ell} \leq t < \gamma_{i,\ell}))\mu_{i,j,k,3}(t;\, \psi_{i,j,k,3}(E_{t}-)),\end{aligned} \tag{8.22}$$

where, for $i \in \{1, 2, 3, 4, 5\}$, $\mu_{i,j,k,r}(t;\, \psi_{i,j,k,r}(E_{t}-)) = 0$ for $k > j$, $r = 1, 2$ and $\mu_{i,j,k,3}(t;\, \psi_{i,j,k,3}(E_{t}-)) = 0$ for $k < j$.

## 8.7 The data augmentation approach

One reason for computing the likelihood function in the previous sections is to use it in order to update the prior distribution, $\pi(\boldsymbol{\theta})$, for the parameter vector $\boldsymbol{\theta}$ to a posterior distribution based on operational data. In this section we describe how a standard simulation procedure, the data augmentation method, (see Tanner and Wong, 1987), can be implemented and used to obtain a sample from $\pi$, without even calculating the likelihood explicitly. The approach is not dependent on assuming either cause control or transition rate control, the idea being to extend the history $E_{t^-}$ to the full history $F_{t^-}$ by means of simulation. If we were able to generate alternately variables from the distribution of the 'missing data' given $\boldsymbol{\theta}$ and the observed data $E_{t^-}$, and from the distribution of $\boldsymbol{\theta}$ given $F_{t^-}$, we would obtain a sample from $\pi$ by means of Gibbs sampling (see Casella and George, 1992). Instead, we describe in some detail a Metropolis–Hastings within Gibbs procedure (see Chib and Greenberg, 1995), where drawing from the conditional distributions is replaced by Metropolis–Hastings steps for some or all variables to be simulated.

We need to introduce some new notation. A history process $F_{t^-}(i)$ for component $i$, containing only those events that concern this component, is defined as follows. First, define $T_0(i) = 0$. Let $\ell \geq 1$ be arbitrary and suppose by induction that $T_{\ell-1}(i)$ has been defined. Let $r = \min\{k : T_k > T_{\ell-1}(i), I_k = i\}$ and let $T_\ell(i) = T_r$ and $U_\ell(i) = U_r = (I_r, J_r, K_r)$. Then $F_{t^-}(i) = \{(T_\ell(i), U_\ell(i)); T_\ell(i) < t\}$. Also define $F_{t^-}(-i) = F_{t^-} - F_{t^-}(i)$, the history process with the events concerning component $i$ being removed.

For simplicity, we will only treat the case of one operational period in detail. In this case the missing data just before $t$ are given by $\{F_{(t\wedge\tau_i)^-}(i); i \in C\}$ and we have $\sigma(\{F_{t^-}\}) = \sigma(\{F_{(t\wedge\tau_i)^-}(i); i \in C, E_{t^-}\}) = \sigma(\{F_{(t\wedge\tau_i)^-}(i), F_{t^-}(-i), E_{t^-}\})$. In addition, we restrict the detailed treatment to the case when no information is acquired at the system level, i.e. $\mathcal{C}_t = \mathcal{E}_t$.

The procedure consists of drawing iteratively variables $F^*_{(t\wedge\tau_i)^-}(i)$ from a proposal distribution $p_i$ to be accepted or rejected according to the usual Metropolis–Hastings criterion. The target distribution is $\pi(F_{(t\wedge\tau_i)^-}(i)|E_{t^-}, F_{t^-}(-i), \boldsymbol{\theta})$. Let $F^{**}_{(t\wedge\tau_i)^-}(i)$ denote the old value of $F_{(t\wedge\tau_i)^-}(i)$ and let $F^*_{t^-}$ and $F^{**}_{t^-}$ denote the complete histories arising from extending the data $\{F_{t^-}(-i), E_{t^-}\}$ with, respectively, $F^*_{(t\wedge\tau_i)^-}(i)$ and $F^{**}_{(t\wedge\tau_i)^-}(i)$. The Metropolis–Hastings ratio is then given by

$$[p_i(F^{**}_{(t\wedge\tau_i)^-}(i))/p_i(F^{*}_{(t\wedge\tau_i)^-}(i))]$$

$$\times\,[\pi(F^{*}_{(t\wedge\tau_i)^-}(i)|E_{t^-}, F_{t^-}(-i), \boldsymbol{\theta})/\pi(F^{**}_{(t\wedge\tau_i)^-}(i)|E_{t^-}, F_{t^-}(-i), \boldsymbol{\theta})]$$
$$= [p_i(F^{**}_{(t\wedge\tau_i)^-}(i))/p_i(F^{*}_{(t\wedge\tau_i)^-}(i))][\pi(F^{*}_{t^-}|\boldsymbol{\theta})/\pi(F^{**}_{t^-}|\boldsymbol{\theta})]$$
$$= [p_i(F^{**}_{(t\wedge\tau_i)^-}(i))/p_i(F^{*}_{(t\wedge\tau_i)^-}(i))][L(\boldsymbol{\theta};\, F^{*}_{t^-})/L(\boldsymbol{\theta};\, F^{**}_{t^-})]. \tag{8.23}$$

The proposal $F^{*}_{(t\wedge\tau_i)^-}(i)$ replaces the old value $F^{**}_{(t\wedge\tau_i)^-}(i)$ with probability equal to the minimum of this ratio and 1. This procedure is repeated for each $i \in C$. We then draw samples from each component of $\boldsymbol{\theta}$ given the other components of $\boldsymbol{\theta}$ and $F_{t^-}$. For some of the components of $\boldsymbol{\theta}$ we may have to introduce a Metropolis–Hastings step if we are unable to draw directly from the conditional distributions.

We will now describe in more detail how to handle the drawing of $F_{(t\wedge\tau_i)^-}(i)$. Depending on what kind of information $E_{t^-}$ contains about $F_{(t\wedge\tau_i)^-}(i)$, we have to distinguish between four cases.

**Case 1.** If component $i$ is monitored from time 0, the history $F_{t^-}(i)$ is observed and no drawing is necessary.

**Case 2.** Suppose that $\tau_i > t$, i.e. the component is not monitored at all in $[0, t)$. Let $j_1$ be the known initial state of component $i$. Generate independently $Z(j_1, k)$ as the transition time from $j_1$ to $k$ according to Equation (8.1) for those $k \in S_i$ for which the transition intensity is positive. If $\min_k\{Z(j_1, k)\} = Z(j_1, k_0)$, let $T_1^*(i) = Z(j_1, k_0)$, $J_1^*(i) = j_1$, $K_1^*(i) = k_0$. Proceed inductively until $T_L^*(i) > t$ or $K_{L-1}^*(i) = 0$. Then $(T_1^*(i), J_1^*(i), K_1^*(i), \ldots, T_{L-1}^*(i), J_{L-1}^*(i), K_{L-1}^*(i))$ is the proposed value $F^*_{t^-}(i)$. Noting that $T_0^{**}(i) = T_0^*(i) = 0$, $t \wedge \tau_i = t$, the ratio of proposal densities is chosen to be given by the ratio of the corresponding likelihoods

$$p_i(F^{**}_{(t\wedge\tau_i)^-}(i))/p_i(F^{*}_{(t\wedge\tau_i)^-}(i))$$
$$= \Bigg[ \prod_{\{\ell:0<T_\ell^{**}(i)<(t\wedge\tau_i)\}} \lambda_{i,J_\ell^{**}(i),K_\ell^{**}(i)}(T_\ell^{**}(i);\, F^{**}_{T_\ell^{**}(i)^-})$$
$$\times \exp\Bigg( - \sum_{\{\ell:0\le T_\ell^{**}(i)<(t\wedge\tau_i)\}} \sum_{k<J_{\ell+1}^{**}(i)} \int_{T_\ell^{**}(i)}^{\min\{t\wedge\tau_i, T_{\ell+1}^{**}(i)\}} \lambda_{i,J_{\ell+1}^{**}(i),k}(s;\, F_{s^-}^{**})\mathrm{d}s\Bigg)\Bigg]$$
$$\times \Bigg[ \prod_{\{\ell:0<T_\ell^{*}(i)<(t\wedge\tau_i)\}} \lambda_{i,J_\ell^{*}(i),K_\ell^{*}(i)}(T_\ell^{*}(i);\, F^{*}_{T_\ell^{*}(i)^-})$$

$$\times \exp\Bigg(- \sum_{\{\ell:0\le T^*_\ell(i)<(t\wedge\tau_i)\}} \sum_{k<J^*_{\ell+1}(i)} \int_{T^*_\ell(i)}^{\min\{t\wedge\tau_i, T^*_{\ell+1}(i)\}} \lambda_{i,J^*_{\ell+1}(i),k}(s;\, F^*_{s-})\mathrm{d}s\Bigg)\Bigg]^{-1} \tag{8.24}$$

**Case 3.** Suppose that $\tau_i \le t$ and $X_i(\tau_i) = 0$. Then proceed as in case 2 until $T^*(i, 0)$ is obtained. If $T^*(i, 0) > \tau_i$, start over again. Otherwise, accept the generated $F^*_{\tau_i-}(i)$ as a proposal from $p_i$. Since we are conditioning on the same event $X_i(\tau_i) = 0$ both for the proposal and the old value, Equation (8.24) still holds.

**Case 4.** Suppose that $\tau_i \le t$ and $X_i(\tau_i) = j > 0$. We proceed as in case 2 until $T^*(i, j)$ is obtained. If $T^*(i, j) \le \tau_i$ and $X_i(T^*(i, j)) = j$, we suggest as an approximation to accept this history as a proposal from $p_i$. Otherwise, start over again. Parallel to case 3, since we are conditioning on $T^*(i, j) \le \tau_i$ and $X_i(T^*(i, j)) = j$ both for the proposal and the old value, Equation (8.24) still holds.

We conclude that the cases 2, 3 and 4 differ in the procedure for generating proposal values, but that the ratio of proposal densities is the same given by Equation (8.24). Inserting Equation (8.24) into Equation (8.23) gives the common Metropolis–Hastings ratio for all three cases.

# Appendix A

# Remaining proofs of bounds given in Chapter 3

Since dealing with modular decompositions in this appendix, we will here augment the notation introduced in Theorem 2.15 for the $k$th minimal cut vector to level $j$ for an MMS $(C, \phi)$. We now write $z^j_{\phi k} = (z^j_{\phi 1k}, \ldots, z^j_{\phi nk})$.

## A.1 Proof of the inequalities 14, 7 and 8 of Theorem 3.12

We start by considering the special case that $\psi$ has only one single minimal cut vector to each level. For this special case we prove an equality which leads to inequality 14.

**Lemma A.1:** Make the assumptions of Theorem 3.12. In addition assume that $\psi$ has only one single minimal cut vector to level $j$, which we denote $z^j_\psi = (z^j_{\psi 1}, \ldots, z^j_{\psi r})$. Then, for $j = 1, \ldots, M$

$$\bar{\ell}''^{j(I)}_{\phi} = \prod_{k \in D^j_\psi(z^j_\psi)} \bar{\ell}''^{(z^j_{\psi k}+1)(I)}_{\chi_k}.$$

*Multistate Systems Reliability Theory with Applications* Bent Natvig

**Proof:** We have, from (ii) of Theorem 2.15

$$I_j(\phi(\boldsymbol{x})) = \boldsymbol{J}_{\boldsymbol{z}^j_\psi}(\boldsymbol{\chi}(\boldsymbol{x})) = \coprod_{k\in D^j_\psi(\boldsymbol{z}^j_\psi)} I_{z^j_{\psi k}+1}(\chi_k(\boldsymbol{x}^{A_k}))$$

$$= \coprod_{k\in D^j_\psi(\boldsymbol{z}^j_\psi)} \prod_{\ell=1}^{m_{\chi_k}^{z^j_{\psi k}+1}} \boldsymbol{J}_{\boldsymbol{z}_{\chi_k \ell}^{z^j_{\psi k}+1}}(\boldsymbol{x}^{A_k}).$$

Now the minimal cut parallel structures of $(C, \phi)$ to level $j$ are given by

$$\coprod_{k\in D^j_\psi(\boldsymbol{z}^j_\psi)} \boldsymbol{J}_{\boldsymbol{z}_{\chi_k \ell_k}^{z^j_{\psi k}+1}}(\boldsymbol{x}^{A_k}), \tag{A.1}$$

where for each $k \in D^j_\psi(z^j_\psi)$ we can choose $1 \leq \ell_k \leq m_{\chi_k}^{z^j_{\psi k}+1}$. Hence, using the fact that the modules are totally independent in $I$

$$\bar{\ell}_\phi^{\prime\prime j(I)}$$

$$= \max_{1\leq \ell_k \leq m_{\chi_k}^{z^j_{\psi k}+1},\ k\in D^j_\psi(\boldsymbol{z}^j_\psi)} P[\forall s \in \tau(I),\ \forall k \in D^j_\psi(z^j_\psi)$$

$$\boldsymbol{J}_{\boldsymbol{z}_{\chi_k \ell_k}^{z^j_{\psi k}+1}}(\boldsymbol{X}^{A_k}(s)) = 0]$$

$$= \max_{1\leq \ell_k \leq m_{\chi_k}^{z^j_{\psi k}+1},\ k\in D^j_\psi(\boldsymbol{z}^j_\psi)} \prod_{k\in D^j_\psi(\boldsymbol{z}^j_\psi)} P[\forall s \in \tau(I)\ \boldsymbol{J}_{\boldsymbol{z}_{\chi_k \ell_k}^{z^j_{\psi k}+1}}(\boldsymbol{X}^{A_k}(s)) = 0]$$

$$= \prod_{k\in D^j_\psi(\boldsymbol{z}^j_\psi)} \max_{1\leq \ell \leq m_{\chi_k}^{z^j_{\psi k}+1}} P[\forall s \in \tau(I)\ \boldsymbol{J}_{\boldsymbol{z}_{\chi_k \ell}^{z^j_{\psi k}+1}}(\boldsymbol{X}^{A_k}(s)) = 0]$$

$$= \prod_{k\in D^j_\psi(\boldsymbol{z}^j_\psi)} \bar{\ell}_{\chi_k}^{\prime\prime(z^j_{\psi k}+1)(I)},$$

and the proof of the lemma is completed.

By applying Theorem 2.17 and Lemma A.1, we get

$$\bar{\ell}_\phi^{\prime\prime j(I)} \leq \max_{1\leq \ell \leq m^j_\psi} \max_{1\leq m \leq m^j_{\phi^j_\ell}} P[\boldsymbol{J}_{\boldsymbol{z}^j_{\phi^j_\ell m}}(\boldsymbol{X}(s)) = 0\ \forall s \in \tau(I)]$$

$$= \max_{1 \le \ell \le m_{\psi}^{j}} \bar{\ell}_{\phi_{\ell}^{j}}^{\prime\prime j(I)} = \max_{1 \le \ell \le m_{\psi}^{j}} \prod_{k \in D_{\psi}^{j}(\boldsymbol{z}_{\psi\ell}^{j})} \bar{\ell}_{\chi_k}^{\prime\prime(z_{\psi k\ell}^{j}+1)(I)}$$

$$= \bar{\ell}_{\psi}^{\prime j(I)}(\bar{\boldsymbol{\ell}}_{\psi}^{\prime\prime(I)}), \tag{A.2}$$

and inequality 14 is readily proved.

To prove the inequalities 7 and 8 we first prove, following Gåsemyr (2010), that the inequality in Equation (A.2) actually is an equality. Hence, we prove the opposite inequality. Choose $1 \le \ell \le m_{\psi}^{j}$ and $1 \le m \le m_{\phi_{\ell}^{j}}^{j}$ such that the right-hand side of the inequality in Equation (A.2) is maximized. Remembering the proof of Theorem 2.17, consider the corresponding cut vector $z_{\phi_{\ell}^{j} m}^{j}$ to level $j$ of $(C, \phi)$. In case this is not a minimal cut vector, construct one, $z_{\phi}^{j}$, satisfying $z_{\phi} \ge z_{\phi_{\ell}^{j} m}^{j}$. Then, using the proof of Equation (A.2), we get

$$\begin{aligned}
\bar{\ell}_{\phi}^{\prime\prime j(I)} &= \max_{1 \le k \le m_{\phi}^{j}} P[\cap_{i \in D_{\phi}^{j}(\boldsymbol{z}_{k}^{j})}(X_i(s) \le z_{ik}^{j}) \ \forall s \in \tau(I)] \\
&\ge P[\cap_{i \in D_{\phi}^{j}(\boldsymbol{z}_{\phi}^{j})}(X_i(s) \le z_{\phi i}^{j}) \ \forall s \in \tau(I)] \\
&\ge P[\cap_{i \in D_{\phi_{\ell}^{j}}^{j}(\boldsymbol{z}_{\phi_{\ell}^{j} m}^{j})}(X_i(s) \le z_{\phi_{\ell}^{j} im}^{j}) \ \forall s \in \tau(I)] \\
&= \max_{1 \le \ell \le m_{\psi}^{j}} \max_{1 \le m \le m_{\phi_{\ell}^{j}}^{j}} P[\boldsymbol{J}_{\boldsymbol{z}_{\phi_{\ell}^{j} m}^{j}}(\boldsymbol{X}(s)) = 0 \ \forall s \in \tau(I)] \\
&= \bar{\ell}_{\psi}^{\prime j(I)}(\bar{\boldsymbol{\ell}}_{\psi}^{\prime\prime(I)}).
\end{aligned} \tag{A.3}$$

Now, combining Equations (A.2) and (A.3) for $I = [t, t]$, we get

$$\begin{aligned}
u_{\phi}^{\prime\prime j(I)} &\le \inf_{t \in \tau(I)} u_{\phi}^{\prime\prime j([t,t])} = \inf_{t \in \tau(I)} [1 - \bar{\ell}_{\phi}^{\prime\prime j([t,t])}] \\
&= \inf_{t \in \tau(I)} [1 - \bar{\ell}_{\psi}^{\prime j([t,t])}(\bar{\boldsymbol{\ell}}_{\psi}^{\prime\prime([t,t])})] \\
&\le 1 - \bar{\ell}_{\psi}^{\prime j(I)}(\bar{\boldsymbol{\ell}}_{\psi}^{\prime\prime(I)}),
\end{aligned}$$

and the inequalities 7 and 8 are proved.

## A.2 Proof of inequality 14 of Theorem 3.13

We start by proving an equality, for the special case that $\psi$ has only one single minimal cut vector to each level, which leads to inequality 14.

**Lemma A.2:** Let $(C, \phi)$ be an MMS with modular decomposition given by Definition 1.6. In addition assume that $\psi$ has only one single minimal cut vector to level $j$, which we denote $z^j_\psi = (z^j_{\psi 1}, \ldots, z^j_{\psi r})$. Then, for $j = 1, \ldots, M$

$$\bar{\ell}'^{j(I)}_{\phi}(\boldsymbol{Q}^{(I)}_{\phi}) = \prod_{k \in D^j_\psi(\boldsymbol{z}^j_\psi)} \bar{\ell}'^{(z^j_{\psi k}+1)(I)}_{\chi_k}(\boldsymbol{Q}^{(I)}_{\phi}).$$

**Proof:** From Equation (A.1) we have

$$\begin{aligned}
\bar{\ell}'^{j(I)}_{\phi}(\boldsymbol{Q}^{(I)}_{\phi}) &= \max_{1 \le \ell_k \le m^{z^j_{\psi k}+1}_{\chi_k},\ k \in D^j_\psi(\boldsymbol{z}^j_\psi)} \prod_{k \in D^j_\psi(\boldsymbol{z}^j_\psi)} \prod_{i \in D^{(z^j_{\psi k}+1)}_{\chi_k}(\boldsymbol{z}^{(z^j_{\psi k}+1)}_{\chi_k \ell_k})} q_{\phi i}^{z^{(z^j_{\psi k}+1)}_{\chi_k i \ell_k}+1(I)} \\
&= \prod_{k \in D^j_\psi(\boldsymbol{z}^j_\psi)} \max_{1 \le \ell \le m^{z^j_{\psi k}+1}_{\chi_k}} \prod_{i \in D^{(z^j_{\psi k}+1)}_{\chi_k}(\boldsymbol{z}^{(z^j_{\psi k}+1)}_{\chi_k \ell})} q_{\phi i}^{z^{(z^j_{\psi k}+1)}_{\chi_k i \ell}+1(I)} \\
&= \prod_{k \in D^j_\psi(\boldsymbol{z}^j_\psi)} \bar{\ell}'^{(z^j_{\psi k}+1)(I)}_{\chi_k}(\boldsymbol{Q}^{(I)}_{\phi}),
\end{aligned}$$

and the proof of the lemma is completed.

Now, remembering the main part of the proof from A.1, we get, by applying Theorem 2.17 and Lemma A.2

$$\begin{aligned}
\bar{\ell}'^{j(I)}_{\phi}(\boldsymbol{Q}^{(I)}_{\phi}) &\le \max_{1 \le \ell \le m^j_\psi} \max_{1 \le m \le m^j_{\phi^j_\ell}} \prod_{i \in D^j_{\phi^j_\ell}(\boldsymbol{z}^j_{\phi^j_\ell m})} q_{\phi i}^{z^j_{\phi^j_\ell i m}+1(I)} \\
&= \max_{1 \le \ell \le m^j_\psi} \bar{\ell}'^{j(I)}_{\phi^j_\ell}(\boldsymbol{Q}^{(I)}_{\phi}) = \max_{1 \le \ell \le m^j_\psi} \prod_{k \in D^j_\psi(\boldsymbol{z}^j_{\psi \ell})} \bar{\ell}'^{(z^j_{\psi k \ell}+1)(I)}_{\chi_k}(\boldsymbol{Q}^{(I)}_{\phi}) \\
&= \bar{\ell}'^{j(I)}_{\psi}(\bar{\boldsymbol{\ell}}'^{(I)}_{\psi}(\boldsymbol{Q}^{(I)}_{\phi})).
\end{aligned} \tag{A.4}$$

Hence, we have

$$\bar{\ell}_{\phi}^{\prime j(I)}(\boldsymbol{Q}_{\phi}^{(I)}) \leq \bar{L}_{\psi}^{*j(I)}(\bar{\boldsymbol{\ell}}_{\psi}^{\prime(I)}(\boldsymbol{Q}_{\phi}^{(I)})) \leq \bar{L}_{\psi}^{*j(I)}(\bar{\boldsymbol{L}}_{\psi}^{(I)}) \leq \bar{L}_{\psi}^{*j(I)}(\bar{\boldsymbol{B}}_{\psi}^{(I)})$$
$$\leq \bar{B}_{\psi}^{*j(I)}(\bar{\boldsymbol{B}}_{\psi}^{(I)}),$$

and inequality 14 is proved.

## A.3 Proof of inequality 10 of Theorem 3.17

We start by proving the following lemma, which is a corrected and slightly generalized version of Lemma 3.2 of Butler (1982).

**Lemma A.3:** Make the same assumptions as in Theorem 3.16. In addition assume that $\psi$ has only one single minimal cut vector to level $j$, $z_{\psi}^{j} = (z_{\psi 1}^{j}, \ldots, z_{\psi r}^{j})$. Then, for $j = 1, \ldots, M$

$$\ell_{\phi}^{*j} \leq \coprod_{k \in D_{\psi}^{j}(\boldsymbol{z}_{\psi}^{j})} \ell_{\chi_k}^{*(z_{\psi k}^{j}+1)}.$$

**Proof:** Consider the structure function $\phi^*$ of an MMS defined by

$$I_j(\phi^*(\boldsymbol{Y}(t))) = \coprod_{k \in D_{\psi}^{j}(\boldsymbol{z}_{\psi}^{j})} \prod_{\ell=1}^{m_{\chi_k}^{z_{\psi k}^{j}+1}} Y_{k\ell}^{z_{\psi k}^{j}+1}(t),$$

where $Y_{k\ell}^{j}(t)$ for $\ell = 1, \ldots, m_{\chi_k}^{j}$, $k = 1, \ldots, r$, $j = 1, \ldots, M$ are binary, mutually independent components at time $t$, and hence associated according to Property 5 of associated random variables. The availabilities to level 1 at this point of time are

$$P(Y_{k\ell}^{j}(t) = 1) = P[\boldsymbol{J}_{\boldsymbol{z}_{\chi_k \ell}^{j}}(\boldsymbol{X}^{A_k}(t)) = 1]. \tag{A.5}$$

We have

$$p_{\phi^*}^{j} = E\left[\coprod_{k \in D_{\psi}^{j}(\boldsymbol{z}_{\psi}^{j})} \prod_{\ell=1}^{m_{\chi_k}^{z_{\psi k}^{j}+1}} Y_{k\ell}^{z_{\psi k}^{j}+1}(t)\right]$$

$$= \coprod_{k\in D_\psi^j(\boldsymbol{z}_\psi^j)} \prod_{\ell=1}^{m_{\chi_k}^{z_{\psi k}^j+1}} P\left[ \boldsymbol{J}_{\boldsymbol{z}_{\chi_k \ell}^{z_{\psi k}^j+1}}(\boldsymbol{X}^{A_k}(t)) = 1 \right] = \coprod_{k\in D_\psi^j(\boldsymbol{z}_\psi^j)} \ell_{\chi_k}^{*(z_{\psi k}^j+1)}. \tag{A.6}$$

Now all the minimal cut parallel structures for $\phi^*$ to level $j$ are of the same form as the ones for $\phi$, given by Equation (A.1), just replacing $\boldsymbol{J}_{\boldsymbol{z}_{\chi_k \ell_k}^{z_{\psi k}^j+1}}(\boldsymbol{x}^{A_k})$ by $y_{k\ell_k}^{z_{\psi k}^j+1}$. Since the modules of $\phi$ are totally independent at time $t$, $\{\boldsymbol{J}_{\boldsymbol{z}_{\chi_k \ell_k}^{z_{\psi k}^j+1}}(\boldsymbol{X}^{A_k}(t))\}_{k\in D_\psi^j(\boldsymbol{z}_\psi^j)}$ are independent, exactly as is true for $\{Y_{k\ell_k}^{z_{\psi k}^j+1}(t)\}_{k\in D_\psi^j(\boldsymbol{z}_\psi^j)}$. Due to Equation (A.5) we then have

$$\ell_{\phi^*}^{*j} = \ell_\phi^{*j}.$$

Hence, from Equation (2.29) of Theorem 2.28 and Equation (A.6) we get

$$\ell_\phi^{*j} = \ell_{\phi^*}^{*j} \le p_{\phi^*}^j = \coprod_{k\in D_\psi^j(\boldsymbol{z}_\psi^j)} \ell_{\chi_k}^{*(z_{\psi k}^j+1)},$$

and the proof of the lemma is completed.

If the minimal cut vectors to level $j$ of $(C, \phi)$ are exactly given by

$$\boldsymbol{z}_{\phi_\ell^j m}^j \quad m = 1, \ldots, m_{\phi_\ell^j}^j, \quad \ell = 1, \ldots, m_\psi^j,$$

then, from Theorem 2.17 and Lemma A.3, we have

$$\begin{aligned}\ell_\phi^{*j} &= \prod_{\ell=1}^{m_\psi^j} \prod_{m=1}^{m_{\phi_\ell^j}^j} P[\boldsymbol{J}_{\boldsymbol{z}_{\phi_\ell^j m}^j}(\boldsymbol{X}(t)) = 1] = \prod_{\ell=1}^{m_\psi^j} \ell_{\phi_\ell^j}^{*j} \le \prod_{\ell=1}^{m_\psi^j} \coprod_{k\in D_\psi^j(\boldsymbol{z}_{\psi\ell}^j)} \ell_{\chi_k}^{*(z_{\psi k\ell}^j+1)} \\ &= \ell_\psi^{**j}(\boldsymbol{\ell}_\psi^*).\end{aligned} \tag{A.7}$$

On the other hand, by applying Equation (A.4) on the dual structure and dual level, specializing $I = [t, t]$ and remembering Equation (3.28), we get

$$\ell_\phi^{\prime j}(\boldsymbol{P}_\phi) \le \ell_\psi^{\prime j}(\boldsymbol{\ell}_\psi^\prime(\boldsymbol{P}_\phi)). \tag{A.8}$$

From Equations (A.7) and (A.8) we now get

$$\begin{aligned} L_\phi^j &= \max[\ell'^j_\phi(\boldsymbol{P}_\phi), \ell^{*j}_\phi] \le \max[\ell'^j_\psi(\boldsymbol{L}_\psi), \ell^{**j}_\psi(\boldsymbol{L}_\psi)] \\ &\le \max[\ell'^j_\psi(\boldsymbol{B}_\psi), \ell^{**j}_\psi(\boldsymbol{B}_\psi)] \\ &= L^{*j}_\psi(\boldsymbol{B}_\psi), \end{aligned}$$

and inequality 10 immediately follows.

# Appendix B

# Remaining intensity matrices in Chapter 4

$$A_{a_2} = \begin{bmatrix} -4 & 0 & 0 & 0 & 4 \\ 0 & -8.3721 & 0 & 0 & 8.3721 \\ 0 & 0 & -9 & 0 & 9 \\ 0 & 0 & 0 & -22.5 & 22.5 \\ 0.00006 & 0.01565 & 0.021 & 0.01736 & -0.05406 \end{bmatrix}$$

$$A_{b_1} = \begin{bmatrix} -1.8 & 0 & 0 & 0 & 0 & 1.8 \\ 0 & -6.207 & 0 & 0 & 0 & 6.207 \\ 0 & 0 & -7.3469 & 0 & 0 & 7.3469 \\ 0 & 0 & 0 & -8.3721 & 0 & 8.3721 \\ 0 & 0 & 0 & 0 & -9 & 9 \\ 0.000066 & 0.000072 & 0.0023 & 0.0082 & 0.02 & -0.03064 \end{bmatrix}$$

$$A_{b_2} = \begin{bmatrix} -22.5 & 0 & 0 & 0 & 22.5 \\ 0 & -51.4286 & 0 & 0 & 51.4286 \\ 0 & 0 & -444.4444 & 0 & 444.4444 \\ 0 & 0 & 0 & -720 & 720 \\ 0.109 & 0.0008 & 0.31 & 0.03 & -0.4498 \end{bmatrix}$$

$$A_{b_3} = \begin{bmatrix} -3.6735 & 0 & 0 & 0 & 0 & 3.6735 \\ 0 & -51.4286 & 0 & 0 & 0 & 51.4286 \\ 0 & 0 & -120 & 0 & 0 & 120 \\ 0 & 0 & 0 & -180 & 0 & 180 \\ 0 & 0 & 0 & 0 & -360 & 360 \\ 0.00099 & 0.00017 & 0.0017 & 0.05 & 0.44 & -0.49286 \end{bmatrix}$$

$$A_{b_4} = \begin{bmatrix} -120 & 120 \\ 0.08 & -0.08 \end{bmatrix}$$

$$A_{b_5} = \begin{bmatrix} -18 & 0 & 0 & 0 & 18 \\ 0 & -51.4286 & 0 & 0 & 51.4286 \\ 0 & 0 & -180 & 0 & 180 \\ 0 & 0 & 0 & -360 & 360 \\ 0.0085 & 0.00089 & 1.6 & 1.06 & -2.66939 \end{bmatrix}$$

$$A_{b_6} = \begin{bmatrix} -180 & 0 & 0 & 180 \\ 0 & -211.7647 & 0 & 211.7647 \\ 0 & 0 & -360 & 360 \\ 1.8 & 1.3 & 0.08 & -3.18 \end{bmatrix}$$

$$A_{b_7} = \begin{bmatrix} -7.3469 & 0 & 0 & 0 & 0 & 7.3469 \\ 0 & -8.3721 & 0 & 0 & 0 & 8.3721 \\ 0 & 0 & -22.5 & 0 & 0 & 22.5 \\ 0 & 0 & 0 & -51.4286 & 0 & 51.4286 \\ 0 & 0 & 0 & 0 & -444.4444 & 444.4444 \\ 0.00154 & 0.00554 & 0.09402 & 0.0008 & 0.176 & -0.2779 \end{bmatrix}$$

$$A_{b_8} = A_{c_1} = \begin{bmatrix} -22.5 & 22.5 \\ 0.009 & -0.009 \end{bmatrix}$$

$$A_{c_2} = \begin{bmatrix} -4 & 0 & 0 & 4 \\ 0 & -8.3721 & 0 & 8.3721 \\ 0 & 0 & -22.5 & 22.5 \\ 0.00003 & 0.00636 & 0.01704 & -0.02343 \end{bmatrix}$$

$$A_{d_1} = \begin{bmatrix} -1 & 0 & 0 & 0 & 0 & 0 & 0 & 1 \\ 0 & -1.2 & 0 & 0 & 0 & 0 & 0 & 1.2 \\ 0 & 0 & -6 & 0 & 0 & 0 & 0 & 6 \\ 0 & 0 & 0 & -12 & 0 & 0 & 0 & 12 \\ 0 & 0 & 0 & 0 & -22.5 & 0 & 0 & 22.5 \\ 0 & 0 & 0 & 0 & 0 & -360 & 0 & 360 \\ 0 & 0 & 0 & 0 & 0 & 0 & -900 & 900 \\ 0.00022 & 0.0004 & 0.00031 & 0.0017 & 0.009 & 0.14 & 1.8 & -1.95163 \end{bmatrix}$$

$$A_{d_2} = \begin{bmatrix} -444.4444 & 444.4444 \\ 0.175 & -0.175 \end{bmatrix}$$

$$A_{d_3} = \begin{bmatrix} -8.3721 & 0 & 8.3721 \\ 0 & -22.5 & 22.5 \\ 0.01276 & 0.00004 & -0.0128 \end{bmatrix}$$

$$A_{d_4} = \begin{bmatrix} -22.5 & 0 & 0 & 0 & 0 & 22.5 \\ 0 & -360 & 0 & 0 & 0 & 360 \\ 0 & 0 & -444.4444 & 0 & 0 & 444.4444 \\ 0 & 0 & 0 & -720 & 0 & 720 \\ 0 & 0 & 0 & 0 & -900 & 900 \\ 0.0026 & 0.0017 & 0.38 & 0.06 & 1.8 & -2.2443 \end{bmatrix}$$

$$A_{d_5} = \begin{bmatrix} -1 & 0 & 0 & 0 & 0 & 0 & 1 \\ 0 & -1.2 & 0 & 0 & 0 & 0 & 1.2 \\ 0 & 0 & -2 & 0 & 0 & 0 & 2 \\ 0 & 0 & 0 & -3 & 0 & 0 & 3 \\ 0 & 0 & 0 & 0 & -6 & 0 & 6 \\ 0 & 0 & 0 & 0 & 0 & -12 & 12 \\ 0.00022 & 0.00022 & 0.00007 & 0.00041 & 0.00036 & 0.00111 & -0.00239 \end{bmatrix}$$

$$A_{d_6} = \begin{bmatrix} -444.4444 & 0 & 444.4444 \\ 0 & -720 & 720 \\ 0.165 & 0.02 & -0.185 \end{bmatrix}$$

$$A_e = \begin{bmatrix} -12 & 12 \\ 0.5 & -0.5 \end{bmatrix}$$

$$A_{f_2} = \begin{bmatrix} -12 & 0 & 0 & 12 \\ 0 & -360 & 0 & 360 \\ 0 & 0 & -444.4444 & 444.4444 \\ 0.0014 & 0.008 & 0.311 & -0.3204 \end{bmatrix}$$

$$A_{g_1} = \begin{bmatrix} -1 & 0 & 0 & 0 & 0 & 0 & 0 & 0 & 1 \\ 0 & -1.2 & 0 & 0 & 0 & 0 & 0 & 0 & 1.2 \\ 0 & 0 & -4 & 0 & 0 & 0 & 0 & 0 & 4 \\ 0 & 0 & 0 & -6 & 0 & 0 & 0 & 0 & 6 \\ 0 & 0 & 0 & 0 & -12 & 0 & 0 & 0 & 12 \\ 0 & 0 & 0 & 0 & 0 & -22.5 & 0 & 0 & 22.5 \\ 0 & 0 & 0 & 0 & 0 & 0 & -444.4444 & 0 & 444.4444 \\ 0 & 0 & 0 & 0 & 0 & 0 & 0 & -720 & 720 \\ 0.00066 & 0.00026 & 0.0001 & 0.00062 & 0.003 & 0.008 & 0.301 & 0.02 & -0.33364 \end{bmatrix}$$

$$A_{g_2} = \begin{bmatrix} -1 & 0 & 0 & 0 & 0 & 1 \\ 0 & -6 & 0 & 0 & 0 & 6 \\ 0 & 0 & -12 & 0 & 0 & 12 \\ 0 & 0 & 0 & -360 & 0 & 360 \\ 0 & 0 & 0 & 0 & -900 & 900 \\ 0.00007 & 0.0001 & 0.0001 & 0.14 & 1.8 & -1.94027 \end{bmatrix}$$

$$A_{g_3} = A_{i_1} = \begin{bmatrix} -2 & 0 & 0 & 0 & 0 & 2 \\ 0 & -6 & 0 & 0 & 0 & 6 \\ 0 & 0 & -8.3721 & 0 & 0 & 8.3721 \\ 0 & 0 & 0 & -22.5 & 0 & 22.5 \\ 0 & 0 & 0 & 0 & -360 & 360 \\ 0.0004 & 0.0003 & 0.00506 & 0.018 & 0.14 & -0.16376 \end{bmatrix}$$

$$A_{g_4} = A_{i_2} = \begin{bmatrix} -360 & 0 & 0 & 360 \\ 0 & -444.4444 & 0 & 444.4444 \\ 0 & 0 & -720 & 720 \\ 0.008 & 0.107 & 0.02 & -0.135 \end{bmatrix}$$

$$A_{h_1} = A_{j_1} = \begin{bmatrix} -1212.9 & 0 & 1140.9 & 72 & 0 & 0 & 0 \\ 0 & -1711.35 & 0 & 1711.35 & 0 & 0 & 0 \\ 9.6 & 0 & -652.05 & 0 & 570.45 & 72 & 0 \\ 0 & 9.6 & 0 & -1150.5 & 0 & 1140.9 & 0 \\ 0 & 0 & 19.2 & 0 & -91.2 & 0 & 72 \\ 0 & 0 & 0 & 19.2 & 0 & -589.65 & 570.45 \\ 0 & 0 & 0 & 0 & 15 & 28.8 & -43.8 \end{bmatrix}$$

$$A_{h_2} = A_{j_2} = \begin{bmatrix} -12 & 0 & 0 & 0 & 0 & 12 \\ 0 & -360 & 0 & 0 & 0 & 360 \\ 0 & 0 & -444.4444 & 0 & 0 & 444.4444 \\ 0 & 0 & 0 & -450 & 0 & 450 \\ 0 & 0 & 0 & 0 & -720 & 720 \\ 0.0014 & 0.104 & 0.197 & 0.18 & 0.33 & -0.8124 \end{bmatrix}$$

$$A_k = \begin{bmatrix} -8.3721 & 0 & 0 & 8.3721 \\ 0 & -22.5 & 0 & 22.5 \\ 0 & 0 & -51.4286 & 51.4286 \\ 0.02088 & 0.00002 & 0.00075 & -0.02165 \end{bmatrix}$$

# References


Agrawal, A. and Barlow, R. E. (1984) A survey of network reliability and domination theory. *Operat. Res.* **32**, 478–492.

Apostol, T. M. (1969) *Calculus*, vol. 2, 2nd edn. John Wiley & Sons, Inc.

Arjas, E. (1989) Survival models and martingale dynamics (with discussion). *Scand. J. Statist.* **16**, 177–225.

Aven, T. and Jensen, U. (1999) *Stochastic Models in Reliability*. Springer.

Barlow, R. E. and Proschan, F. (1975a) *Statistical Theory of Reliability and Life Testing. Probability Models*. Holt, Rinehart & Winston.

Barlow, R. E. and Proschan, F. (1975b) Importance of system components and fault tree events. *Stochastic Process. Appl.* **3**, 153–173.

Barlow, R. E. and Wu, A. S. (1978) Coherent systems with multistate components. *Math. Oper. Res.* **4**, 275–281.

Birnbaum, Z. W. (1969) On the importance of different components in a multicomponent system. In *Multivariate Analysis-II*, (ed. Krishnaiah, P. R.), pp. 581–592. Academic Press.

Block, H. W. and Savits, T. H. (1982) A decomposition for multistate monotone systems. *J. Appl. Prob.* **19**, 391–402.

Bodin, L. D. (1970) Approximations to system reliability using a modular decomposition. *Technometrics* **12**, 335–344.

Butler, D. A. (1982) Bounding the reliability of multistate systems. *Operat. Res.* **30**, 530–544.

Casella, G. and George, E. I. (1992) Explaining the Gibbs sampler. *Amer. Statistician* **46**, 167–174.

Chib, S. and Greenberg, E. (1995) Understanding the Metropolis–Hastings algorithm. *Amer. Statistician* **49**, 327–335.

El-Neweihi, E., Proschan, F. and Sethuraman, J. (1978) Multistate coherent systems. *J. Appl. Prob.* **15**, 675–688.

Esary, J. D. and Proschan, F. (1970) A reliability bound for systems of maintained, interdependent components. *J. Amer. Statist. Assoc.* **65**, 329–338.

Feller, W. (1968) *An Introduction to Probability Theory and its Applications*, vol. 1, 3rd edn. John Wiley & Sons, Inc.

Ford, L. R. and Fulkerson, D. R. (1956) Maximal flow through a network. *Canadian J. Math.* **8**, 399–404.

Funnemark, E. and Natvig, B. (1985) Bounds for the availabilities in a fixed time interval for multistate monotone systems. *Adv. Appl. Prob*. **17**, 638–665.

Gåsemyr, J. (2010) Bounds for the availabilities for multistate monotone systems based on decomposition into stochastically independent modules. Submitted.

Gåsemyr, J. and Natvig, B. (2001) Bayesian inference based on partial monitoring of components with applications to preventive system maintenance. *Nav. Res. Logistics* **48**, 551–577.

Gåsemyr, J. and Natvig, B. (2005) Probabilistic modelling of monitoring and maintenance of multistate monotone systems with dependent components. *Methodol. Comp. Appl. Prob*. **7**, 63–78.

Griffith, W. S. (1980) Multistate reliability models. *J. Appl. Prob*. **17**, 735–744.

Høgåsen, G. (1990) *MUSTAFA programs for MUltiSTAte Fault-tree Analysis*, Ph.D thesis, University of Oslo.

Huseby, A. B. and Natvig, B. (2010) Advanced discrete simulation methods applied to repairable multistate systems. In *Reliability, Risk and Safety. Theory and Applications-I*, (eds. Bris, R., Guedes Soares, C. and Martorell, S.), pp. 659–666. London, CRC Press.

Huseby, A. B., Eide, K. A., Isaksen, S. L., Natvig, B. and Gåsemyr, J. (2010) Advanced discrete event simulation methods with application to importance measure estimation in reliability. In *Discrete Event Simulations*, (ed. Goti, A.), pp. 205–222. SCIYO.

Karlin, S. and Taylor, H. M. (1975) *A First Course in Stochastic Processes*, 2nd edn. Academic Press.

Lisnianski, A. and Levitin, G. (2003) *Multi-state System Reliability*. World Scientific.

Lisnianski, A., Frenkel, I. and Ding, Y. (2010) *Multi-state System Reliability Analysis and Optimization for Engineers and Industrial Managers*. Springer.

Mastran, D. V. and Singpurwalla, N. D. (1978) A Bayesian estimation of the reliability of coherent structures. *Operat. Res*. **26**, 663–672.

Mørch, H. W. (1991) *An application of multistate reliability theory on an existing network of gas pipelines*, Cand. scient. thesis, University of Oslo (in Norwegian).

Nature, (1986) Near catastrophe at Le Bugey. *Nature* **321**, 462.

Natvig, B. (1979) A suggestion of a new measure of importance of system components. *Stochastic Process. Appl*. **9**, 319–330.

Natvig, B. (1980) Improved bounds for the availability and unavailability in a fixed time interval for systems of maintained, interdependent components. *Adv. Appl. Prob*. **12**, 200–221.

Natvig, B. (1982a) Two suggestions of how to define a multistate coherent system. *Adv. Appl. Prob*. **14**, 434–455.

Natvig, B. (1982b) On the reduction in remaining system lifetime due to the failure of a specific component. *J. Appl. Prob*. **19**, 642–652. Correction *J. Appl. Prob*. **20**, 713.

Natvig, B. (1985a) Multistate coherent systems. In *Encyclopedia of Statistical Sciences*, vol. 5 (eds Johnson, N. L. and Kotz, S.), pp. 732–735. John Wiley & Sons, Inc.

Natvig, B. (1985b) New light on measures of importance of system components. *Scand. J. Statist*. **12**, 43–54.

Natvig, B. (1986) Improved upper bounds for the availabilities in a fixed time interval for multistate monotone systems. *Adv. Appl. Prob*. **18**, 577–579.

Natvig, B. (1993) Strict and exact bounds for the availabilities in a fixed time interval for multistate monotone systems. *Scand. J. Statist*. **20**, 171–175.

Natvig, B. (2007) Multistate reliability theory. In *Encyclopedia of Statistics in Quality and Reliability*, vol. 3 (eds Ruggeri, F., Kenett, R. S. and Faltin, F. W.), pp. 1160–1164. John Wiley & Sons, Ltd.

Natvig, B. (2011) Measures of component importance in nonrepairable and repairable multistate strongly coherent systems. *Methodol. Comput. Appl. Prob*. **13**.

Natvig, B. and Eide, H. (1987) Bayesian estimation of system reliability. *Scand. J. Statist*. **14**, 319–327.

Natvig, B. and Gåsemyr, J. (2009) New results on the Barlow–Proschan and Natvig measures of component importance in nonrepairable and repairable systems. *Methodol. Comput. Appl. Prob*. **11**, 603–620.

Natvig, B. and Mørch, H. W. (2003) An application of multistate reliability theory to an offshore gas pipeline network. *Int. J. Reliability, Quality and Safety Eng*. **10**, 361–381.

Natvig, B., Gåsemyr, J. and Reitan, T. (2010a) Bayesian assessment of availabilities and unavailabilities of multistate monotone systems. Submitted.

Natvig, B., Huseby, A. B. and Reistadbakk, M. O. (2010b) Measures of component importance in repairable multistate systems–a numerical study. Submitted.

Natvig, B., Sørmo, S., Holen, A. T. and Høgåsen, G. (1986) Multistate reliability theory–a case study. *Adv. Appl. Prob*. **18**, 921–932.

Natvig, B., Eide, K. A., Gåsemyr, J., Huseby, A. B. and Isaksen, S. L. (2009) Simulation based analysis and an application to an offshore oil and gas production system of the Natvig measures of component importance in repairable systems. *Rel. Eng. System Safety* **94**, 1629–1638.

Ross, S. (1979) Multivalued state component reliability systems. *Ann. Prob*. **7**, 379–383.

Satyanarayana, A. (1982) A unified formula for analysis of some network reliability problems. *IEEE Trans. Rel*. **31**, 23–32.

Satyanarayana, A. and Chang, M. K. (1983) Network reliability and the factoring theorem. *Networks* **13**, 107–120.

Satyanarayana, A. and Prabhakar, A. (1978) New topological formula and rapid algorithm for reliability analysis of complex networks. *IEEE Trans. Rel*. **27**, 82–100.

Signoret, J. P. and Clave, N. (2007) SAFERELNET V3: Production Availability Test Case. *TOTAL*, (DGEP/TDO/EXP/SRF 04-013).

Tanner, M. and Wong, W. (1987) The calculation of posterior distributions by data augmentation. *J. Amer. Statist. Assoc*. **82**, 528–550.

# Index

# WILEY SERIES IN PROBABILITY AND STATISTICS



The Wiley Series in Probability and Statistics is well established and authoritative. It covers many topics of current research interest in both pure and applied statistics and probability theory. Written by leading statisticians and institutions, the titles span both state-of-the-art developments in the field and classical methods.

Reflecting the wide range of current research in statistics, the series encompasses applied, methodological and theoretical statistics, ranging from applications and new techniques made possible by advances in computerized practice to rigorous treatment of theoretical approaches.

This series provides essential and invaluable reading for all statisticians, whether in academia, industry, government, or research.

ABRAHAM and LEDOLTER · Statistical Methods for Forecasting
AGRESTI · Analysis of Ordinal Categorical Data
AGRESTI · An Introduction to Categorical Data Analysis
AGRESTI · Categorical Data Analysis, Second Edition
ALTMAN, GILL and McDONALD · Numerical Issues in Statistical Computing for the Social Scientist
AMARATUNGA and CABRERA · Exploration and Analysis of DNA Microarray and Protein Array Data
ANDĚL · Mathematics of Chance
ANDERSON · An Introduction to Multivariate Statistical Analysis, Third Edition
*ANDERSON · The Statistical Analysis of Time Series
ANDERSON, AUQUIER, HAUCK, OAKES, VANDAELE and WEISBERG · Statistical Methods for Comparative Studies
ANDERSON and LOYNES · The Teaching of Practical Statistics
ARMITAGE and DAVID (editors) · Advances in Biometry
ARNOLD, BALAKRISHNAN and NAGARAJA · Records
*ARTHANARI and DODGE · Mathematical Programming in Statistics
*BAILEY · The Elements of Stochastic Processes with Applications to the Natural Sciences
BALAKRISHNAN and KOUTRAS · Runs and Scans with Applications
BALAKRISHNAN and NG · Precedence-Type Tests and Applications
BARNETT · Comparative Statistical Inference, Third Edition

*Now available in a lower priced paperback edition in the Wiley Classics Library.

BARNETT · Environmental Statistics: Methods & Applications
BARNETT and LEWIS · Outliers in Statistical Data, Third Edition
BARTOSZYNSKI and NIEWIADOMSKA-BUGAJ · Probability and Statistical Inference
BASILEVSKY · Statistical Factor Analysis and Related Methods: Theory and Applications
BASU and RIGDON · Statistical Methods for the Reliability of Repairable Systems
BATES and WATTS · Nonlinear Regression Analysis and Its Applications
BECHHOFER, SANTNER and GOLDSMAN · Design and Analysis of Experiments for Statistical Selection, Screening and Multiple Comparisons
BEIRLANT, GOEGEBEUR, SEGERS, TEUGELS and DE WAAL · Statistics of Extremes: Theory and Applications
BELSLEY · Conditioning Diagnostics: Collinearity and Weak Data in Regression
BELSLEY, KUH and WELSCH · Regression Diagnostics: Identifying Influential Data and Sources of Collinearity
BENDAT and PIERSOL · Random Data: Analysis and Measurement Procedures, Third Edition
BERNARDO and SMITH · Bayesian Theory
BERRY, CHALONER and GEWEKE · Bayesian Analysis in Statistics and Econometrics: Essays in Honor of Arnold Zellner
BHAT and MILLER · Elements of Applied Stochastic Processes, Third Edition
BHATTACHARYA and JOHNSON · Statistical Concepts and Methods
BHATTACHARYA and WAYMIRE · Stochastic Processes with Applications
BIEMER, GROVES, LYBERG, MATHIOWETZ and SUDMAN · Measurement Errors in Surveys
BILLINGSLEY · Convergence of Probability Measures, Second Edition
BILLINGSLEY · Probability and Measure, Third Edition
BIRKES and DODGE · Alternative Methods of Regression
BISWAS, DATTA, FINE and SEGAL · Statistical Advances in the Biomedical Sciences: Clinical Trials, Epidemiology, Survival Analysis, and Bioinformatics
BLISCHKE and MURTHY (editors) · Case Studies in Reliability and Maintenance
BLISCHKE and MURTHY · Reliability: Modeling, Prediction and Optimization
BLOOMFIELD · Fourier Analysis of Time Series: An Introduction, Second Edition
BOLLEN · Structural Equations with Latent Variables
BOLLEN and CURRAN · Latent Curve Models: A Structural Equation Perspective
BOROVKOV · Ergodicity and Stability of Stochastic Processes
BOSQ and BLANKE · Inference and Prediction in Large Dimensions
BOULEAU · Numerical Methods for Stochastic Processes
BOX · Bayesian Inference in Statistical Analysis
BOX · R. A. Fisher, the Life of a Scientist
BOX and DRAPER · Empirical Model-Building and Response Surfaces
*BOX and DRAPER · Evolutionary Operation: A Statistical Method for Process Improvement
BOX · Improving Almost Anything *Revised Edition*
BOX, HUNTER and HUNTER · Statistics for Experimenters: An Introduction to Design, Data Analysis and Model Building

*Now available in a lower priced paperback edition in the Wiley Classics Library.

BOX, HUNTER and HUNTER · Statistics for Experimenters: Design, Innovation and Discovery, Second Edition
BOX and LUCEÑO · Statistical Control by Monitoring and Feedback Adjustment
BRANDIMARTE · Numerical Methods in Finance: A MATLAB-Based Introduction
BROWN and HOLLANDER · Statistics: A Biomedical Introduction
BRUNNER, DOMHOF and LANGER · Nonparametric Analysis of Longitudinal Data in Factorial Experiments
BUCKLEW · Large Deviation Techniques in Decision, Simulation and Estimation
CAIROLI and DALANG · Sequential Stochastic Optimization
CASTILLO, HADI, BALAKRISHNAN and SARABIA · Extreme Value and Related Models with Applications in Engineering and Science
CHAN · Time Series: Applications to Finance
CHARALAMBIDES · Combinatorial Methods in Discrete Distributions
CHATTERJEE and HADI · Regression Analysis by Example, Fourth Edition
CHATTERJEE and HADI · Sensitivity Analysis in Linear Regression
CHERNICK · Bootstrap Methods: A Practitioner's Guide
CHERNICK and FRIIS · Introductory Biostatistics for the Health Sciences
CHILÉS and DELFINER · Geostatistics: Modeling Spatial Uncertainty
CHOW and LIU · Design and Analysis of Clinical Trials: Concepts and Methodologies, Second Edition
CLARKE · Linear Models: The Theory and Application of Analysis of Variance
CLARKE and DISNEY · Probability and Random Processes: A First Course with Applications, Second Edition
*COCHRAN and COX · Experimental Designs, Second Edition
CONGDON · Applied Bayesian Modelling
CONGDON · Bayesian Models for Categorical Data
CONGDON · Bayesian Statistical Modelling, Second Edition
CONOVER · Practical Nonparametric Statistics, Second Edition
COOK · Regression Graphics
COOK and WEISBERG · An Introduction to Regression Graphics
COOK and WEISBERG · Applied Regression Including Computing and Graphics
CORNELL · Experiments with Mixtures, Designs, Models and the Analysis of Mixture Data, Third Edition
COVER and THOMAS · Elements of Information Theory
COX · A Handbook of Introductory Statistical Methods
*COX · Planning of Experiments
CRESSIE · Statistics for Spatial Data, Revised Edition
CSÖRGÖ and HORVÁTH · Limit Theorems in Change Point Analysis
DANIEL · Applications of Statistics to Industrial Experimentation
DANIEL · Biostatistics: A Foundation for Analysis in the Health Sciences, Sixth Edition
*DANIEL · Fitting Equations to Data: Computer Analysis of Multifactor Data, Second Edition
DASU and JOHNSON · Exploratory Data Mining and Data Cleaning
DAVID and NAGARAJA · Order Statistics, Third Edition
*DEGROOT, FIENBERG and KADANE · Statistics and the Law

*Now available in a lower priced paperback edition in the Wiley Classics Library.

DEL CASTILLO · Statistical Process Adjustment for Quality Control
DEMARIS · Regression with Social Data: Modeling Continuous and Limited Response Variables
DEMIDENKO · Mixed Models: Theory and Applications
DENISON, HOLMES, MALLICK and SMITH · Bayesian Methods for Nonlinear Classification and Regression
DETTE and STUDDEN · The Theory of Canonical Moments with Applications in Statistics, Probability and Analysis
DEY and MUKERJEE · Fractional Factorial Plans
DILLON and GOLDSTEIN · Multivariate Analysis: Methods and Applications
DODGE · Alternative Methods of Regression
*DODGE and ROMIG · Sampling Inspection Tables, Second Edition
*DOOB · Stochastic Processes
DOWDY, WEARDEN and CHILKO · Statistics for Research, Third Edition
DRAPER and SMITH · Applied Regression Analysis, Third Edition
DRYDEN and MARDIA · Statistical Shape Analysis
DUDEWICZ and MISHRA · Modern Mathematical Statistics
DUNN and CLARK · Applied Statistics: Analysis of Variance and Regression, Second Edition
DUNN and CLARK · Basic Statistics: A Primer for the Biomedical Sciences, Third Edition
DUPUIS and ELLIS · A Weak Convergence Approach to the Theory of Large Deviations
EDLER and KITSOS (editors) · Recent Advances in Quantitative Methods in Cancer and Human Health Risk Assessment
*ELANDT-JOHNSON and JOHNSON · Survival Models and Data Analysis
ENDERS · Applied Econometric Time Series
ETHIER and KURTZ · Markov Processes: Characterization and Convergence
EVANS, HASTINGS and PEACOCK · Statistical Distribution, Third Edition
FELLER · An Introduction to Probability Theory and Its Applications, Volume I, Third Edition, Revised; Volume II, Second Edition
FISHER and VAN BELLE · Biostatistics: A Methodology for the Health Sciences
FITZMAURICE, LAIRD and WARE · Applied Longitudinal Analysis
*FLEISS · The Design and Analysis of Clinical Experiments
FLEISS · Statistical Methods for Rates and Proportions, Second Edition
FLEMING and HARRINGTON · Counting Processes and Survival Analysis
FULLER · Introduction to Statistical Time Series, Second Edition
FULLER · Measurement Error Models
GALLANT · Nonlinear Statistical Models.
GEISSER · Modes of Parametric Statistical Inference
GELMAN and MENG (editors) · Applied Bayesian Modeling and Casual Inference from Incomplete-data Perspectives
GEWEKE · Contemporary Bayesian Econometrics and Statistics
GHOSH, MUKHOPADHYAY and SEN · Sequential Estimation
GIESBRECHT and GUMPERTZ · Planning, Construction and Statistical Analysis of Comparative Experiments

*Now available in a lower priced paperback edition in the Wiley Classics Library.

GIFI · Nonlinear Multivariate Analysis
GIVENS and HOETING · Computational Statistics
GLASSERMAN and YAO · Monotone Structure in Discrete-Event Systems
GNANADESIKAN · Methods for Statistical Data Analysis of Multivariate Observations, Second Edition
GOLDSTEIN · Multilevel Statistical Models, 4th Edition
GOLDSTEIN and WOLFF · Bayes Linear Statistics, Theory & Methods
GOLDSTEIN and LEWIS · Assessment: Problems, Development and Statistical Issues
GREENWOOD and NIKULIN · A Guide to Chi-Squared Testing
GROSS, SHORTLE, THOMPSON and HARRIS · Fundamentals of Queueing Theory, fourth Edition
GROSS, SHORTLE, THOMPSON and HARRIS · Solutions Manual to Accompany Fundamentals of Queueing Theory, fourth Edition
*HAHN and SHAPIRO · Statistical Models in Engineering
HAHN and MEEKER · Statistical Intervals: A Guide for Practitioners
HALD · A History of Probability and Statistics and their Applications Before 1750
HALD · A History of Mathematical Statistics from 1750 to 1930
HAMPEL · Robust Statistics: The Approach Based on Influence Functions
HANNAN and DEISTLER · The Statistical Theory of Linear Systems
HARTUNG, KNAPP and SINHA · Statistical Meta-Analysis with Applications
HEIBERGER · Computation for the Analysis of Designed Experiments
HEDAYAT and SINHA · Design and Inference in Finite Population Sampling
HEDEKER and GIBBONS · Longitudinal Data Analysis
HELLER · MACSYMA for Statisticians
HERITIER, CANTONI, COPT and VICTORIA-FESER · Robust Methods in Biostatistics
HINKELMANN and KEMPTHORNE · Design and Analysis of Experiments, Volume 1: Introduction to Experimental Design
HINKELMANN and KEMPTHORNE · Design and analysis of experiments, Volume 2: Advanced Experimental Design
HOAGLIN, MOSTELLER and TUKEY · Exploratory Approach to Analysis of Variance
HOAGLIN, MOSTELLER and TUKEY · Exploring Data Tables, Trends and Shapes
*HOAGLIN, MOSTELLER and TUKEY · Understanding Robust and Exploratory Data Analysis
HOCHBERG and TAMHANE · Multiple Comparison Procedures
HOCKING · Methods and Applications of Linear Models: Regression and the Analysis of Variance, Second Edition
HOEL · Introduction to Mathematical Statistics, Fifth Edition
HOGG and KLUGMAN · Loss Distributions
HOLLANDER and WOLFE · Nonparametric Statistical Methods, Second Edition
HOSMER and LEMESHOW · Applied Logistic Regression, Second Edition
HOSMER and LEMESHOW · Applied Survival Analysis: Regression Modeling of Time to Event Data
HUBER · Robust Statistics
HUBERTY · Applied Discriminant Analysis
HUBERY and OLEJNKI · Applied MANOVA and Discriminant Analysis, 2nd Edition

*Now available in a lower priced paperback edition in the Wiley Classics Library.

HUNT and KENNEDY · Financial Derivatives in Theory and Practice, Revised Edition
HURD and MIAMEE · Periodically Correlated Random Sequences: Spectral Theory and Practice
HUSKOVA, BERAN and DUPAC · Collected Works of Jaroslav Hajek – with Commentary
HUZURBAZAR · Flowgraph Models for Multistate Time-to-Event Data
IMAN and CONOVER · A Modern Approach to Statistics
JACKMAN · Bayesian Analysis for the Social Sciences
JACKSON · A User's Guide to Principle Components
JOHN · Statistical Methods in Engineering and Quality Assurance
JOHNSON · Multivariate Statistical Simulation
JOHNSON and BALAKRISHNAN · Advances in the Theory and Practice of Statistics: A Volume in Honor of Samuel Kotz
JOHNSON and BHATTACHARYYA · Statistics: Principles and Methods, Fifth Edition
JOHNSON and KOTZ · Distributions in Statistics
JOHNSON and KOTZ (editors) · Leading Personalities in Statistical Sciences: From the Seventeenth Century to the Present
JOHNSON, KOTZ and BALAKRISHNAN · Continuous Univariate Distributions, Volume 1, Second Edition
JOHNSON, KOTZ and BALAKRISHNAN · Continuous Univariate Distributions, Volume 2, Second Edition
JOHNSON, KOTZ and BALAKRISHNAN · Discrete Multivariate Distributions
JOHNSON, KOTZ and KEMP · Univariate Discrete Distributions, Second Edition
JUDGE, GRIFFITHS, HILL, LU TKEPOHL and LEE · The Theory and Practice of Econometrics, Second Edition
JURECKOVÁ and SEN · Robust Statistical Procedures: Asymptotics and Interrelations
JUREK and MASON · Operator-Limit Distributions in Probability Theory
KADANE · Bayesian Methods and Ethics in a Clinical Trial Design
KADANE and SCHUM · A Probabilistic Analysis of the Sacco and Vanzetti Evidence
KALBFLEISCH and PRENTICE · The Statistical Analysis of Failure Time Data, Second Edition
KARIYA and KURATA · Generalized Least Squares
KASS and VOS · Geometrical Foundations of Asymptotic Inference
KAUFMAN and ROUSSEEUW · Finding Groups in Data: An Introduction to Cluster Analysis
KEDEM and FOKIANOS · Regression Models for Time Series Analysis
KENDALL, BARDEN, CARNE and LE · Shape and Shape Theory
KHURI · Advanced Calculus with Applications in Statistics, Second Edition
KHURI, MATHEW and SINHA · Statistical Tests for Mixed Linear Models
*KISH · Statistical Design for Research
KLEIBER and KOTZ · Statistical Size Distributions in Economics and Actuarial Sciences
KLUGMAN, PANJER and WILLMOT · Loss Models: From Data to Decisions
KLUGMAN, PANJER and WILLMOT · Solutions Manual to Accompany Loss Models: From Data to Decisions

*Now available in a lower priced paperback edition in the Wiley – Interscience Paperback Series.

KOSKI and NOBLE · Bayesian Networks: An Introduction
KOTZ, BALAKRISHNAN and JOHNSON · Continuous Multivariate Distributions, Volume 1, Second Edition
KOTZ and JOHNSON (editors) · Encyclopedia of Statistical Sciences: Volumes 1 to 9 with Index
KOTZ and JOHNSON (editors) · Encyclopedia of Statistical Sciences: Supplement Volume
KOTZ, READ and BANKS (editors) · Encyclopedia of Statistical Sciences: Update Volume 1
KOTZ, READ and BANKS (editors) · Encyclopedia of Statistical Sciences: Update Volume 2
KOVALENKO, KUZNETZOV and PEGG · Mathematical Theory of Reliability of Time-Dependent Systems with Practical Applications
KOWALSI and TU · Modern Applied U-Statistics
KROONENBERG · Applied Multiway Data Analysis
KULINSKAYA, MORGENTHALER and STAUDTE · Meta Analysis: A Guide to Calibrating and Combining Statistical Evidence
KUROWICKA and COOKE · Uncertainty Analysis with High Dimensional Dependence Modelling
KVAM and VIDAKOVIC · Nonparametric Statistics with Applications to Science and Engineering
LACHIN · Biostatistical Methods: The Assessment of Relative Risks
LAD · Operational Subjective Statistical Methods: A Mathematical, Philosophical and Historical Introduction
LAMPERTI · Probability: A Survey of the Mathematical Theory, Second Edition
LANGE, RYAN, BILLARD, BRILLINGER, CONQUEST and GREENHOUSE · Case Studies in Biometry
LARSON · Introduction to Probability Theory and Statistical Inference, Third Edition
LAWLESS · Statistical Models and Methods for Lifetime Data, Second Edition
LAWSON · Statistical Methods in Spatial Epidemiology, Second Edition
LE · Applied Categorical Data Analysis
LE · Applied Survival Analysis
LEE · Structural Equation Modelling: A Bayesian Approach
LEE and WANG · Statistical Methods for Survival Data Analysis, Third Edition
LEPAGE and BILLARD · Exploring the Limits of Bootstrap
LEYLAND and GOLDSTEIN (editors) · Multilevel Modelling of Health Statistics
LIAO · Statistical Group Comparison
LINDVALL · Lectures on the Coupling Method
LIN · Introductory Stochastic Analysis for Finance and Insurance
LINHART and ZUCCHINI · Model Selection
LITTLE and RUBIN · Statistical Analysis with Missing Data, Second Edition
LLOYD · The Statistical Analysis of Categorical Data
LOWEN and TEICH · Fractal-Based Point Processes
MAGNUS and NEUDECKER · Matrix Differential Calculus with Applications in Statistics and Econometrics, Revised Edition
MALLER and ZHOU · Survival Analysis with Long Term Survivors

*Now available in a lower priced paperback edition in the Wiley Classics Library.

MALLOWS · Design, Data and Analysis by Some Friends of Cuthbert Daniel
MANN, SCHAFER and SINGPURWALLA · Methods for Statistical Analysis of Reliability and Life Data
MANTON, WOODBURY and TOLLEY · Statistical Applications Using Fuzzy Sets
MARCHETTE · Random Graphs for Statistical Pattern Recognition
MARKOVICH · Nonparametric Analysis of Univariate Heavy-Tailed Data: Research and practice
MARDIA and JUPP · Directional Statistics
MARKOVICH · Nonparametric Analysis of Univariate Heavy-Tailed Data: Research and Practice
MARONNA, MARTIN and YOHAI · Robust Statistics: Theory and Methods
MASON, GUNST and HESS · Statistical Design and Analysis of Experiments with Applications to Engineering and Science, Second Edition
MCCULLOCH and SERLE · Generalized, Linear and Mixed Models
MCFADDEN · Management of Data in Clinical Trials
MCLACHLAN · Discriminant Analysis and Statistical Pattern Recognition
MCLACHLAN, DO and AMBROISE · Analyzing Microarray Gene Expression Data
MCLACHLAN and KRISHNAN · The EM Algorithm and Extensions
MCLACHLAN and PEEL · Finite Mixture Models
MCNEIL · Epidemiological Research Methods
MEEKER and ESCOBAR · Statistical Methods for Reliability Data
MEERSCHAERT and SCHEFFLER · Limit Distributions for Sums of Independent Random Vectors: Heavy Tails in Theory and Practice
MICKEY, DUNN and CLARK · Applied Statistics: Analysis of Variance and Regression, Third Edition
*MILLER · Survival Analysis, Second Edition
MONTGOMERY, JENNINGS and KULAHCI · Introduction to Time Series Analysis and Forecasting Solutions Set
MONTGOMERY, PECK and VINING · Introduction to Linear Regression Analysis, Fourth Edition
MORGENTHALER and TUKEY · Configural Polysampling: A Route to Practical Robustness
MUIRHEAD · Aspects of Multivariate Statistical Theory
MULLER and STEWART · Linear Model Theory: Univariate, Multivariate and Mixed Models
MURRAY · X-STAT 2.0 Statistical Experimentation, Design Data Analysis and Nonlinear Optimization
MURTHY, XIE and JIANG · Weibull Models
MYERS and MONTGOMERY · Response Surface Methodology: Process and Product Optimization Using Designed Experiments, Second Edition
MYERS, MONTGOMERY and VINING · Generalized Linear Models. With Applications in Engineering and the Sciences
†NELSON · Accelerated Testing, Statistical Models, Test Plans and Data Analysis
†NELSON · Applied Life Data Analysis
NEWMAN · Biostatistical Methods in Epidemiology

*Now available in a lower priced paperback edition in the Wiley Classics Library.
†Now available in a lower priced paperback edition in the Wiley - Interscience Paperback Series.

OCHI · Applied Probability and Stochastic Processes in Engineering and Physical Sciences
OKABE, BOOTS, SUGIHARA and CHIU · Spatial Tesselations: Concepts and Applications of Voronoi Diagrams, Second Edition
OLIVER and SMITH · Influence Diagrams, Belief Nets and Decision Analysis
PALTA · Quantitative Methods in Population Health: Extentions of Ordinary Regression
PANJER · Operational Risks: Modeling Analytics
PANKRATZ · Forecasting with Dynamic Regression Models
PANKRATZ · Forecasting with Univariate Box-Jenkins Models: Concepts and Cases
PARDOUX · Markov Processes and Applications: Algorithms, Networks, Genome and Finance
PARMIGIANI and INOUE · Decision Theory: Principles and Approaches
*PARZEN · Modern Probability Theory and Its Applications
PEÑA, TIAO and TSAY · A Course in Time Series Analysis
PESARIN and SALMASO · Permutation Tests for Complex Data: Theory, Applications and Software
PIANTADOSI · Clinical Trials: A Methodologic Perspective
PORT · Theoretical Probability for Applications
POURAHMADI · Foundations of Time Series Analysis and Prediction Theory
POWELL · Approximate Dynamic Programming: Solving the Curses of Dimensionality
PRESS · Bayesian Statistics: Principles, Models and Applications
PRESS · Subjective and Objective Bayesian Statistics, Second Edition
PRESS and TANUR · The Subjectivity of Scientists and the Bayesian Approach
PUKELSHEIM · Optimal Experimental Design
PURI, VILAPLANA and WERTZ · New Perspectives in Theoretical and Applied Statistics
PUTERMAN · Markov Decision Processes: Discrete Stochastic Dynamic Programming
QIU · Image Processing and Jump Regression Analysis
RAO · Linear Statistical Inference and its Applications, Second Edition
RAUSAND and HØYLAND · System Reliability Theory: Models, Statistical Methods and Applications, Second Edition
RENCHER · Linear Models in Statistics
RENCHER · Methods of Multivariate Analysis, Second Edition
RENCHER · Multivariate Statistical Inference with Applications
RIPLEY · Spatial Statistics
RIPLEY · Stochastic Simulation
ROBINSON · Practical Strategies for Experimenting
ROHATGI and SALEH · An Introduction to Probability and Statistics, Second Edition
ROLSKI, SCHMIDLI, SCHMIDT and TEUGELS · Stochastic Processes for Insurance and Finance
ROSENBERGER and LACHIN · Randomization in Clinical Trials: Theory and Practice
ROSS · Introduction to Probability and Statistics for Engineers and Scientists
ROSSI, ALLENBY and MCCULLOCH · Bayesian Statistics and Marketing
ROUSSEEUW and LEROY · Robust Regression and Outline Detection
ROYSTON and SAUERBREI · Multivariable Model - Building: A Pragmatic Approach to Regression Anaylsis based on Fractional Polynomials for Modelling Continuous Variables

*Now available in a lower priced paperback edition in the Wiley Classics Library.

RUBIN · Multiple Imputation for Nonresponse in Surveys
RUBINSTEIN · Simulation and the Monte Carlo Method, Second Edition
RUBINSTEIN and MELAMED · Modern Simulation and Modeling
RYAN · Modern Engineering Statistics
RYAN · Modern Experimental Design
RYAN · Modern Regression Methods
RYAN · Statistical Methods for Quality Improvement, Second Edition
SALEH · Theory of Preliminary Test and Stein-Type Estimation with Applications
SALTELLI, CHAN and SCOTT (editors) · Sensitivity Analysis
*SCHEFFE · The Analysis of Variance
SCHIMEK · Smoothing and Regression: Approaches, Computation and Application
SCHOTT · Matrix Analysis for Statistics
SCHOUTENS · Levy Processes in Finance: Pricing Financial Derivatives
SCHUSS · Theory and Applications of Stochastic Differential Equations
SCOTT · Multivariate Density Estimation: Theory, Practice and Visualization
*SEARLE · Linear Models
SEARLE · Linear Models for Unbalanced Data
SEARLE · Matrix Algebra Useful for Statistics
SEARLE and WILLETT · Matrix Algebra for Applied Economics
SEBER · Multivariate Observations
SEBER and LEE · Linear Regression Analysis, Second Edition
SEBER and WILD · Nonlinear Regression
SENNOTT · Stochastic Dynamic Programming and the Control of Queueing Systems
*SERFLING · Approximation Theorems of Mathematical Statistics
SHAFER and VOVK · Probability and Finance: Its Only a Game!
SILVAPULLE and SEN · Constrained Statistical Inference: Inequality, Order and Shape Restrictions
SINGPURWALLA · Reliability and Risk: A Bayesian Perspective
SMALL and MCLEISH · Hilbert Space Methods in Probability and Statistical Inference
SRIVASTAVA · Methods of Multivariate Statistics
STAPLETON · Linear Statistical Models
STAUDTE and SHEATHER · Robust Estimation and Testing
STOYAN, KENDALL and MECKE · Stochastic Geometry and Its Applications, Second Edition
STOYAN and STOYAN · Fractals, Random and Point Fields: Methods of Geometrical Statistics
STREET and BURGESS · The Construction of Optimal Stated Choice Experiments: Theory and Methods
STYAN · The Collected Papers of T. W. Anderson: 1943–1985
SUTTON, ABRAMS, JONES, SHELDON and SONG · Methods for Meta-Analysis in Medical Research
TAKEZAWA · Introduction to Nonparametric Regression
TAMHANE · Statistical Analysis of Designed Experiments: Theory and Applications
TANAKA · Time Series Analysis: Nonstationary and Noninvertible Distribution Theory
THOMPSON · Empirical Model Building
THOMPSON · Sampling, Second Edition

*Now available in a lower priced paperback edition in the Wiley Classics Library.

THOMPSON · Simulation: A Modeler's Approach
THOMPSON and SEBER · Adaptive Sampling
THOMPSON, WILLIAMS and FINDLAY · Models for Investors in Real World Markets
TIAO, BISGAARD, HILL, PEÑA and STIGLER (editors) · Box on Quality and Discovery: with Design, Control and Robustness
TIERNEY · LISP-STAT: An Object-Oriented Environment for Statistical Computing and Dynamic Graphics
TSAY · Analysis of Financial Time Series
UPTON and FINGLETON · Spatial Data Analysis by Example, Volume II: Categorical and Directional Data
VAN BELLE · Statistical Rules of Thumb
VAN BELLE, FISHER, HEAGERTY and LUMLEY · Biostatistics: A Methodology for the Health Sciences, Second Edition
VESTRUP · The Theory of Measures and Integration
VIDAKOVIC · Statistical Modeling by Wavelets
VINOD and REAGLE · Preparing for the Worst: Incorporating Downside Risk in Stock Market Investments
WALLER and GOTWAY · Applied Spatial Statistics for Public Health Data
WEERAHANDI · Generalized Inference in Repeated Measures: Exact Methods in MANOVA and Mixed Models
WEISBERG · Applied Linear Regression, Second Edition
WELSH · Aspects of Statistical Inference
WESTFALL and YOUNG · Resampling-Based Multiple Testing: Examples and Methods for p-Value Adjustment
WHITTAKER · Graphical Models in Applied Multivariate Statistics
WINKER · Optimization Heuristics in Economics: Applications of Threshold Accepting
WONNACOTT and WONNACOTT · Econometrics, Second Edition
WOODING · Planning Pharmaceutical Clinical Trials: Basic Statistical Principles
WOODWORTH · Biostatistics: A Bayesian Introduction
WOOLSON and CLARKE · Statistical Methods for the Analysis of Biomedical Data, Second Edition
WU and HAMADA · Experiments: Planning, Analysis and Parameter Design Optimization
WU and ZHANG · Nonparametric Regression Methods for Longitudinal Data Analysis: Mixed-Effects Modeling Approaches
YANG · The Construction Theory of Denumerable Markov Processes
YOUNG, VALERO-MORA and FRIENDLY · Visual Statistics: Seeing Data with Dynamic Interactive Graphics
ZACKS · Stage-Wise Adaptive Designs
*ZELLNER · An Introduction to Bayesian Inference in Econometrics
ZELTERMAN · Discrete Distributions: Applications in the Health Sciences
ZHOU, OBUCHOWSKI and McCLISH · Statistical Methods in Diagnostic Medicine

*Now available in a lower priced paperback edition in the Wiley Classics Library.